Lecture Notes in Civil Engineering

Volume 289

Lecture Notes in Civil Engineering (LNCE) publishes the latest developments in Civil Engineering—quickly, informally and in top quality. Though original research reported in proceedings and post-proceedings represents the core of LNCE, edited volumes of exceptionally high quality and interest may also be considered for publication. Volumes published in LNCE embrace all aspects and subfields of, as well as new challenges in, Civil Engineering. Topics in the series include:

- Construction and Structural Mechanics
- Building Materials
- Concrete, Steel and Timber Structures
- Geotechnical Engineering
- Earthquake Engineering
- Coastal Engineering
- Ocean and Offshore Engineering; Ships and Floating Structures
- Hydraulics, Hydrology and Water Resources Engineering
- Environmental Engineering and Sustainability
- Structural Health and Monitoring
- Surveying and Geographical Information Systems
- Indoor Environments
- Transportation and Traffic
- Risk Analysis
- Safety and Security

To submit a proposal or request further information, please contact the appropriate Springer Editor:

- Pierpaolo Riva at pierpaolo.riva@springer.com (Europe and Americas);
- Swati Meherishi at swati.meherishi@springer.com (Asia–except China, and Australia, New Zealand);
- Wayne Hu at wayne.hu@springer.com (China).

All books in the series now indexed by Scopus and EI Compendex database!

Simone Pascuzzi · Francesco Santoro
Editors

Farm Machinery and Processes Management in Sustainable Agriculture

XI International Scientific Symposium 2022

Springer

Editors
Simone Pascuzzi
DiSAAT
University of Bari Aldo Moro
Bari, Italy

Francesco Santoro
DiSAAT
University of Bari Aldo Moro
Bari, Italy

ISSN 2366-2557 ISSN 2366-2565 (electronic)
Lecture Notes in Civil Engineering
ISBN 978-3-031-13089-2 ISBN 978-3-031-13090-8 (eBook)
https://doi.org/10.1007/978-3-031-13090-8

This Springer imprint is published by the registered company Springer Nature Switzerland AG
The registered company address is: Gewerbestrasse 11, 6330 Cham, Switzerland

Preface

The Russian-Ukrainian war is the most serious geopolitical conflict since the Second World War and will have certainly more significant global consequences than the previous financial or health crises suffered in the last twenty years.

In this particular critical moment, it is necessary to carefully analyze and evaluate the foreseeable repercussions on food and energy supplies and the potential impact on the international scene. At the same time, it appears indispensable and urgent to respond flexibly to current dynamics and make strategic choices consistent with long-term interests.

The direct and indirect criticalities generated by the difficult situation that has developed in recent times have already determined negative impacts on the current management of agricultural production and food products, with high-risk consequences for the stability of numerous agri-food chains.

These critical issues immediately have a different impact depending on the sector such as, for example:

- for cereal crops, protein, and oil crops, there is the sudden unavailability of raw materials due to the closure of important markets, the consequent sharp rise in prices also due to limitations in logistics and procurement;
- in the fruit and vegetable sector, there may be an oversupply on domestic markets and a drop in the remuneration of producers;
- in the nursery and horticultural sector in hot greenhouses, the interruption of production processes may occur due to the increase in fuel prices for heating;
- in breeding, the unavailability of basic raw materials for the production of feed and therefore for nutrition, also exacerbated by an unsustainable increase in energy costs, could lead to the impossibility of completing the production cycles, forcing farmers to resort to the slaughter of heads in full production.

In the medium to long terms, these criticalities can aggravate some structural problems of agricultural systems such as, for example:

- the growing dependence on foreign countries for genetic materials (varieties and hybrids) used for national crops, such as corn and wheat;

- the continuing difficult situation for many companies linked to the scarce availability of necessary manpower and production costs that often exceed the remuneration of the products;
- the increase in fuel costs, with an immediate impact on fruit processing and conditioning plants and on transport to markets.

In this difficult and cyclical phase of instability, the eleventh edition of the "Farm Machinery and Process Management in Sustainable Agriculture" International Symposium was an opportunity to present and discuss some proposals based on objective criteria to be implemented in the short and medium terms to address the main problems and possible solutions for agriculture and for the entire agri-food system.

The FMPMSA International Symposium, in large part, in responding to the ambitious goal of proposing scientific and technological solutions as well as structural measures useful for increasing the resilience of agri-food systems.

Bari, Italy

Simone Pascuzzi
Francesco Santoro

Contents

The Structure of Engine Power of Agricultural Tractors Registered in Poland Between 2010 and 2020

Artur Przywara, Artur Kraszkiewicz, Milan Koszel, Atanas Zdravkov Atanasov, and Stanisław Parafiniuk

Abstract The aim of this study was to analyse changes on the new farm tractors market in Poland. The work covered the period between 2010 and 2020, and was based on new farm-tractor sales data measured by the number of registrations. The paper presents the semi-annual sales figures, the demand for selected tractor brands, and the sales of new farm tractors representing various engine power segments. New tractors were considered to be those registered for the first time in Poland with either the same or preceding year of manufacture. Between 2010 and 2020, over 144,000 new farm tractors were registered in Poland, with the highest number of tractors sold being recorded in 2012, and the lowest in 2016. Unfortunately, no Polish brand made its way to the top 10 of the top-selling tractors. The Polish market was largely dominated by tractors produced in Western Europe and the USA. The analysis revealed that New Holland tractors (23,780 vehicles) accounted for the largest number of registrations, followed by John Deere (19,453), Zetor (16,398) and Deutz-Fahr (10,508). The total share of these four manufacturers between 2010 and 2020 was 48.6%. In the reference period, vehicles representing the 51–70 kW engine power segment had the largest share (35.17%), followed by 71–100 (24.44%), 31–50 (14.00%), <30 (7.68%), 101–120 (7.4%), 121–140 (4.85%), >161 (3.69%), and 141–160 kW (2.76%). Among the farm tractors representing lower power segments, i.e., up to 50 kW, Kubota was the most frequently purchased brand. As regards higher engine power segments, i.e., 50–140 kW, New Holland tractors held market dominance, while in the >140 kW engine power segment, John Deere tractors were prevalent.

Keywords Farm tractor · Tractors market · Sales of farm tractors in Poland · Registration · Engine power

A. Przywara (✉) · A. Kraszkiewicz · M. Koszel · S. Parafiniuk
Department of Machinery Exploitation and Management of Production Processes, University of Life Sciences in Lublin, Lublin, Poland
e-mail: artur.przywara@up.lublin.pl

A. Z. Atanasov
Department of Agricultural Machinery, University of Ruse Angel Kanchev, Ruse, Bulgaria

S. Pascuzzi and F. Santoro (eds.), *Farm Machinery and Processes Management in Sustainable Agriculture*, Lecture Notes in Civil Engineering 289,
https://doi.org/10.1007/978-3-031-13090-8_1

1 Introduction

In recent years, the agricultural sector in Poland has undergone profound changes as a result of the country's systemic transformation and accession to the European Union. In order to remain competitive, agricultural holdings must continue to modernise their machines. EU funds are currently one of the main sources of financing for the purchases of new machinery. The replacement of technical fixed assets in an average Polish family farm usually takes place over one generation, i.e., every 25–30 years [1]. This is attributable not only to the low use of machinery and the corresponding low technical wear and tear, but also to some financial constraints. Moreover, the relationship between machinery prices and agricultural-product prices is becoming more and more unfavourable.

The prices of machinery and equipment are gradually increasing in line with technological progress and novel solutions introduced to the market. Prices are also influenced by the economic and political situation. Poland's accession to the European Union in 2004 triggered an increase in gross prices to include VAT (22%), which had been suspended until that year. The implementation of the Common Agricultural Policy (CAP), under which farmers receive various subsidies and grants, is also reflected in how producers shape prices [1]. The investments made under the "Modernisation of agricultural holdings" measure of the Rural Development Programmes for 2007–2013 and 2014–2020 involved the purchase of equipment. Investments in machinery, tools and equipment for crop production had the largest share. In the first Programme period, the purchase of farm tractors accounted for 37% of all investments in agricultural equipment [2], and in the second period – for 29.6% [3].

According to one study, when buying a new tractor, 91.2% of all farmers believe that the most important technical parameter is engine power [4]. Before making the purchase, Polish farmers also attach much importance to the tractor's failure-free performance (54.7%), and to fast access to maintenance services and spare parts (41.4%). Regional variations on the tractors market are similar to the national variations on this market [5]. Lorencowicz found that in the western regions of Poland, where agricultural holdings are two or three times larger than in the rest of the country, the tractors purchased had higher power outputs, younger age and higher prices (81–95 kW, 14–18 years old, PLN 168,000–186,000) [6]. In contrast, tractors purchased in south-eastern Poland, where agricultural holdings below 10 ha are prevalent, have worse parameters (76–86 kW, 14–18 years old, PLN 76,000–86,000).

Analyses of changes in the spatial variations in the use of funds under the "Modernisation of agricultural holdings" measure of the Rural Development Programmes for 2007–2013 and 2014–2020 revealed that the use of these funds in the financing periods was significantly higher in the provinces with larger agricultural holdings [7]. In the first financing period, the provinces in which small agricultural holdings were prevalent were characterised by the highest share of holdings benefiting from investment subsidies in the whole group of agricultural holdings eligible for such support (above 20%). However, in the next financing period, the activity of farmers in these provinces was lower than in others, while high activity rates were observed in the provinces with large agricultural holdings, where the share of holdings benefiting from EU aid was around 35–40%, compared to less than 20% in the 2007–2013 financing period.

Changes taking place in the area structure of agricultural holdings have an indirect impact on labour inputs, as well as on the provision of machinery and equipment in agriculture. The latter, in turn, has a decisive influence on the production and economic effects of agriculture. The development of agricultural technology in agricultural holdings generally leads to positive changes in the organisation of such holdings. Among other factors, this is achieved through production simplification and increase having influence upon the concentration of agricultural land [8].

The economic situation on the domestic market of means of agricultural mechanisation, including tractors, reflects the income situation of farmers [9]. Choosing the right engine power appears crucial, as it has a direct bearing on the tractor price. Therefore, the demand for tractors is an important aspect taken into consideration when analysing the market of means of agricultural mechanisation. Changes in the market conditions affecting the profitability of agricultural holdings create the need for constantly monitoring the situation on the farm tractors market. It is, therefore, necessary to obtain information on tractor sales that accounts for their evolution over time [10].

The aim of this publication is to analyse the variability of the new farm tractors market, with a particular focus on the engine power of tractors registered in Poland between 2010 and 2020.

2 Research Method

The analysis of the new farm tractors market was based on their sales measured by the number of registrations. Data obtained from the Central Registry of Vehicles and Drivers (Polish abbreviation: CEPiK) were used in the study. To establish a uniform definition, new tractors were considered to be those registered in the year under analysis, with either the same or preceding year of manufacture. The data on tractor registrations may not be identical to the values previously reported in other sources due to database updates and, as additionally noted by Pawlak [9], they do not equal to the value of domestic supply, calculated as the sum of the number of manufactured and imported tractors less exports. Nonetheless, they can still be used to analyse the demand for new tractors, primarily in terms of the farmers' interest in particular manufacturers.

The analysis covered the seasonal nature of sales presented on a semi-annual basis, the demand for specific brands (manufacturers), and the breakdown of engine power outputs in new farm tractors registered in Poland between 2010 and 2020. Based on the engine power shown in the collected data and available literature [10–13], tractors were grouped into the following classes: <30, 31–50, 51–70, 71–100, 101–120, 121–140, 141–160, and >161 kW.

3 Results and Discussion

The analysis between 2010 and 2020, over 144,000 new farm tractors were registered in Poland. Between 2010 and 2012, tractor sales increased from 13,163 to 18,484 vehicles, with 2012 being the best year in terms of sales of new tractors. A gradual decline in the sales figures was then observed, and the lowest number (8,429 vehicles) of registrations was recorded in 2016. This number was lower compared to the best-selling year by 54.4%. There were several reasons for this decline, including the termination of support under the Rural Development Programme for 2007–2013, the unfavourable situation on the milk and pork markets, which had a significant impact on the investment climate and the purchasing power of agricultural holdings, and a temporary lack of new active programmes to support the modernisation of holdings in terms of purchasing new agricultural machinery. Market saturation, resulting from increased investments supported by EU funds, also contributed to the drop in the sales of new tractors [14]. However, a slow rise in the sales of tractors was observed from 2017 to the end of the period under analysis. This resulted mostly from the transition period between successive Rural Development Programmes (for 2007–2013 and 2014–2020), which caused the suspension of investments by farmers who decided to wait for the funds to be provided under the next programme and who could not purchase a second (auxiliary) tractor under the new programme (launched by the Agency for Restructuring and Modernisation of Agriculture, ARMA).

The farm tractors sold on the Polish market mainly come from Western Europe and the USA. Polish farmers can also purchase vehicles produced in Belarus or Southern Asia (India) and South-Western Asia (Japan, South Korea and the People's Republic of China).

The analysis of tractor sales between 2010 and 2015, presented on a semi-annual basis, showed that the first six months of the year were characterised by a higher number of registrations. In contrast, from 2016 to the end of the period under analysis, the sales pattern changed, with the first six months recording lower sales than the second half of the year.

The numbers of tractor registrations in each half-year period between 2010 and 2020 are presented in Fig. 1.

The highest number of registrations in the period under analysis was recorded in the first half of 2012 (10,331 vehicles), and the lowest in the first half of 2016 (4,022 vehicles). The year 2020 is also worth mentioning as it saw the outbreak of the COVID-19 pandemic, which involved a huge number of restrictions disrupting production and sales processes. In spite of that difficult situation, 14,379 farm tractors were registered in that year.

As regards new vehicles, New Holland tractors (23,780 vehicles) accounted for the highest number of registrations between 2010 and 2020, followed by John Deere (19,453), Zetor (16,398) and Deutz-Fahr tractors (10,508). The combined share of these manufacturers in the total tractor sales in the period under analysis was 48.58%. Tractors offered by foreign manufacturers were definitely prevalent. Registrations of Polish tractor brands, including Ursus, Farmer and Pronar, were at low levels,

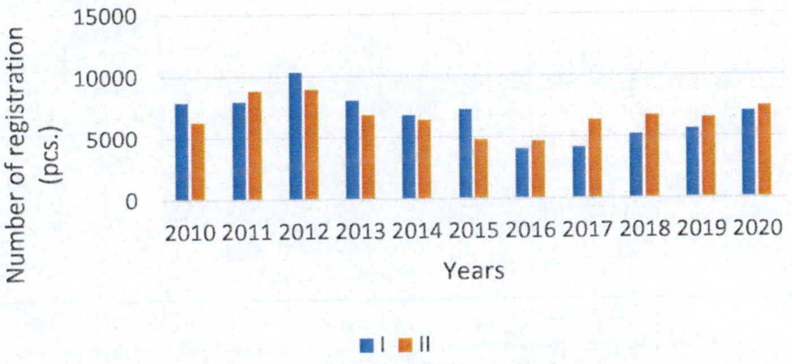

Fig. 1 The number of registrations of new farm tractors in Poland between 2010 and 2020, presented on a semi-annual basis

amounting to 3.14, 0.51 and 0.42%, respectively. In addition, Case IH could certainly be classified among the most popular farm tractors, as its sales between 2010 and 2012 made it the fifth most commonly purchased tractor. From 2013, its place was taken by the Japanese manufacturer Kubota, which, over the following years until 2020, achieved significantly better sales even when compared to the German Deutz-Fahr. From 2018, Kubota even managed to exceed the Czech Zetor, which recorded 2,137 registrations by 2020, compared to 2,756 Kubota tractors. The Belarusian farm tractor manufacturer (named Belarus) also attained a satisfactory result in the period under analysis, with 5,221 registrations, which in 2011, 2019 and 2020 translated into better figures than those recorded for its predecessor, Case IH. Further down the rankings, Class, Farmtrac, Massey Ferguson and Ursus recorded average registration numbers of between 4,825 and 4,547 vehicles over the analysed decade. The numbers of registrations of Poland's best-selling tractors were consistent with the overall trend of sales increases and decreases for all manufacturers between 2010 and 2020, with the highest sales recorded in 2012 and the lowest in 2016. Kubota, Belarus and Ursus were the only exceptions, with peaking sales in 2014, 2012 and 2017, respectively (Table 1).

In terms of engine power, the largest share in the registration breakdown was recorded for tractors in the 51–70 kW power engine segment, which accounted for 35.17% of all sales between 2010 and 2020. A smaller share was attributable to 71–100 kW tractors (24.44%), followed by 31–50 (14%), <30 (7.68%), 101–120 (7.40%), 121–140 (4.85%), and >161 kW (3.69%). The smallest group (2.76%) comprised tractors in the 141–160 kW segment. The shares of tractor registrations by engine power in the past decade are presented in Fig. 2.

As regards the sales of tractors in individual years in terms of engine power, the registration of new farm tractors exhibited some variations (Fig. 3). In the first segment, i.e. below 30 kW, no correlation ($R = -0.14$) was observed with respect to the total number of registrations. For this power range, the lowest sales were recorded in 2014, and then they gradually increased until 2020 to achieve the highest result in this class. Similar trends were observed for the following engine power

Table 1 The numbers of new farm tractors registered between 2010 and 2020 by manufacturer

Brand	Year											Total
	2010	2011	2012	2013	2014	2015	2016	2017	2018	2019	2020	
New Holland	2383	2639	3725	2578	2374	1984	1210	1505	1920	1713	1749	23,780
John Deere	2101	2328	2707	2442	1753	1794	1041	1282	1530	1252	1223	19,453
Zetor	2173	1843	2701	2085	1695	1556	1081	1127	741	680	716	16,398
Deutz-Fahr	1059	1245	1609	899	878	1034	626	1020	641	774	723	10,508
Kubota	130	308	306	604	1128	1045	626	771	869	881	1006	7674
Case IH	715	829	1072	452	592	546	488	665	770	583	751	7463
Belarus	700	991	341	102	86	217	174	272	411	971	956	5221
Class	242	446	843	613	461	452	360	413	274	345	376	4825
Farmtrac	427	596	787	468	599	424	259	501	204	196	329	4790
Massey Ferguson	417	587	768	480	375	322	143	284	213	385	601	4575
Ursus	347	241	464	439	293	403	894	936	323	139	68	4547
Valtra	357	373	379	334	364	372	239	337	352	315	352	3774
Fendt	460	363	297	179	207	131	63	54	104	167	203	2228
Same	371	363	408	469	68	86	45	36	3	2	2	1853
Steyr	10	58	157	125	162	202	220	249	225	183	249	1840
Other	1271	1922	1920	2415	2089	1350	960	1839	3198	3400	5075	25,439
Total	13,163	15,132	18,484	14,684	13,124	11,918	8429	11,291	11,778	11,986	14,379	144,368

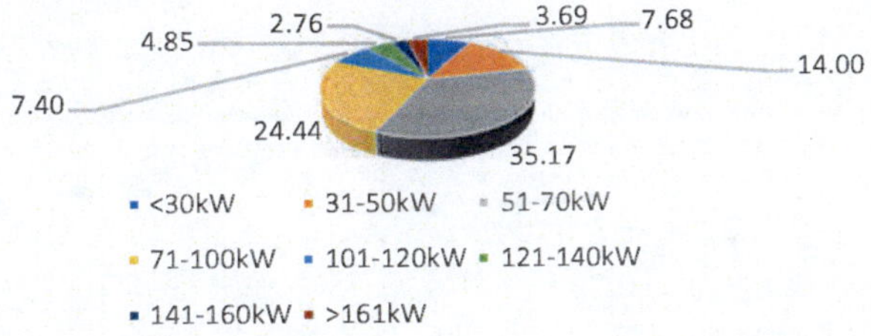

Fig. 2 Breakdown of engine power outputs in new farm tractors registered in Poland between 2010 and 2020

ranges: 121–140, >161, 141–160 and 101–120 kW. The highest positive correlation of registrations by engine power with the total number of registrations in the years under analysis was observed for 71–100 kW (R = 0.91). As regards this power range, most tractors were sold in 2012 (5,669 vehicles) and the fewest in 2016 (2,061 vehicles). A similar pattern concerned the 51–70 kW engine power segment (R = 0.87), with 7,661 tractors sold in the peak year 2012, accounting for 40.86% of all tractors sold in that year. This was also the highest sales figure among all groups. In contrast, 3,117 tractors were sold in 2016. In the third group (R = 0.56), 2,216 tractors were sold in 2012, and slightly fewer (1,716 vehicles) in 2016. Among all engine power segments, the lowest number of registrations (209 vehicles) was recorded in 2014 for tractors below 30 kW, accounting for only 1.6% of all tractors sold. It is also worth noting that after 2016, with sales being the lowest over the whole decade, only the 101–120 kW tractors did not record a higher number of registrations.

The following section presents the share of the top ten best-selling farm tractors in relation to the engine power ranges under analysis. The first segment concerned tractors below 30 kW, with Kubota (58.34%), John Deere (35.16%) and Belarus (6.88%) recording the highest numbers of registrations (Fig. 4), followed by Farmtrac

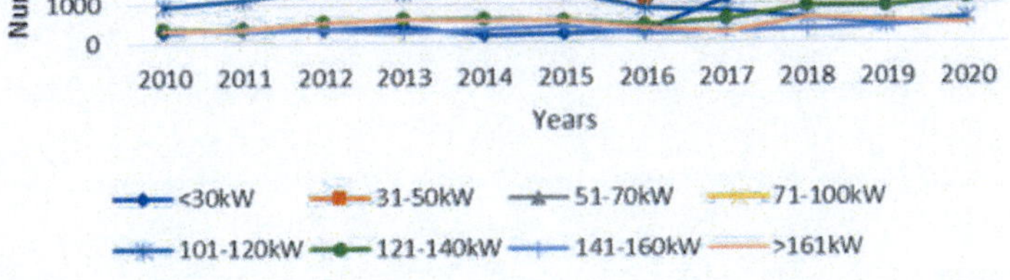

Fig. 3 The number of new farm tractors registered between 2010 and 2020 by engine power range

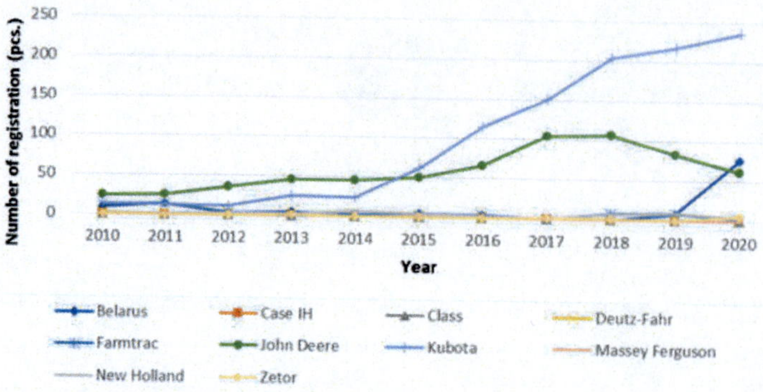

Fig. 4 The number of tractors by selected manufacturers, registered in Poland between 2010 and 2020, in the <30 kW engine power range

(3.14%) and New Holland (2.33%). The share of the remaining manufacturers was below 0.5%.

As regards the second segment, i.e., 31–50 kW (Fig. 5), New Holland recorded the highest number of tractors sold (26.32%), followed by Kubota (22.86%), Deutz-Fahr (12.69%), Farmtrac (12.19%) and John Deere (10.22%). Lower shares were attributable to the following manufacturers: Zetor (7.90%), Case IH (6.13%) and Massey Ferguson (1.67%), while the overall share of the remaining manufacturers did not exceed 1%.

In the third power segment, i.e., 51–70 kW, which comprised the most numerous group in terms of registrations, the best-selling tractor manufacturer was Zetor, with a share of 24.16% (Fig. 6). New Holland was rated second (22.26%) with the best sales result recorded in 2012 (1,577 vehicles). Johne Deere (14.53%), Deutz-Fahr (10.47%), Belarus (9.92%), Farmtrac (9.06%), Case IH (8.24%) and Kubota (5.65%) were next, while the overall share of the remaining manufacturers did not exceed 5%.

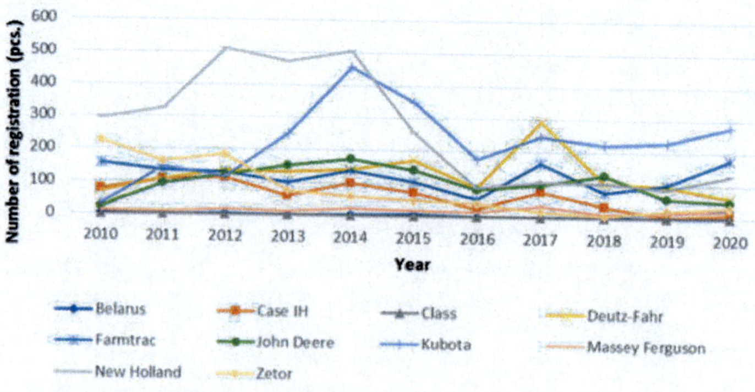

Fig. 5 The number of tractors by selected manufacturers, registered in Poland between 2010 and 2020, in the 31-50 kW engine power range

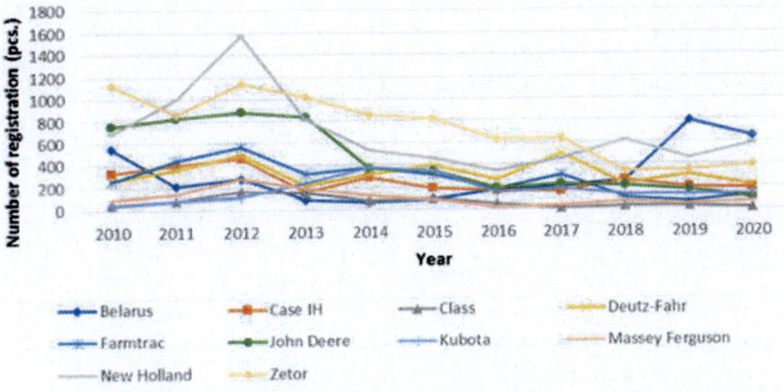

Fig. 6 The number of tractors by selected manufacturers, registered in Poland between 2010 and 2020, in the 51–70 kW engine power range

The fourth important group in terms of registrations was composed of tractors in the 71–100 kW engine power segment (Fig. 7), in which the largest seller was the US-based New Holland (24.07%) with 6,707 vehicles registered in the entire period under analysis. John Deere, also from the USA, was rated second (21.86%), followed by Zetor (14.42%), Deutz-Fahr (13.71%), Class (6.92%), and Case IH (6.86%). The remaining manufacturers contributed to less than 5% of the registered vehicles.

In the next engine power segment, i.e., 101–120 kW, the best outcome was recorded by John Deere (32.58%), with 2,524 vehicles sold, attaining the highest sales figures in 2012 amounting to 438 vehicles (Fig. 8). This was the only group that did not record a sales growth between 2016 and 2020. Another favourable result was recorded by New Holland (25.76%), with 1,996 vehicles sold over the decade. Deutz-Fahr (14.07%), Class (12.88%), Massey Ferguson (7.53%), and Kubota (5.64%) came next, while the overall share of the remaining manufacturers did not exceed 5%.

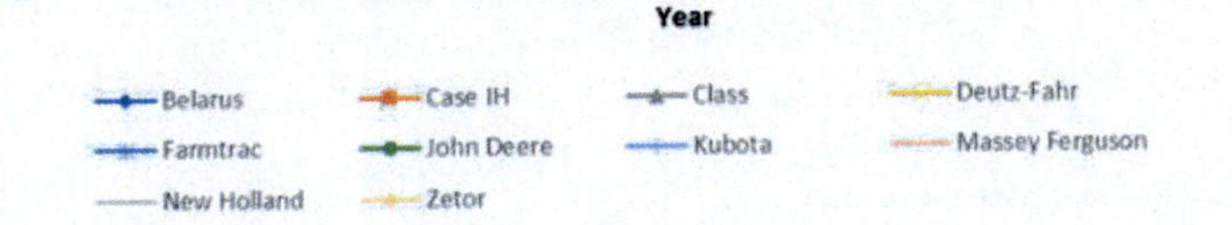

Fig. 7 The number of tractors by selected manufacturers, registered in Poland between 2010 and 2020, in the 71–100 kW engine power range

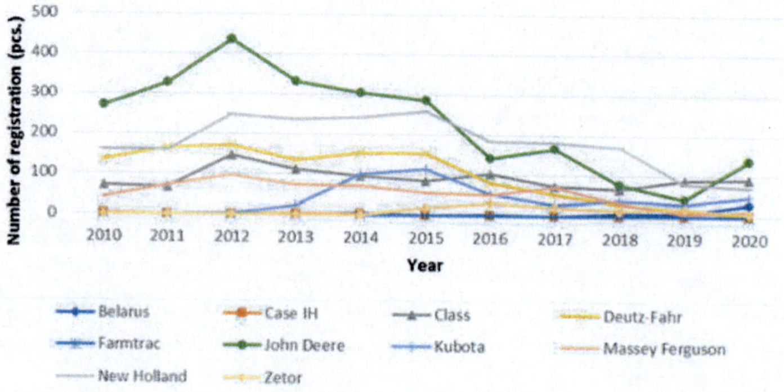

Fig. 8 The number of tractors by selected manufacturers, registered in Poland between 2010 and 2020, in the 101–120 kW engine power range

Among tractors in the 121–140 kW power range, New Holland proved to be the unrivalled manufacturer (48.70%), with 2,208 vehicles sold in whole registration period under analysis (Fig. 9). It recorded its best sales in 2020, amounting to 281 tractors. In this group, five manufacturers had a dominant position on the Polish market, including John Deere (24.02%), Massey Ferguson (8.40%), Deutz-Fahr (7.76%) and Class (7.06%).

In the last but one segment, i.e., 141–160 kW, which comprised the lowest number of registered tractors, only three tractor brands recorded meaningful figures (Fig. 10). The first of these was John Deere (40.21%), which recorded a total of 895 registrations, attaining the best result in 2020 with 178 vehicles sold. It was followed by New Holland (35.44%) with 789 vehicles sold, and by Class (11.14%) with 248 registrations.

The most powerful segment (over 161 kW) was dominated by John Deere (44.97%) with total sales of 1,789 vehicles, out of a total of 3978 vehicles in the

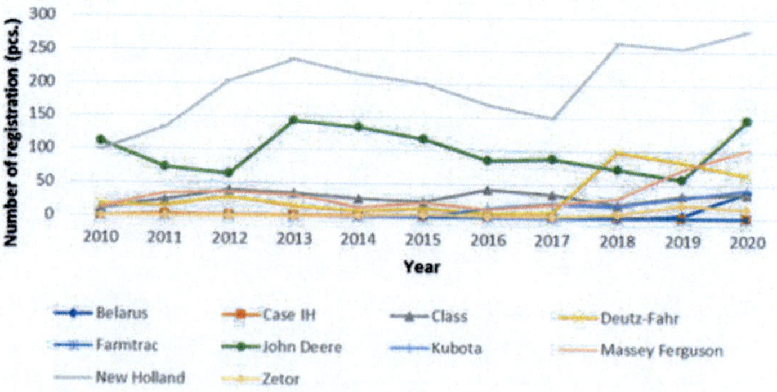

Fig. 9 The number of tractors by selected manufacturers, registered in Poland between 2010 and 2020, in the 121–140 kW engine power range

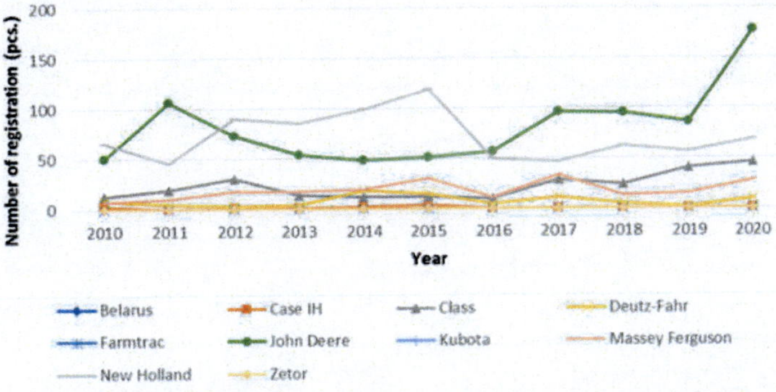

Fig. 10 The number of tractors by selected manufacturers, registered in Poland between 2010 and 2020, in the 141–160 kW engine power range

whole group. Tractors representing this engine power range showed almost no correlation (R = 0.03) to the overall pattern of registrations over the period under analysis. Despite their small share in total sales, the best-selling brands, along with the above-mentioned John Deere, once again included New Holland (25.84%) with 1,028 vehicles sold, and Class (15.21%) with 605 vehicles sold. Some manufacturers, such as Belarus, Farmtrac, Kubota and Zetor, recorded no sales of tractors representing this engine power segment due to the fact that they were not marketed in Poland or were not produced.

Of the eight engine power ranges of farm tractors identified, the following three segments had the highest shares: 31-50, 51–70 and 71–100 kW (Fig. 11). Among these, the US-based New Holland was rated first. As regards the first segment, it had a share of 13.99% in the entire Polish farm tractor market, advancing to the leader position. In the second segment, its share of 14.52% put it in the second position, and in the third segment it was also in the lead, with a share of 18.66%. With 108,251 registered tractors in these three segments, the manufacturer sold 17,095 vehicles, maintaining its share at the level of 15.79%. The second largest manufacturer, as regards the share in sales, was Zetor (12.03%) with 13,028 vehicles sold, and the third was John Deere (11.18%) with 12,105 vehicles sold (Fig. 12).

As revealed by the analyses, the farm tractors market between 2010 and 2020 was characterised by high volatility. The demand for farm tractors undoubtedly reflected changes in Polish agricultural holdings in the period under analysis. Thanks to EU funds, the demand for farm tractors was high, especially for those with engine power of 31 to 100 kW. Young farmers, when purchasing farm tractors as part of the subsidies for young farmers, not only took into account the applicable selection rules, but also the machinery park they own. Research conducted by Buchowski found that incomplete saturation of agricultural land would contribute to purchases of tractors with greater power, which farmers are normally reluctant to do, mainly because they are afraid they would have to replace most machines collaborating with the tractors. In this respect, tractor purchases are rational [15]. Nonetheless, the demand for high-powered farm tractors has been on the rise. As the western and northern provinces of

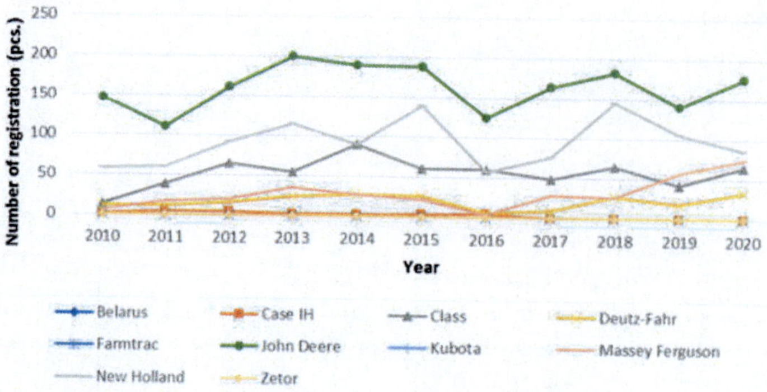

Fig. 11 The number of tractors by selected manufacturers, registered in Poland between 2010 and 2020, in the >161 kW engine power range

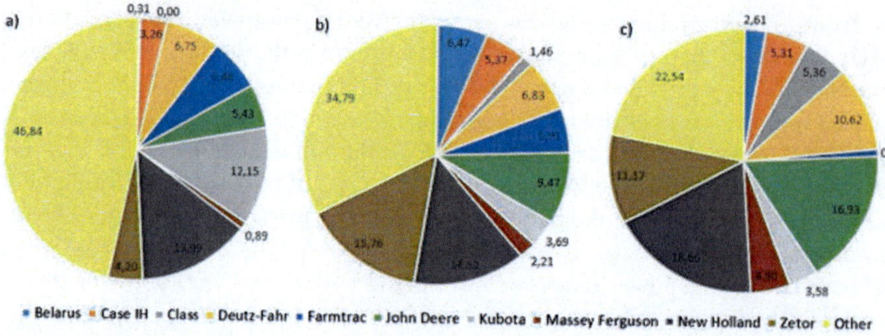

Fig. 12 The percentage shares of registrations of selected manufacturers in Poland between 2010 and 2020 in the following engine power segments: **a** 31–50, **b** 51–70, and **c** 71–100 kW

Poland are characterised by agricultural holdings occupying larger areas, they require more powerful tractors. In addition, holdings situated in these regions have both greater income and more opportunities to obtain EU funds. Changes in production technology, attempts to increase efficiency, the improving economic situation, as well as the gradual specialisation and concentration of production in development-oriented holdings were other factors contributing to such tendencies.

4 Conclusions

Between 2010 and 2020, more than 144,000 farm tractors were registered in Poland. The highest sales were recorded in 2012 with 18,484 vehicles registered, while the lowest was in 2016, with 8429 vehicles registered – this was 54.4% less compared to the best-selling year. Tractor purchases were influenced by access to EU funds and support programmes, as well as by the unfavourable situation on the agricultural crops market. Of note is also the fact that in 2020, despite the difficult situation related to the outbreak of the COVID-19 pandemic, tractor registrations continued to display an upward trend and were about 20% higher than in the previous year, with 14,379 vehicles registered.

Between 2010 and 2020, most tractors registered in Poland were those representing the 51–70 kW engine power segment, accounting for 35.17% of all tractors sold. The second most popular power segment was 71–100 (24.44%), followed by 31–50 (14%), <30 (7.68%), 101–120 (7.40), 121–140 (4.85%), and > 161 kW (3.69%). The smallest group (2.76%) included tractors representing the 141-160 kW engine power segment.

Among the three most popular engine power segments, the best sales were recorded by the following manufacturers: New Holland, Zetor and John Deere. In terms of the highest number of registrations in all segments, New Holland proved to be the leader (23,780 vehicles), followed by John Deere (19,453 vehicles), Zetor (16,398 vehicles) and Deutz-Fahr (10,508 vehicles). The combined share of these four manufacturers between 2010 and 2020 was 48.6%.

References

1. Lorencowicz E (2017) Zmiany cen ciągników rolniczych w Polsce w latach 2000–2014 (Changes in prices of agricultural tractors in Poland between 2000 and 2014). Roczniki Naukowe. Stowarzyszenie Ekonomistów Rolnictwa i Agrobiznesu 18(4):163–166
2. Hornowski A (2015) Evaluation of economic effectiveness of financial investments from the measure modernization of agricultural holdings under RDP 2007–2013 in Polish agricultural holdings. Pol Assoc Agric Agribus Econ Ann 17(3):131–137
3. Final Report: Evaluation of the results of the implementation and impact of the Rural Development Programme 2014–2020: Competitiveness, innovation, knowledge transfer. Ministry of Agriculture and Rural Development (MAaRD), Warsaw (2019)
4. Centrum Prasowe PAP (Polskiej Agencji Prasowej). https://www.kalendarzrolnikow.pl/6244/przy-zakupie-traktora-liczy-sie-moc. Accessed 11 Jan 2022
5. Zalewski A (2017) Uwarunkowania regionalnego zróżnicowania wyposażenia gospodarstw rolnych w ciągniki rolnicze w Polsce (Determinants of regional diversification of farm equipment with agricultural tractors in Poland). RN SERiA 19(6):219–224
6. Lorencowicz E, Uziak J (2020) Regional structure of tractor market in Poland. Agric Eng 24(3):51–62
7. Wicki L, Pietrzykowski R (2018) Zróżnicowanie przestrzenne wykorzystania środków na modernizację gospodarstw rolnych z Programu Rozwoju Obszarów Wiejskich (Spatial differentiation of the use of funds for modernization of agricultural holdings from the Rural Development Programme). Zeszyty Naukowe SGGW – Ekonomika i Organizacja Gospodarki Żywnościowej 124:93–108
8. Kluba M, Rudnicki R, Wiśniewski Ł (2016) Intensity of agrilutural production organization and the mechanization of agriculture level in Poland in the light of the national agricultural census (PSR) 2010. Studia KPZK, pp 214–230
9. Pawlak J (2015) Podaż krajowa ciągników a ich rejestracja (Domestic supply of the tractors and their registration). Problemy Inżynierii Rolniczej 2(88):29–36

10. Skudlarski J (2017) Registration of farm tractors in Poland in 2015. Probl Agric Econ 2(351):164–177
11. Skudlarski J (2017) Rynek nowych i używanych ciągników rolniczych w Polsce w latach 2012–2016 (New and used agricultural tractors market in Poland between 2012 and 2016). Roczniki Naukowe Stowarzyszenia Ekonomistów Rolnictwa i Agrobiznesu 19(1):165–169
12. Zając S, Izdebski W, Skudlarski J (2012) The market for farm tractors in Poland in 2001–2012. Pol Assoc Agric Agribus Econ Ann 16(3):318–323
13. Przywara A, Koszel M, Krawczuk A, Anifantis AS (2022) The analysis of the new farm tractor market in Poland in 2010–2020 in context of income generated by farmer's households. Agric Eng 26(1):65–79
14. Kaczmarek M (2016) Rynek maszyn rolniczych (The market of agricultural machinery). http://analizybranzowe.pl/?p=121. Accessed 28 Feb 2021
15. Buchowski P (2020) Evaluation of machinery investment decisions undertaken by young farmers in Poland's Suwalki County (in Polish). Probl World Agric 20(3):16–25

Tillage Quality and Draught Force of a Novel Plough Concept with Cambered Disc

Gerhard Moitzi, Christian Rechberger, Franz-Ferdinand Huber, Matthias Trimmel, and Helmut Wagentristl

Abstract A prototype of a novel plough system ("hybrid plough"), invented by Huber Solution GmbH was tested on a silty loam soil according tillage quality (in-soil mixing of residual plants, soil loosening, soil penetration resistance). Reference system was a mouldboard plough. The draught force was measured for light and heavy soil conditions additionally.

Results show, that the working principle of the "hybrid plough" has a better in-soil mixing of the residual plants, where residual plants were also detected on the soil surface. Soil loosening efficiency was higher, indicated by a significant lower mean-weight diameter of soil aggregates than mouldboard ploughing. There were no significant differences of the soil penetration resistance between "hybrid plough" and mouldboard plough. At heavy soil conditions, the "hybrid plough" had more than 20% significant lower draught force requirement compared to mouldboard plough. This potential energy saving effect is explained by a lower pull-, push and shear forces for soil loosening.

Keywords Hybrid disc plough · Soil loosening · Soil aggregate size distribution · Soil penetration resistance · Draught force

G. Moitzi (✉) · H. Wagentristl
Department of Crop Sciences, University of Natural Resources and Life Sciences (BOKU), Experimental Farm Groß-Enzersdorf, Schloßhofer Straße 31, 2301 Groß-Enzersdorf, Austria
e-mail: gerhard.moitzi@boku.ac.at

C. Rechberger
HBLFA Francisco Josephinum – BLT Wieselburg, Rottenhauser Strasse 1, 3250 Wieselburg, Austria

F.-F. Huber
Huber Soil Solution GmbH, Franz Nabl Gasse 15, 8435 Wagna, Austria

M. Trimmel
Josephinum Research, Rottenhauser Strasse 1, 3250 Wieselburg, Austria

S. Pascuzzi and F. Santoro (eds.), *Farm Machinery and Processes Management in Sustainable Agriculture*, Lecture Notes in Civil Engineering 289,
https://doi.org/10.1007/978-3-031-13090-8_2

1 Introduction

The machinery stock, fuel consumption and work time are crucial economic factors for the profit potential in the arable farming sector. Mouldboard ploughing is still a widely used tillage method in Europe under difficult conditions (heavy soils in humid regions). It is based on high intensity of soil engagement and inversion of the soil. Mouldboard generates large aggregates and a rough soil surface. The soil is over-loosened and the residual plants are buried to a deeper soil horizon. A follow-up by secondary tillage in order to prepare a seedbed, where the soil is stratified re-compacted is necessary. The main advantage of the system is to produce controlled soil conditions, a thorough loosening and a field clear of plant residues. Plant residues, weed seeds or shattered seeds are buried totally. Furthermore, tractors running in the furrow and there is the risk of ploughpan compaction [1].

An alternative to the mouldboard plough is the disc plough, which is widely used in arid climates as it is less susceptible to the obstacles in the soil and to rough conditions [2, 3]. According CIGR [1], the quality of inverting and crumbling the soil is lower than with the moldboard plough, and great amounts of organic residue can be a problem. In comparison to mouldboard plough, the bottom of the furrow ground generated by disc plough is not level. It has a "moutainous" shape.

A replacement of the mouldboards plough through a cultivator reduced the work time and fuel consumption for soil tillage as well as the energy consumption per moved soil matter about more than 50% [5]. The geometry of the tools in the tillage implement has a great influence on specific draught force [3–5].

Previous studies have shown, that disc ploughs can be advantageous on heavier soils and show up potential for lower specific draught force requirements [7, 8]. In contrast to mouldboard ploughs, disc ploughs have a better incorporation of crops residues through mixing effects [9]. In addition, rotating cutting tools usually show less wear compared to rigid ploughshares. Disc harrows, on the other hand, are not suitable for greater working depths, especially on heavier soils.

From the point of tillage quality and energy saving, a novel ploughing concept was developed (Fig. 1). The aim was to combine the functional advantages of the mouldboard plough (e.g. equal loosening to an even furrow level) with disc plough (e.g. lower draught force requirement, better mixing of plant residues into the soil). In serveral functional field experiments in Austria, a prototype of a hybrid plough (Figs. 1 and 4) was tested according tillage quality (soil coverage with residual plants, soil aggregate distribution and soil penetration resistance) and draught force.

Fig. 1 Hybrid plough (cambered disc and disc for horizontal soil cutting)

2 Materials and Methods

2.1 Site Description

The field experiments for tillage quality assessment were carried out at the Experimental Station of the University of Natural Resources and Life Sciences, Vienna (BOKU) in Groß-Enzersdorf (48°12′1.79″ N, 16°33′33.75″ E) in calcaric chernozem with silty loam (36% sand, 44% silt, 20% clay). The soil organic matter is 16.3 g kg^{-1}.

The moisture content at the experiment was measured with soil moisture meter (Delta – T Devices HH2 moisture meter with WET Sensor Wet-2, Cambridge, England) and was for the tillage quality measurement between 15% and 16%, which were well suitable for tillage.

The draught force measurements were conducted at the research site of the Higher Federal Education and Research Institute of Agriculture, Agricultural Engineering and Food Technology (HBLFA) Francisco-Josephinum in Wieselburg (48°07′28.6″ N, 15°08′48, 7 E). The soil type of the field plot was an alluvial loess/loamy silt. The gravimetric water content was 19.4% with a standard deviation of 0.8% and the bulk density 1,384.7 kg m^{-3} with a standard deviation of 30.7 kg m^{-3} (n = 6).

2.2 Design of the Hybrid Plough

The hybrid plough (Fig. 1) is a result of a long-term development process driven by the Huber soil solution GmbH, Austria. The aim was to develop a plough with rotational tools to increase the loosening efficiency with low draught force and energy requirement. A four reversible furrow mouldboard plough was modified to a hybrid

Fig. 2 Cambered disc with knife blade for horizontal soil cutting

plough, with cambered disc and tools for horizontal soil cutting. For horizontal soil cutting, two kind of tools were equipped: disc (see Fig. 1) and knife blade (see Fig. 2).

2.3 Experimental Design

Tillage quality test was carried out on an arable field after harvest of soybean at the experimental site Groß-Enzersdorf and nearby site Raasdorf (same soil condition). The hybrid plough was compared with a reversible four furrow mouldboard plough. The working width and working speed were set to 20–25 cm and 6–7 km h^{-1}, respecitively. The plots were 400 m long. The tillage quality was assessed with three measured parameters after tillage: soil coverage with residual plants, soil aggregate distribution and soil penetration resistance.

For the draught force measurement at Wieselburg, the plot was divided into sections according soil texture (Fig. 3). Each section had a length of approximately 70 m in direction of tillage.

2.4 Tillage Quality Assessment

The soil coverage with residual plants were detected with optical image proccessing approach, developed by Josephinum Research, Austria (https://soilcover.josephinum.at/). Eight digital images were taken with a compact camera from each treatment and analysed with SoilCover online.

The soil samples for the soil aggregate distribution measurement were taken carefully with a spade (25 × 25 cm) from the loosend soil (0–10/15 cm) and filled in a paper bag. Before sieving, the soil samples were air-dried in a glasshouse. Seven samples were taken of each treatment. The dry sieving analysis (mesh size: 40, 20,

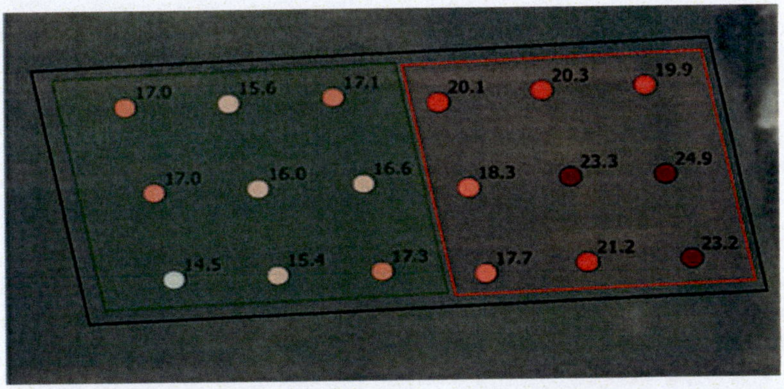

Fig. 3 The field trial plot at Wieselburg with the clay contents in percent. The plot was divided into two parts: "Light soils" (green rectangle on the left) and "Heavy soils" (red rectangle on the right)

Fig. 4 Experimental setup for comparative draught force measurements

10, 5, 2.5, 1.25 und <1.25 mm) was carried out with a vibratory sieve shaker (Retsch AS 200, Haan, Germany) with the adjusted amplitude of 1.3 mm and acceleration of gravity of 18.75 g). One sieving analysis took 2 × 15 s. The sieving fractions were weighted with a precision scale (0.1 g).

The mean-weight diameter (MWD) was calculated with the seven sieving fractions according Eq. 1.

$$MWD = \frac{\sum(n_i \times d_i)}{\sum n_i} \tag{1}$$

MWD: mean-weight diameter (mm).

n_i: weight of the sieving fraction in the mesh size i (g).

d_i: mean diameter of the mesh size i (mm).

Soil penetration resistance was measured with a cone penetrometer (Eijkelkamp, The Netherlands) (60° cone and 1 cm^2) to a depth of 40 cm. For each treatment, seven cone penetrations were recorded and presented in mean soil penetration resistance (MPa) for the soil depth 0–5, 6–20, 21–25 and 26–40 cm.

2.5 *Draught Force Measurement*

For the comparative draught force measurements, a three furrowed reversible mouldboard plough was equipped with disc type modules on one side (Fig. 2). The forces at the three-point-linkage were determined with the in-house developed triaxial force measuring device in combination with an IMC CRONOS-SL-4 data acquisition system.

The tillage depth was set to 22 cm and could kept constant within small deviations for both ploughs. Due to limitations in the adjustment range of the hybrid plough, the total working width could not be set to the same value as for the mouldboard plough. Hence the effective total working width was 1.35 m for the mouldboard plough and 0.96 m for the hybrid plough. Target forward speed was chosen to be 2.2 m s^{-1} (8 km h^{-1}). From the raw data of the force measurement, a mean value was calculated for each field length of 10 m. Three repetitions per treatment resulted in a sample of 21 values per treatment for statistical evaluation. For consideration of different working widths of both implements, the drought force was related to the tilled cross-sectional area (specific drought force in kN m^{-2}).

2.6 *Statistical Analysis*

All analyses were conducted using IBM® SPSS® Statistics 21. The requirements for analysis of variance were tested with the Levene test and assessment of normal distribution of residues. One-factorial analysis of variance (ANOVA) was carried out for soil coverage with residual plants, soil aggregate distribution and soil penetration resistance testing treatments ("hybrid plough" with knife blade for horizontal soil cutting, "hybrid plough" with disc for horizontal soil cutting, mouldboard plough-reference). Draught force affected by soil type (heavy soil, light soil) and plough (hybrid plough, mouldboard plough) was tested with two-factorial ANOVA. Multiple comparison tests to separate means were carried out with the Student–Newman–Keuls procedure ($p < 0.05$).

3 Results and Discussion

3.1 Tillage Quality

As expected, no residual plants of soybeans were detected after mouldboard ploughing (Figs. 5 and 6). After ploughing with a hybrid plough, residual soybean plants were found on the soil surface.

At the experimental site Groß-Enzersdorf (Fig. 5), the mean coverage of residual plants was 5.4% (hybrid plough with knife blade) and 6.4% (hybrid plough with disc) and at the experimental site Raasdorf (Fig. 6) 6.6 and 6.7%, respectively.

During ploughing, different forces are applied to the soil, such as shearing, compression, deformation, impact and acceleration [1]. The results are a loosened soil with different soil aggregate fractions (Tables 1 and 2), as an effect of cutting, crumbling, inversion, mixing, leveling and shaping in a particular section profile of the soil.

At the two experimental sites, in the soil samples from the loosened top soil (0–10/15 cm) plant residues were found (Tables 1 and 2) in the hybrid plough treatment, indicated that the plant residues were better mixed with the soil. This better incorporation effect promotes biological degradation and humification of the plant residues. Totally buried plant residues, which is carried out by mouldboard plough, generates dry stabilized plant residues layer in arid and semi-arid regions. In humid regions, there is a risk of anaerob decomposition with greenhouse gas emissions (N_2O and CH_4).

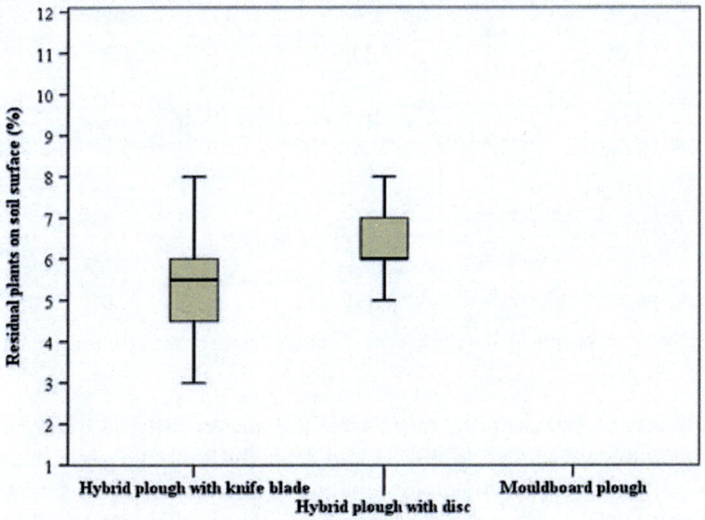

Fig. 5 Soil coverage with residual plants in different tillage treatments. Boxplots with median, Experimental site: Groß-Enzersdorf

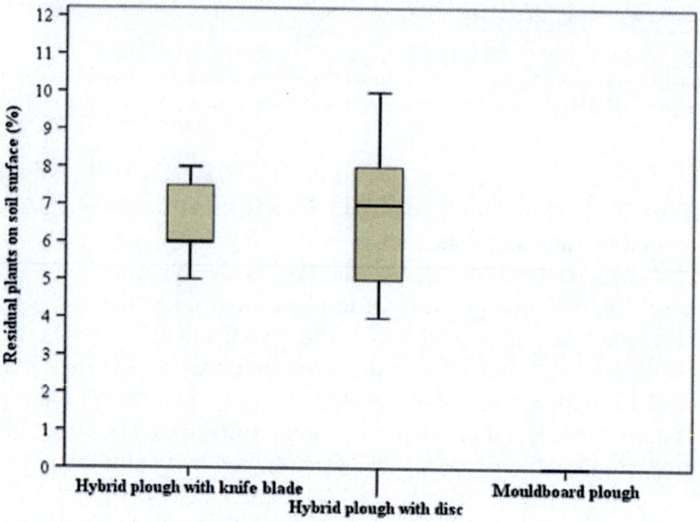

Fig. 6 Soil coverage with residual plants in different tillage treatments. Boxplots with median. Experimental site: Raasdorf

Table 1 Soil aggregate fractions (%) and mean-weight diameter as well as plant residues (%), experimental site: Groß-Enzersdorf

	Hybrid plough		Mouldboard plough
	Knife blade for horizontal soil cutting	Disc for horizontal soil cutting	
>40 mm	6,7^{a*}	18,0^{a}	39,1^{b}
20–40 mm	3,0^{a}	13,9^{c}	6,3^{b}
10–20 mm	11,8	14,7	11,0
5–10 mm	17,0^{b}	12,8^{ab}	10,6^{a}
2.5–5 mm	17,0^{b}	11,4^{a}	9,2^{a}
1.25–2.5 mm	14,8^{b}	9,1^{a}	7,9^{a}
<1.25 mm	29,6^{b}	19,8^{a}	15,9^{a}
MWD (mm)	**8,9^{a}**	**18,8^{b}**	**28,3^{c}**
Plant residues	**0,1^{b}**	**0,2^{b}**	**0,0^{a}**

* Significant differences for the tillage implement effect are shown with small superscript letters

Soil aggregate fractions and the mean-weight diameter differed between hybrid plough and mouldboard plough (Tables 1 and 2). Hybrid plough generates smaller soil aggregates than mouldboard plough, indicated that acceleration forces on the cambered disc induced additional soil breaking. Similar results were also determined in a previous work by visual detection of the soil roughness using a stereo camera [10, 11].

Table 2 Soil aggregate fractions (%) and mean-weight diameter as well as plant residues (%), experimental site: Raasdorf

	Hybrid plough		Mouldboard plough
	Knife blade for horizontal soil cutting	Disc for horizontal soil cutting	
>40 mm	21,0[a*]	15.3[a]	44,4[b]
20–40 mm	8,4	7,9	8,6
10–20 mm	11,3	12,4	9,7
5–10 mm	12,4[a]	17,1[b]	9,8[a]
2.5–5 mm	12,2	13,8	8,8
1.25–2.5 mm	10,7[b]	12,0[b]	6,1[a]
<1.25 mm	23,5[b]	21,0[b]	12,8[a]
MWD (mm)	**18,5[a]**	**15,5[a]**	**31,8[b]**
Plant residues	**0,5[b]**	**0,6[b]**	**0,0[a]**

* Significant differences for the tillage implement effect are shown with small superscript letters

The statistically significant effect between the knife blade and disc for horizontal soil cutting of the hybrid plough at the experimental site Groß-Enzersdorf (Table 1) could be explained by the lower working depth whereby the loosening effect of the soil on the cambered disc is more intensive.

Soil penetration resistance (Table 3) in the topsoil (0–25 cm) was not affected by the three ploughing treatments. Hybrid plough with knife blade tended to higher soil penetration in the subsoil (26–40 cm) and resulted in significant higher mean soil penetration in whole soil depth of 0–40 cm. The reasons could not be clearly explained.

Table 3 Mean soil penetration resistance (MPa), experimental site: Raasdorf

	Hybrid plough		Mouldboard plough
	Knife blade for horizontal soil cutting	Disc for horizontal soil cutting	
0–5 cm	0,3	0,3	0,2
6–20 cm	0,4	0,5	0,4
21–25 cm	1,4	0,9	1,0
26–40 cm	4,6	3,6	3,6
0–40 cm	2.2[b*]	1,3[a]	1,3[a]

* Significant differences for the tillage implement effect are shown with small superscript letters

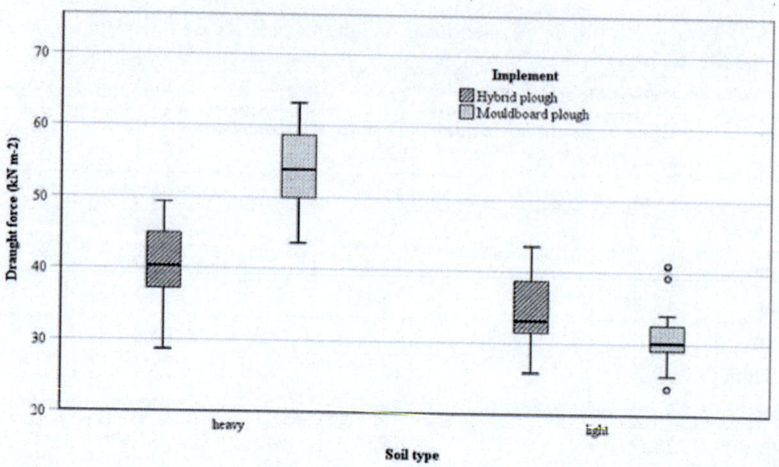

Fig. 7 Boxplots of the specific drought force for both plough types at heavy and light soil conditions

3.2 *Draught Force*

The mean measured specific draught force was on the heavy soil 48.6 kN m^{-2} and on light soil 32.5 kN m^{-2}, respectively, which was significant different ($p < 0.001$). On heavy soils, the hybrid plough showed a significant lower specific draught force than the mouldboard plough (40.7 vs. 53.9 kN m^{-2}, $p < 0.001$). On light soils, the hybrid plough and the mouldboard plough did not differ significantly (34.4 kN m^{-2} vs. 31.2 kN m^{-2}, $p = 0.059$) (Fig. 7).

These results confirm the findings of previous studies, where disc ploughs have potential advantages in terms of draft requirements, especially on heavier soils.

References

1. CIGR Handbook of Agricultural Engineering Volume III Plant Production Engineering (1999): Edited by CIGR–The Internation Commission of Agricultural Engineering. Published by the American Society of Agricultural Engineers. https://cigr.org/node/640. Accessed 1 Apr 2022
2. Ojeniyi SO (1986) Effect of zero-tillage and disc ploughing on soil water, soil temperature and growth and yield of maize. Soil Tillage Res 7(1–2):173–182
3. Olatunji O (2009) Effect of weight and draught on the performance of disc plough on sandy–loam soil. Res J Appl Sci Eng Technol 1:22–26. ISSN: 2040-7467
4. Khaliullin D, Belinsky A, Valiev A, Lukmanov R, Bourges G (2020) Optimization of plow adjustment. In: BIO Web Conference, vol 27, p 00103.https://doi.org/10.1051/bioconf/20202700103
5. Moitzi G et al (2013) Effects of tillage systems and mechanization on work time, fuel and energy consumption for cereal cropping in Austria. Agric Eng Int CIGR J 15(4):94–101
6. Arvidsson J, Keller T, Gustafsson K (2004) Specific draught for mouldboard plough, chisel plough and disc harrow at different water contents. Soil Tillage Res 79:221–231
7. Söhne W (1951) Die Scheibenpflüge. Grundlagen der Landtechnik Heft 1/1951

8. Groth H-J (1964) Auch der Scheibenpflug leistet gute Pflugarbeit. Deutsche Agrartechnik Bd. 14, Nr. 7
9. Groth H-J (1971) Über den Mischeffekt des Pfluges. Deutsche Agrartechnik Bd. 21, Nr. 12
10. Rechberger C, Trimmel M, Huber F (2019) Hybrid-Disc-Plough–An energy saving plough concept for heavier soils in: LAND.TECHNIK AgEng, pp 503–508. https://doi.org/10.51202/9783181023617-503
11. Riegler-Nurscher P et al (2020) Machine vision for soil roughness measurement and control of tillage machines during seedbed preparation. Soil Tillage Res 196:104351. https://doi.org/10.1016/j.still.2019.104351

Theoretical Study of the Trajectory of Movement of a Ploughing Aggregate with a Reversible Plough on the Headlands

Volodymyr Bulgakov, Simone Pascuzzi, Valerii Adamchuk, Volodymyr Nadykto, Semjons Ivanovs, Aivars Aboltins, Hryhoriy Kaletnik, Francesco Santoro, Francesco Vicino, and Janusz Nowak

Abstract Ploughing with reversible ploughs is increasingly used. Their main advantage is that the aggregates based on them allow ploughing without formation of dump ridges and breakup furrows (i.e. ensure formation of a smooth field surface). The trajectory of the movement of the ploughing aggregate with a reversible plough on the headlands has its own specifics and length due to the fact that the aggregate moves across the field in a shuttle way (i.e., processing is constantly carried out at the butt in relation to the previous pass). This trajectory depends on the length of the cultivated field and the working width of the plough. In comparison with the conventional aggregates the total length of the path of movement of the ploughing aggregates with reversible ploughs on the headland under certain conditions may be much larger. As estimated indicators in the work there are taken the total lengths of the trajectories of movement of the ploughing aggregates with reversible and conventional ploughs on the headland. On the basis of theoretical studies comparative calculated graphic

V. Bulgakov
National University of Life and Environmental Sciences of Ukraine, Kyiv, Ukraine

S. Pascuzzi (✉) · F. Santoro · F. Vicino
Department of Agricultural and Environmental Science, University of Bari Aldo Moro, Bari, Italy
e-mail: simone.pascuzzi@uniba.it

V. Adamchuk
Institute of Mechanics and Automation of Agricultural Production of the National Academy of Agrarian Sciences of Ukraine, Glevakha, Vasylkiv Disctrict, Kyiv Region, Ukraine

V. Nadykto
Dmytro Motornyi Tavria State Agrotechnological University, Melitopol, Ukraine

S. Ivanovs · A. Aboltins
Latvia University of Life Sciences and Technologies, Jelgava, Latvia

H. Kaletnik
Vinnytsia National Agrarian University, 3 Soniachna Str., Vinnytsia UA 21008, Ukraine

J. Nowak
Department of Operation of Machines and Management of Production Processes, University of Life Sciences in Lublin, 13, Akademicka Str., 20-950 Lublin, Poland

S. Pascuzzi and F. Santoro (eds.), *Farm Machinery and Processes Management in Sustainable Agriculture*, Lecture Notes in Civil Engineering 289,
https://doi.org/10.1007/978-3-031-13090-8_3

dependences of the length of the trajectory of the movement of the ploughing aggregates were constructed for various lengths of the track (length of the cultivated field) and various plough widths. It will be an obvious advantage of the ploughing aggregate as part of a wheeled tractor with a minimum turning radius of not more than 5.3 m with a field length of 1400 m.

Keywords Turn · Headland · Ploughing width · Length of the field · Shuttle-like movement

1 Introduction

Reversible ploughs of various designs are widely used in the world [1–4].

The developers of these machines, together with the researchers, pay great attention to the improvement of the plough bodies and other elements [5–8]. In agrotechnical terms, they are interested to improve the running stability and performance quality of reversible ploughs both in a horizontal and a longitudinal-vertical plane [9, 10].

It should be noted that during the main operations the mode and trajectory of movement of the ploughing aggregates with reversible and conventional ploughs have no fundamental difference. This difference manifests itself when running of these ploughing aggregates on the headland.

The main advantage of the reversible ploughs is their ability to perform ploughing without tailing ridges and back furrows (Fig. 1) [11]. In this case the machine-and-tractor aggregate moves in a shuttle-like manner. In the process of moving on the headland, it makes loopy and, as a rule, pear-shaped turns. But the execution of such a maneuver is more complicated than, for instance, a loopless turn. And this complexity is the higher, the smaller is the width of the ploughing aggregate [12, 13].

When ploughing with a conventional plough, the ploughing aggregate with a working width B_p also performs loopy pear-shaped turns. But it does this only at the initial stage of work. And this stage continues until, after n passes of the aggregate, the width of the ploughed area (S_p) becomes equal to, or greater than twice, the minimum turning radius (R_o) of the tractor [14]: $S_p = n \cdot B_p \geq 2\,R_o$. After that such a ploughing aggregate makes ordinary loopless turns with a rectilinear section.

As a result, it may turn out that, under certain conditions of work on the headlands of the field, the total length of the path of movement of the ploughing aggregates with reversible and conventional ploughs will be different. Because of this, the time spent by them for non-useful work will also be different. An aggregate with a large value of such time will have a lower performance and worse indicators of the specific fuel consumption. A priori assuming an increase in the non-productive time, spent by the ploughing aggregate with a reversible plough, the researchers propose certain solutions to this problem. One of them is the equipment of the ploughing aggregate with a reversible plough of special navigation systems [15]. Another solution is to

Fig. 1 Ploughing with a reversible plough

complicate the design of the tractor itself [16, 17]. It is a special device in the form of an additional support wheel which, after the unit has entered the headland, raises the front steering wheels of the tractor. As a result, turning of the aggregate takes place relative to the support point of the left or right rear wheel of the tractor that is braked during the turn. According to research data such a change in the design of the tractor reduces the width of the turning lane of the aggregate by 50%. In addition, the reduction in turning time reaches 35%. It should be emphasized that in practice the arable plots quite often have a shape that differs to varying degrees from the rectangular one. The choice of the optimal trajectory for turning the ploughing aggregate with a reversible plough, when working under such conditions, is a subject of research [18–22].

However, the authors of this article did not find studies that would allow giving a comparative assessment of the modes of movement of the ploughing aggregate on the headland, based on a reversible and a simple plough. Moreover, there are practically no such data for individual ploughing aggregate, gained on the basis either of a reverse plough or a conventional plough. As a result, this state of affairs does not allow one to decide more or less unequivocally on the expediency of using a reversible plough instead of a conventional one. In view of the foregoing, the present article is devoted to solving this problem)".

2 Materials and Methods

As indicators for estimation here were selected the total lengths of the trajectories (paths) of movement the ploughing aggregates with reversible ($\sum L_{xp}$) and conventional ($\sum L_{xs}$) ploughs on the headland.

For their calculation, we used our previously obtained theoretical assumptions (Bulgakov et al., 2016):

$$\sum L_{xp} = (7R_o + 2E) \cdot \left[Integer\left(\frac{\sqrt{2 \cdot (8R_o^2 + B_p L_p)}}{B_p} \right) - 1 \right] \quad (1)$$

$$\sum L_{xs} = (7R_o + 2E) \cdot \left[Integer\left(\frac{2R_o}{B_p} \right) - 1 \right] + (1.7R_o + 2E) \cdot \left[Integer\left(\frac{\sqrt{2 \cdot (8R_o^2 + B_p L_p)}}{B_p} \right) - Integer\left(\frac{2R_o}{B_p} \right) \right] + B_p \sum_{i=0}^{n_{bp}} i \quad (2)$$

where E – the path of movement of the ploughing aggregate on the headland before it begins to turn, m; L_p – the length of the arable area of the field, m; n_{bp} – the number of the loopless turns, performed by the ploughing aggregate with a conventional plough. Their number is determined from the following equation:

$$n_{bp} = Integer\left(\frac{\sqrt{2 \cdot (8R_o^2 + B_p L_p)}}{B_p} \right) - Integer\left(\frac{2R_o}{B_p} \right) \quad (3)$$

The designations of the remaining parameters, included in Eqs. (1)–(3), are indicated above in the text of the article.

In the research there were evaluated ploughing aggregates the ploughs of which are equipped with 0.35 m wide bodies. The smallest working width of the compared ploughing aggregates $B_p = 1.75$ m, and the largest - $B_p = 3.15$ m. These values correspond to the plough designs with 5 and 9 bodies, respectively.

A tractor was selected as the base one the minimum turning radius of which was changed within a 4–7 m range. The following values of parameter E corresponded to the ploughing aggregates composed on the basis of such a tractor were: 8.5; 9.5; 10.5; 11.5; 12.5 m.

The value of the length of the arable plot in the theoretical calculations was changed within a range $L_p = 600$–1400 m.

3 Results and Discussion

Dependences (1) and (2) were obtained by comparing the operation of the ploughing aggregates with reversible and conventional ploughs on an arable plot of optimal width (C_{opt}, m). The value of this indicator is determined from the following relationship [23, 24]:

$$C_{opt} = \sqrt{2 \cdot \left(8R_o^2 + B_p L_p\right)} \tag{4}$$

The results of calculations using dependencies (1) and (2) show that, when the operating width B_p of the compared ploughs changes from 1.75 to 3.15 m, the values of the optimal width of the arable plots C_{opt} change from 50 to about 95 m (Fig. 2).

The smallest range of variation of this parameter (from 50 to 63 m) occurs at the field length $L_p = 600$ m (the red line 1, Fig. 2) but the largest (from 72 to 93 m) at $L_p = 1400$ m (red line 3). This fact indicates that the intensity of growth of the optimal width of the arable land is directly dependent on the increase in the length of the field. This fact indicates that the intensity of the growth of the optimal width of the arable land is directly dependent on the increase in the length of the field.

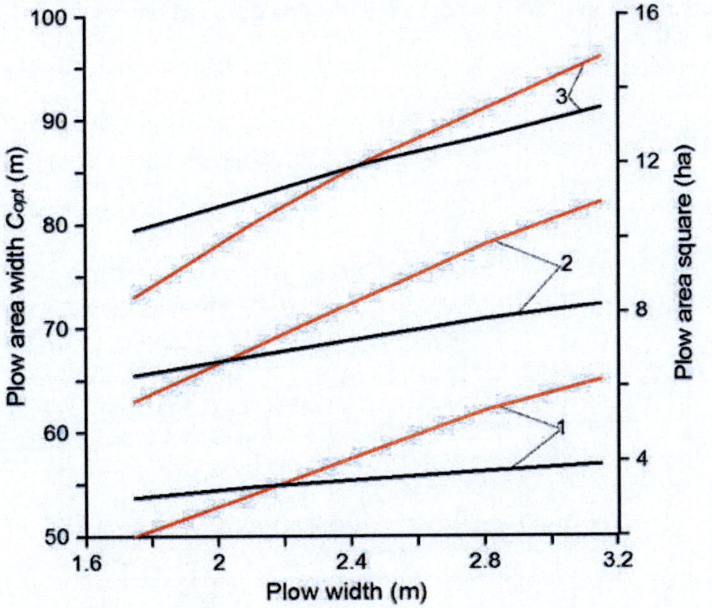

Fig. 2 Dependence of the optimal width of the arable plot (—) and its area (—) upon the operating width of the plough for various lengths of the field: 1 – $L_p = 600$ m; 2 – $L_p = 1000$ m; 3 – $L_p = 1400$ m

A quantitatively different but qualitatively similar nature of the change corresponds to the dynamics of the growth of the arable land area as parameters B_p and L_p increase (the black lines, Fig. 2).

When processing plots with such optimal values of the width and area, it turned out that expedience of using a ploughing aggregate with a reversible plough depends significantly on the length of the field. With the value of this parameter at the level of 600 m, the path travelled by the ploughing aggregate with a reversible plow on the headland for all values of the machine-tractor unit operating width is greater (curves 1, Fig. 3).The equality of the values of the indicators ΣL_{xp} (the reversible plough) and ΣL_{xs} (the ordinary plough) takes place at point *A*. This corresponds to the maximum value of parameter $B_p = 3.15$ m. At the minimum value of parameter $B_p = 1.75$ m indicator ΣL_{sp} is by 13% higher than the similar indicator ΣL_{xs}.

Analysis of these results leads to the conclusion that, when the length of the working area of the cultivated field is 600 m, the use of the ploughing aggregate with a reversible plow is not expedient.

When the field length is 1000 m and the plough working width is 1.75 m, the conventional ploughing aggregate has again an advantage, though a minimal one. The equality of the values of indicators ΣL_{xp} and ΣL_{xs} occurs at point *B* (Curves 2, Fig. 3), which corresponds to the plough width of 1.9 m. With a further increase in the value of this design parameter, the advantage passes to the ploughing aggregate with a reversible plough. At $B_p = 3.15$ m indicator ΣL_{xp} becomes less by 13% than the similar indicator ΣL_{xs}.

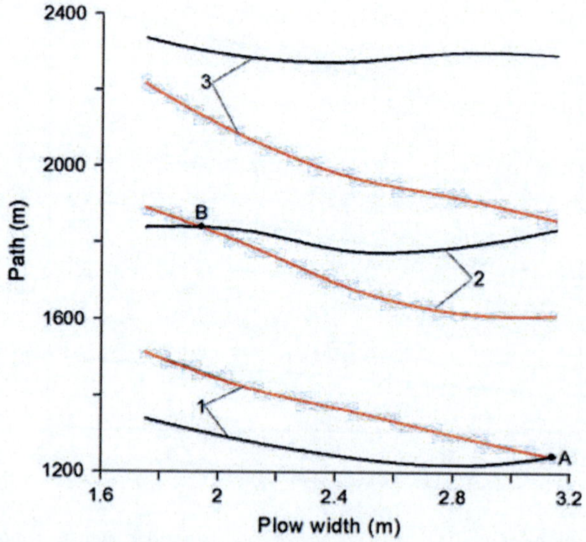

Fig. 3 The path of movement of the ploughing aggregate with a reversible (—) and a conventional (—) plough on the headland upon the width of the plough at different values of the length of the field: 1 – $L_p = 600$ m; 2 – $L_p = 1000$ m; 3 – $L_p = 1400$ m

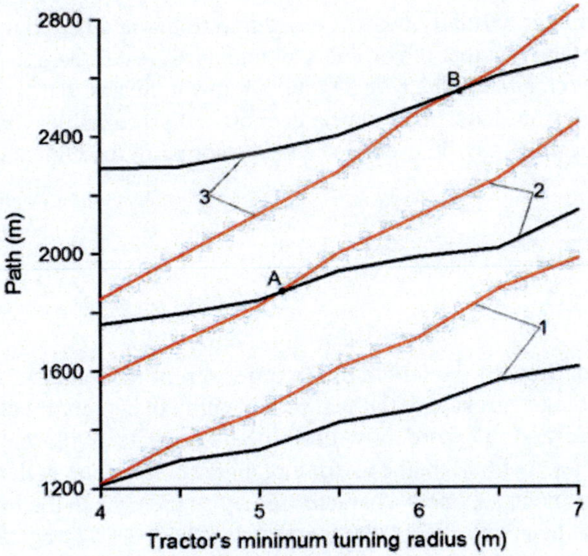

Fig. 4 The path of movement of the ploughing aggregate with a reversible (—) and a conventional (—) plough on the headland upon the minimum turning radius of the tractor at various values of field length: 1 – $L_p = 600$ m; 2 – $L_p = 1000$ m; 3 – $L_p = 1400$ m

The ploughing aggregate with a reversible plough gains full advantage when working on a field the length of which is 1400 m (Curves 3, Fig. 3). In this case the decrease in indicator ΣL_{xp} as compared to the similar indicator ΣL_{xs} increases from 5% (at $B_p = 1.75$ m) to 20% (at $B_p = 3.15$ m).

When the length of the field is 600 m, the length of the trajectory of the aggregate with a reversible plough will be longer than usual for any value of the minimum turning radius (Curves 1, Fig. 4). When the length of the field is 1000 m, the situation depends on the value of parameter R_o. At $R_o \leq 5.3$ m, the aggregate with a reversible plow has the best performance. In this case the value of indicator ΣL_{xp} is less than the value of indicator ΣL_{xs} (Curves 2, Fig. 4). At point B, which corresponds to the value of parameter R_o approximately 5.3 m, the difference between the values of these indicators is equal to zero. With further increase in the minimum turning radius of the tractor (and in this case, of the entire aggregate), a ploughing aggregate with a conventional plough becomes more efficient. A qualitatively similar result takes place when the length of the arable field is 1400 m. The only difference is that the preference of the ploughing aggregate with the reversible plough is maintained when the value of parameter R_o is increased to about 6.6 m (Curves 3, Fig. 4). At $L_p = 1400$ m the efficiency of the ploughing aggregate with a reversible plough has the greatest value when parameter R_o has a minimum value, i.e., 4 m. In this case, the value of the estimation indicator ΣL_{xp} is by 20% less than the value of indicator ΣL_{xs}. It should be added to the above analysis that the practitioners are aware of the technological methods that allow the operation of the ploughing aggregate with a conventional plough virtually without any dump ridges and breakup furrows. One should not forget the fact that the ordinary plough is almost three times cheaper and much less complicated than the reverse one. All this indicates that without justified analysis the use of the latter may not be economically profitable. That is why the above research results may be useful when substantiating the choice of the type of the plough.

4 Conclusions

From the point of view of the length of the trajectory of movement the expediency of using a ploughing aggregate with the reversible plough essentially depends on the length of the field and the working width of the aggregate. With the value of this parameter at the level of 600 m, the use of a ploughing aggregate with the reversible plough instead of the usual one is characterised by an increase in the total path of its movement on the headland. When the working width of the aggregate is increased from 1.75 to 3.15 m, this growth takes place at any (from those considered in the calculations) minimum turning radius of the tractor, and it reaches 13%.

When the length of the field is 1000 m, the advantage of the ploughing aggregate with a reversible plough over the conventional one takes place at a working width of more than 1.9 m, and with a minimum tractor turning radius of less than 5.3 m.

Full advantage of the ploughing aggregate as part of a wheeled tractor with a minimum turning radius of not more than 6.3 m and a reversible plough takes place at the length of the field 1400 m. In comparison with the usual one, the total path of movement of such an aggregate on the headland, depending on its working width, is reduced by 5–20%.

References

1. Balwani SN, Gulhane UD, Sahu AR (2016) Design, modification and analysis of sha of two furrow reversible plough. Int J Sci Technol Eng 3:108–116
2. Chandra Mouli K, Arunkumar S, Satwik B, Ram SB, Rushi Tej J, Chaitanya AS (2018) Design of reversible plough attachment. Mater Today Proc 5:23702–23709. https://doi.org/10.1016/J.MATPR.2018.10.160
3. Yin Y, Guo S, Meng Z, Qin W, Li B, Luo C (2018) Method and system of plowing depth online sensing for reversible plough. IFAC-PapersOnLine 51:326–331. https://doi.org/10.1016/j.ifacol.2018.08.199
4. Bulgakov V, Pascuzzi S, Nadykto V, Ivanovs S, Adamchuk V (2021) Experimental study of the implement-and-tractor aggregate used for laying tracks of permanent traffic lanes inside controlled traffic farming systems. Soil Tillage Res 208:104895. https://doi.org/10.1016/j.still.2020.104895
5. Belousov SV, Trubilin EI (2013) The development of a reversible plough design for heavy soil treatment. Sci Kuban 1:31–37

6. Pascuzzi S, Anifantis AS, Santoro F (2020) The concept of a compact profile agricultural tractor suitable for use on specialised tree crops. Agriculture 10:123. https://doi.org/10.3390/agriculture10040123
7. Guerrieri AS, Anifantis AS, Santoro F, Pascuzzi S (2019) Study of a large square baler with innovative technological systems that optimize the baling effectiveness. Agriculture 9(5):86
8. Khalaf AAS, Salem NM, Khalaf AAS (2013) The dynamic properties of vertisols as affected by different types of plough under dry farming conditions. Int Res J Agric Sci Soil Sci 3:322–327. https://doi.org/10.14303/irjas.2013.082
9. Bulgakov V, Pascuzzi S, Adamchuk V, Kuvachov V, Nozdrovicky L (2019) Theoretical study of transverse offsets of wide span tractor working implements and their influence on damage to row crops. Agriculture 9:144
10. Trendafilov K (2010) Theoretical determination of the width of strip for turning when ploughing with trailed and semi-mounted reversible ploughs in a field with irregular shape. Agric Sci Technol 2:218–220
11. Pascuzzi S, Santoro F (2017) Analysis of the almond harvesting and hulling mechanization process. A case study. Agriculture (Switzerland) 7(12):100
12. Manetto G, Cerruto E, Pascuzzi S, Santoro F (2017) Improvements in citrus packing lines to reduce the mechanical damage to fruit. Chem Eng Trans 58:391–396
13. Bulgakov V et al (2016) Theoretical premises for enhancement of efficiency in the use of reversible ploughs. Mech Agric Conserv Resour 62:30–34
14. Baldoin C, Balsari P, Cerruto E, Pascuzzi S, Raffaelli M (2008) Improvement in Pesticide application on greenhouse crops: results of a survey about greenhouse structures in Italy. Acta Hortic 801:609–614. https://doi.org/10.17660/ActaHortic.2008.801.69
15. Heiß A, Paraforos DS, Griepentrog HW (2019) Determination of cultivated area, field boundary and overlapping for a plowing operation using ISO 11783 communication and D-GNSS position data. Agriculture 9:1–17. https://doi.org/10.3390/agriculture9020038
16. Zeleniy PV, Scherbakova OK (2016) Mathematical model and methodology for calculation of minimization on turning radius of tractor unit with replaceable supporting and maneuvering device. Sci Tech 15:216–224. https://doi.org/10.21122/2227-1031-2016-15-3-216-224
17. Bulgakov V, Pascuzzi S, Anifantis AS, Santoro F (2019) Oscillations analysis of front-mounted beet topper machine for biomass harvesting. Energies 12:2774
18. Trendafilov K (2017) Analysis of non-working moves performed by semi-mounted reversible plough in a headland. Appl Res Tech Technol Educ 5:79–85. https://doi.org/10.15547/artte.2017.02.001
19. Bulgakov V et al (2021) Study of the steering of a wide span vehicle controlled by a local positioning system. J Agric Eng LII:1144. https://doi.org/10.4081/jae.2021.1144
20. Bulgakov V, Pascuzzi S, Ivanovs S, Nadykto V, Nowak J (2020) Kinematic discrepancy between driving wheels evaluated for a modular traction device. Biosys Eng 196:88–96. https://doi.org/10.1016/j.biosystemseng.2020.05.017
21. Pascuzzi S, Blanco I, Anifantis AS, Scarascia Mugnozza G (2016) Hazards assessment and technical actions due to the production of pressured hydrogen within a pilot photovoltaic-electrolyzer-fuel cell power system for agricultural equipment. J Agric Eng XLVII:507:89–93
22. Bulgakov V, Pascuzzi S, Adamchuk V, Ivanovs S, Pylypaka S (2019) A theoretical study of the limit path of the movement of a layer of soil along the plough mouldboard. Soil Tillage Res 195:104406. https://doi.org/10.1016/j.still.2019.104406
23. Bulgakov V, Pascuzzi S, Beloev H, Ivanovs S (2019) Theoretical investigations of the headland turning agility of a trailed asymmetric implement-and-tractor aggregate. Agriculture 9:224. https://doi.org/10.3390/agriculture9100224
24. Pascuzzi S, Santoro F, Manetto G, Cerruto E (2018) Study of the correlation between foliar and patternator deposits in a Tendone vineyard. Agric Eng Int CIGR J 20(3):97–107

Research of the Possibilities to Improve the Quality of Potatoes Harvesting by Including an Experimental Heap Leveler-Distributor in the Design of Harvesting Machines

Semjons Ivanovs, Volodymyr Bulgakov, Simone Pascuzzi, Valerii Adamchuk, Viktor Kornuchin, Zinovii Ruzhylo, Oleksandra Trokhaniak, Janusz Nowak, Zbigniew Kiernicki, Francesco Santoro, and Oksana Bulgakova

Abstract Improving the quality indicators of the operation of potato harvesters under difficult soil and climatic conditions is an urgent scientific and practical task. The paper reflects a research of the impact upon the quality of operation of two kinds of potato diggers of an experimental design for additional levelling of a heap (mass of soil, potato tubers, plant residues, etc.) on the conveyor. This device evenly distributes the potato heap along the plane of the main rod elevator, improves the separation conditions, reduces the total loss of tubers from 2.5 to 0.8%, that is, 3.1 times. A rational mode of operation of the digger has been determined, which, under these conditions, provides acceptable values for the losses and damage to the tubers.

Keywords Potatoes · Harvesting · Tuber purity · Damage · Additional devices

---

S. Ivanovs
Latvia University of Life Sciences and Technologies, Jelgava, Latvia

V. Bulgakov · Z. Ruzhylo · O. Trokhaniak · O. Bulgakova
National University of Life and Environmental Sciences of Ukraine, Kyiv, Ukraine

S. Pascuzzi · F. Santoro (✉)
Department of Agricultural and Environmental Science, University of Bari Aldo Moro, Bari, Italy
e-mail: francesco.santoro@uniba.it

V. Adamchuk · V. Kornuchin
Institute of Mechanics and Automation of Agricultural Production of the National Academy of Agrarian Sciences of Ukraine, Kyiv, Ukraine

J. Nowak
Department of Operation of Machines and Management of Production Processes, University of Life Sciences in Lublin, Lublin, Poland

Z. Kiernicki
Lublin University of Technology, Lublin, Poland

S. Pascuzzi and F. Santoro (eds.), *Farm Machinery and Processes Management in Sustainable Agriculture*, Lecture Notes in Civil Engineering 289,
https://doi.org/10.1007/978-3-031-13090-8_4

1 Introduction

Potato growing is an important branch of agricultural production in many countries of the world [1–3]. Potato harvesting is the most labor-intensive technological process in potato production [4–6]. Experience of the operation of potato harvesters shows that they do not always ensure high-quality potato harvesting; and they do not provide sufficient reliability of the technological process [7, 8]. The shortcomings are especially pronounced when working on heavy loamy and wet soils. In many countries, for harvesting potatoes in relatively small areas or on waterlogged soils potato diggers are used, which, compared to the expensive potato harvesters, have a smaller number of conveyors and separating devices [9, 10]. Even though such machines are mainly used for harvesting potatoes, intended for consumption in the local market (i.e., not on the large commercial potato farms), they have a large numerical distribution; and improvement of their designs is of current importance. To reduce losses and obtain tubers of the required purity under difficult working conditions on such machines, it is necessary to work at low speeds and carry out post-cleaning of the crop at the potato sorting stations, which leads to increased cost of potato production since it requires significant additional labour, including manual labor [11–13].

There are works of many scientists, devoted to the problem how to improve the quality and productivity of the potato harvesters [14–17]. However, the problem of improving the quality of potato harvesting in many countries is still relevant. One of the promising types of potato diggers that provide a higher quality of harvesting under difficult conditions are machines equipped with additional active rod rotors [18]. Yet in the process of their operation the supply of a mass of soil with tubers (heap) is observed in the central part of the rod separator-conveyor; and a layer of heap, uneven in thickness, is created (a mixture of soil, tubers, leaves of tops and weeds, etc.), (i.e. in the central parts of the thickness of this layer is greater. Obviously, due to this uneven layer the soil separation will not be optimal. Therefore, a device is needed that ensures the creation of a more uniform layer of soil with the tubers on the conveyor. At the same time, introduction of additional elements of impact upon the potato tubers into the structure may increase injuries to the tubers. Therefore, in such cases, a wider study of the impact of an additional device upon the quality indicators of potato harvesting is required.

The purpose of this work is to study the impact upon the quality of potato harvesting using an experimental design of a heap leveler-distributor, located above the supplying conveyor.

Fig. 1 An experimental potato digger KTH type from impurities during its operation

2 Materials and Methods

Investigations of digging and cleaning potatoes from impurities were conducted with two kinds of experimental potato harvesters: a two-row mounted potato digger (variant A) (Fig. 1), and a trailed two-row potato digger-loader (variant B) (Fig. 2).

Fig. 2 Advanced design of a double-row experimental potato digger-loader (equipped with a heap leveler-distributor)

In both cases, the machines had bar cleaning drums with a diameter of 570 mm (frequency of rotation – 57 rpm), which improved the separation of soil from tubers; yet they created an uneven layer of the potato heap on the inclined conveyor (UA 63436A). In both versions the technological scheme included an experimental heap leveler-distributor, installed above the supply conveyor of the potato harvester (at a height of 15 cm) (Fig. 3).

This device was easy-to-remove, and it allowed conducting comparative experiments on harvesting potatoes both with and without a leveler-distributor.

The mounted experimental potato digger of the KTH type has a working width of 1.4 m (2 rows). Its operating speed is – 2.0…2.5 km h^{-1}, the maximum travel depth of digging shares is 0.27 m, productivity is 0.28…0.0.35 ha h^{-1}, its mass is 1500 kg.

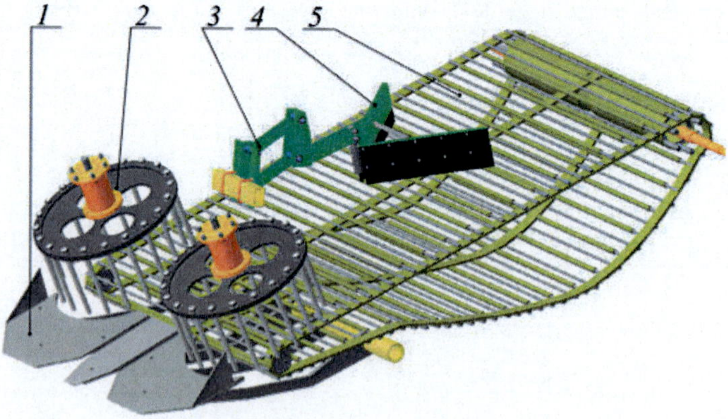

Fig. 3 General view of the experimental digging-separating body of the potato digger: 1 – the digging shares; 2 – the bar separating drums; 3 – a mechanism for installing the heap distributor; 4 – the distributor of a V-shape type; 5 – the supplying bar separating conveyor

The potato digger consists of a frame, track rollers, plough shares, bar drums, the main supply conveyor, a leveler-distributor, a haulm-separating device, a cascade elevator, a mechanism for driving the working bodies and the cutting discs. After the passage of such a digger the potato tubers remain on the surface of the field, and then they need to be manually picked, or a special picking machine must be used for this purpose.

The trailed double-row experimental potato digger-loader has a similar design at the stage of digging and primary separation, but then, at the exit, it has a system of conveyors for the supply of tubers directly into the body of a transport trailer (moving along it). Considering that this option has more conveyors, the process of separation from the soil in it is more intense, and therefore it was important to evaluate the efficiency of the leveler-distributor and to determine whether it is needed in this option. A brief description of the trailed double-row potato digger-loader: it harvests potatoes from two rows (the working width 1.4 m) and ensures loading the tubers into the body of the nearby vehicle; the operating speed – 1...6 km h^{-1}, the maximum travel depth of the digging shares – 0.25 m, efficiency – 0.45...0.0.65 ha h^{-1}, weight 3500 kg. When conducting experimental studies of the new cleaner, potatoes of the Nevsky variety were harvested, cultivated on ridges (beds) with a row spacing of 70 cm. When harvesting the tubers, the tops were previously mowed, the average height of the ridges was 19.2 cm, the average depth of the lower tuber was 18.2 cm, and the width of the ridges was on average 22 cm. The mass ratio of tubers by fractions was: 52.9% up to 50 mm including, 17.6% from 50 to 80 mm including, and 29.5% more than 80 mm. The conditions under which the tests were carried out were typical for heavy soils, the soil hardness in the layer of the tubers was from $1.4...3.0 \cdot 10^6$ N m^{-2}. The type of soil is loamy chernozem with a flat relief. Soil moisture content in the layers was 16.1...17.0%, The quality characteristics of work were determined at the operating speeds of 3.2...4.8 km h^{-1}, the air temperature during the potato harvesting was 10° C. The plot was covered with weeds and haulm residues (1.56 t ha^{-1}). Overall, during the experimental studies the potatoes were harvested on an area of 10.1 ha. When the leveler-distributor is on, the impurities amount in the second elevator is much less than when the leveler-distributor is off. This leads to an improvement in the purity of tubers in the containers. Standard methods were used during the investigations for testing the potato harvesters [19–21], as well as additionally developed methods for assessing the influence of various factors. The quality of harvesting the potatoes by machines in Ukraine and other countries of Eastern Europe is regulated by the so-called agricultural requirements [14] which define the criteria and their minimum values that the machines must satisfy from an agronomic point of view.

3 Results and Discussion

The quality indicators of work of the potato digger with a heap leveler-distributor were determined at an operating speed of 2.1 km h^{-1}. Table 1 shows the results of studies of an experimental sample.

Figure 4 shows the results of investigations of the loss of the potato tubers and the degree of their damage when digging out with a potato digger, equipped with the new cleaner from impurities.

When changing the depth of the plough shares h at a working speed $V = 2.1$ m s^{-1} from 20 to 22.8 cm, the loss L of the potato tubers increases by 16…18%, and their damage D decreases by 8…10%. When the plough share travel depth is 21.4 cm and the tuber depth is 18.3 cm, the loss of tubers was 2.5%, and the damage was 4.8%, which is within the acceptable limits. The quality of work was significantly affected by the heap leveler-distributor, which, when put into operation, evenly distributed the potato heap along the plane of the main rod elevator, improved the conditions for separating the heap, reduced soiling of the tubers, which made it possible to reduce the total loss of tubers from 2.5 to 0.8%, that is, in 3.1 times. Sprinkling of tubers with

Table 1 Quality indicators of tuber digging, using a potato digger (variant A) with an experimental heap leveler-distributor

Name of the indicator	Value or characteristic according to agricultural requirements		
	With a leveler-distributor	Without a leveler-distributor	According to requirements
Working speed, km h^{-1}	2.1	2.1	2…4
Depth of the share travel, cm	21.4	21.4	up to 25
Losses of tubers, %	0.8	2.5	up to 3%
Damage to tubers by mass (weight), %	4.8	4.8	up to 7%
Number of the damaged per 100 tubers, pcs.:			
– the peel torn off from 1/4 to 1/2 of the surface	2	2	–
– the peel torn off more than 1/2 of the surface	0	0	–
– the mass torn out more than 5 mm	2	2	–
– cracks longer than 20 mm	1	1	–
– compressed tubers	2	2	–
– cut tubers	1	1	up to 1.5%
Strip width of the dug out tubers, cm	110.0	58.0	–

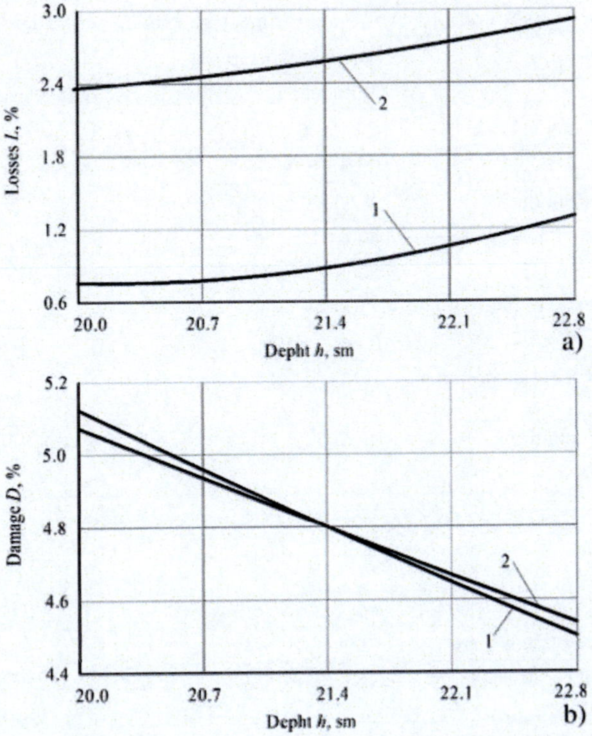

Fig. 4 Graphical dependences of the degree of losses of the potato tubers *L* **a** and their damage *D* **b** during digging by a potato digger (variant A) at the depth of the plough shares *h* at a working speed V = 2.1 m s^{-1}: 1 – with a heap leveler-distributor; 2 – without a heap leveler-distributor

the soil is reduced, which leads to an increase in the efficiency of workers-potato diggers. A rational mode of operation of the digger has been determined, which, under these conditions, provides acceptable values for the losses and damage to the tubers. At a speed of the aggregate 2.1 km h^{-1} the labour efficiency was 0.29 ha h^{-1}. To the quality of work, the experimental potato digger basically meets the agricultural requirements (Table 2).

Figure 5 presents the results of studies of the degree of the loss of the potato tubers, their damage and the purity of potatoes when dug up by a two-row potato digger-loader.

Table 2 Performance indicators of a two-row experimental potato digger-loader (variant B)

Name of the indicator	Value of indicators	
	Without a leveler-distributor	With a leveler- distributor
Working speed of travel: km h^{-1}	2.8…3.6	3.2…4.8
Digging depth, cm	19.3	19.3
Purity of potato cleaning, %:		
– Tubers	85.8	97.4
– Free soil	2.2	0.9
– Soil clods (up to 50 mm in size)	11.8	1.7
– Vegetable admixtures	0.2	0
Completeness of collection of the tubers, %:		
– Collected in a container	98.6	99.4
– Left on the surface	1.4	0.6
– Total loss, % kg	1.4 340	0.6 146
Damage to the potato tubers, %:		
– Total	4.6	4.6
– Tubers cut by the plough shares	1.6	1.6
– Cracks on the tuber surface more than 20 mm long	0.8	0.8
– Turn out crumbs, more than 5 mm deep	2.2	2.2
– Piece by piece (from 100 tubers)	6.0	6.0

When changing the depth of the plough shares h at a working speed $V = 3.2$ m•s^{-1} from 17.9 to 20.7 cm, the loss L of potato tubers increases by 12–15%, and their damage D decreases by 6–8%. A positive effect of the heap leveler-distributor on the quality of the machine: the purity of tubers in the container reached 97.4% (without the leveler-distributor it was 85.8%) at a digging depth h = 19.3 cm. That is, the use of a leveler-distributor increases such an important indicator of the operation of the potato harvester as the purity of tubers in the container by at least 10%.

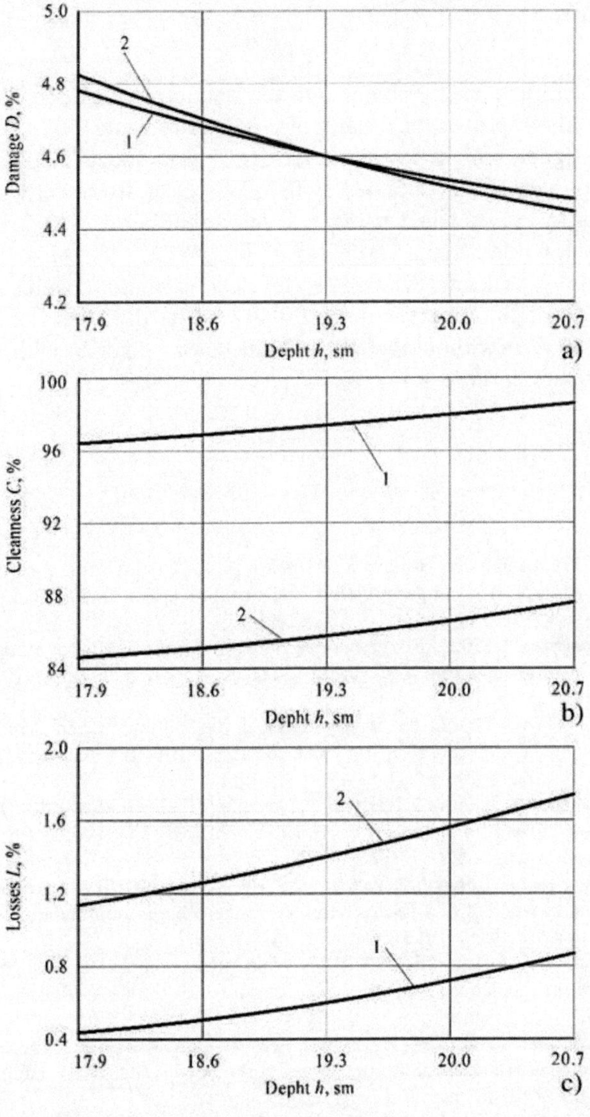

Fig. 5 Dependences of the degree of losses L of the potato tubers **a**, their damage D **b** and the purity of potatoes C **c** when digging by potato digger-loader (variant B), equipped with a new cleaner, from the depth of digging h at a working speed $V = 3.2\ m\ s^{-1}$: 1 – with a heap leveler-distributor; 2 – without a heap leveler-distributor

4 Conclusions

Inclusion of a heap leveler-distributor into the potato harvesters of the considered design significantly improves the quality of potato harvesting.

The potato digger with a heap leveler-distributor, located above the supply conveyor, when harvesting potatoes with a yield of 30.6 t h^{-1}, reduces the losses of tubers by 31%, when using a leveler-distributor, and 2.5% without a leveler-distributor, while damage to the tubers was 4.8%.

Experimental digger-loader with a heap leveler-distributor ensures the completeness of collection of the tubers – up to 99.4%, while the purity of the tubers in the container is 97.4% (without the leveler-distributor – 85.8%) with an acceptable damage to the tubers of 4.6%.

References

1. Adamchuk V, Prysyazhnyi V, Ivanovs S, Bulgakov V (2016) Investigations in technological method of growing potatoes under mulch of straw and its effect on the yield. Eng Rural Dev 15:1098–1103
2. Bulgakov V, Pascuzzi S (2021) Mathematical model of the movement of a potato body along the surface of a spiral separator. Acta Hortic 1311:447–454. https://doi.org/10.17660/ActaHortic.2021.1311.57
3. Pascuzzi S, Bulgakov V (2021) The theory of sifting the soil mass when cleaning potatoes on a spiral separator. Acta Hortic 1311:417–424. https://doi.org/10.17660/ActaHortic.2021.1311.53
4. Bulgakov V, Pascuzzi S, Ivanovs S, Nadykto V, Nowak J (2020) Kinematic discrepancy between driving wheels evaluated for a modular traction device. Biosys Eng 196:88–96. https://doi.org/10.1016/j.biosystemseng.2020.05.017
5. Bulgakov V, Pascuzzi S, Adamchuk V, Ivanovs S, Pylypaka S (2019) A theoretical study of the limit path of the movement of a layer of soil along the plough mouldboard. Soil Tillage Res 195:104406. https://doi.org/10.1016/j.still.2019.104406
6. Pascuzzi S, Anifantis AS, Santoro F (2020) The concept of a compact profile agricultural tractor suitable for use on specialised tree crops. Agriculture 10:123. https://doi.org/10.3390/agriculture10040123
7. Bulgakov V, Pascuzzi S, Ivanovs S, Ruzhylo Z, Fedosiy I, Santoro F (2020) A new spiral potato cleaner to enhance the removal of impurities and soil clods in potato harvesting. Sustainability (Switzerland) 12(23):9788. https://doi.org/10.3390/su12239788
8. Bulgakov V et al (2021) Study of the steering of a wide span vehicle controlled by a local positioning system. J Agric Eng LII:1144. https://doi.org/10.4081/jae.2021.1144
9. Bulgakov V, Ivanovs S, Adamchuk V, Ihnatiev Y (2017) Investigation of the influence of the parameters of the experimental spiral potato heap separator on the quality of work. Agron Res 15(1):44–54
10. Vereshchagin N, Pshechenkov K, Gerasimov V (1983) Harvesting potatoes in difficult conditions, p 208. Kolos, Moscow
11. Hevko RB, Tkachenko IG, Synii SV, Flonts IV (2016) Development of design and investigation of operation processes of small-sclale root crop and potato harvesters. INMATEH Agric Eng 49(2):53–60
12. Zavgorodniy AF, Shumilo MM (1986) Combined digging-separating body of potato harvesters. Tech Agric 10:60–61

13. Bulgakov V, Pascuzzi S, Nadykto V, Ivanovs S, Adamchuk V (2021) Experimental study of the implement-and-tractor aggregate used for laying tracks of permanent traffic lanes inside controlled traffic farming systems. Soil Tillage Res 208:104895. https://doi.org/10.1016/j.still.2020.104895
14. Petrov G (2004) Potato Harvesting Machines, p 320. Mashinostroenije, Moskow
15. Ichiki H, Van Nguyen N, Yoshinaga K (2013) Stone-clod separation and its application to potato cultivation in Hokkaido. Eng Agric Environ Food 6(2):77–85
16. Bulgakov V, Pascuzzi S, Beloev H, Ivanovs S (2019) Theoretical investigations of the headland turning agility of a trailed asymmetric implement-and-tractor aggregate. Agriculture 9:224. https://doi.org/10.3390/agriculture910ZS0224
17. Bulgakov V, Pascuzzi S, Adamchuk V, Kuvachov V, Nozdrovicky L (2019) Theoretical study of transverse offsets of wide span tractor working implements and their influence on damage to row crops. Agriculture 9:144
18. UA 63436A: (2004) Digging and separating device of potato harvester. Ukrainian patent
19. Standard DSTU 7794: (2015) Potato harvesters. Test methods. Kyiv, Ukraine, p 14
20. Standard DSTU 7572: (2014) Potato harvesters. General technical requirements. Kyiv, Ukraine, p 8
21. Pascuzzi S, Santoro F, Manetto G, Cerruto E (2018) Study of the correlation between foliar and patternator deposits in a "Tendone" vineyard. Agric Eng Int CIGR J 20(3):97–107

Theoretical Studies of the Relationship Between the Parameters of the Bridge Tractor and the Engineering Area of the Field

Volodymyr Bulgakov, Stanislav Nikolaenko, Simone Pascuzzi, Valerii Adamchuk, Volodymyr Kuvachov, Valentyna Krutyakova, Francesco Santoro, and Janusz Nowak

Abstract The planning of the field under the bridge system of agriculture should be carried out taking into account the size of both the bridge tractor, which moves on specially created tracks of the permanent technological track, and auxiliary means of mechanization. The purpose of the research is to study the processes of field planning and organization of the movement of bridge tractors, considering the prospects for further automation of all technological processes that they perform, including transport. Theoretical research was carried out by modeling the conditions of the bridge tractor operation on the PC using the provisions of theoretical mechanics and tractor theory. Experimental studies of the bridge tractor were carried out according to both generally accepted and developed methods and involved the use of modern and specially designed equipment. As a result of research, mathematical models and algorithms were developed to allow interdependent choice of construction parameters of bridge tractors, in particular, the width of its track and propulsion, stable traffic conditions (taking into account the value of technological tolerance) and parameters of the field for its permanent technological track. Mathematical analysis of the obtained models substantiates the rational track width of these bridge tractors at the

V. Bulgakov · S. Nikolaenko
National University of Life and Environmental Sciences of Ukraine, Kyiv, Ukraine

S. Pascuzzi (✉) · F. Santoro
Department of Agricultural and Environmental Science, University of Bari Aldo Moro, Bari, Italy
e-mail: simone.pascuzzi@uniba.it

V. Adamchuk
Institute of Mechanics and Automation of Agricultural Production of the National Academy of Agrarian Sciences of Ukraine, Glevakha, Ukraine

V. Kuvachov
Dmytro Motornyi Tavria State Agrotechnological University, Melitopol, Ukraine

V. Krutyakova
Engineering and Technological Institute "Biotechnica" NAASU, Khlibodarske, Ukraine

J. Nowak
Department of Operation of Machines and Management of Production Processes, University of Life Sciences in Lublin, Lublin, Poland

S. Pascuzzi and F. Santoro (eds.), *Farm Machinery and Processes Management in Sustainable Agriculture*, Lecture Notes in Civil Engineering 289,
https://doi.org/10.1007/978-3-031-13090-8_5

level of 7.5...9 m, according to which the loss of field area under the engineering zone when using them is not more than 5...6%.

Keywords Bridge tractor · Permanent technological track · Engineering and technological zone of the field · Theoretical and experimental research

1 Introduction

In track and bridge agriculture, part of the total field area is allocated for the engineering zone, which reduces the productive part of the field – agro-technical zone [1–6]. The engineering zone includes a transport system for the movement of all means of mechanization, communication for energy and water supply, telecommunication channels and orientation system [7, 8]. The area of the engineering zone depends mainly on the parameters of the transport system, which, in turn, are determined by the parameters of technological and transport machines [9, 10]. According to the requirements of automation, mechanization should be subordinated to the principles of the coordinate transport system, in which machines can move only in two mutually perpendicular directions and for the implementation of which the field must have strictly defined dimensions [11–13].

When analyzing the use of productive field area in bridge farming, it is necessary to consider, first of all, the issues of planning and organization of machine movement, taking into account the prospects of further automation of all technological processes, including transport [14–20].

It is obvious that with the increase of the track width of the bridge tractor the losses of the fertile area of the field under the engineering zone decrease [21]. The nature of this pattern depends on many factors, including the design parameters of the bridge tractor and the stability and controllability of its movement, organization and mode of its rotation on the turning lane, etc. [22]. From the standpoint of efficient use of the productive field area, it is advisable to have an accessible method and be able to determine the rational track width of the bridge tractor under specific conditions of its use.

Study of regularities of influence of parameters of agricultural bridge systems and conditions of their operation on the character of field planning and organization of movement of bridge tractors, taking into account prospects for further automation of all technological processes performed by them, including transport.

2 Materials and Methods

Theoretical research was carried out by modeling the conditions of the bridge tractor operation on the PC using the provisions of theoretical mechanics and tractor theory. Experimental studies of the bridge tractor (Fig. 1) were carried out according to both generally accepted and developed methods and involved the use of modern and specially designed equipment. Processing of experimental data was performed on a PC using regression and correlation-spectral analyzes. The research methods are based on the foundations of theoretical mechanics using the Mathcad package.

Experimental studies were conducted in a specially equipped laboratory for testing a bridge tractor with a test length of 50 m.

During the experimental studies after the working movement of the bridge tractor, the amplitude of its transverse deviation from the axis of symmetry of the constant technological track was measured. For this purpose, the amplitude of the transverse deviation from the axis of symmetry of the constant technological track of the tracks of its front x_s and rear x_r wheels from one side of the bridge tractor was determined with a step of 0.2 m (Fig. 2).

From the obtained implementations, statistical characteristics such as standard deviations and normalized spectral densities were determined according to the method presented in [23, 24].

The experimental amplitude-frequency characteristic of the oscillations of the linear transverse displacement of the bridge tractor as part of the harrowing unit was calculated from the expression [23, 24]:

Fig. 1 Experimental studies of a bridge tractor of new construction

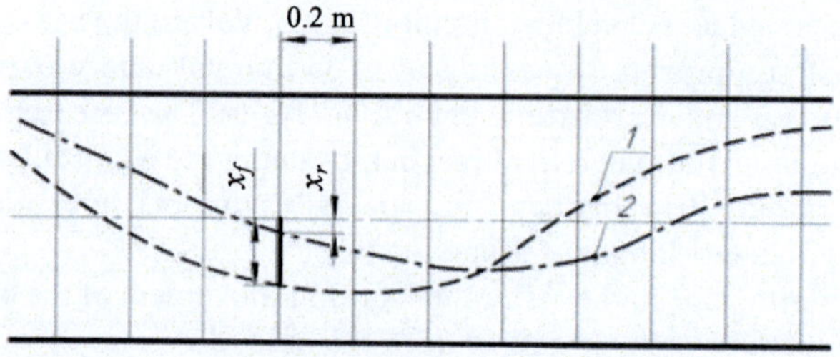

Fig. 2 Scheme for determining the transverse displacement of the bridge tractor: 1, 2 – respectively, the middle of the trajectories of its front and rear wheels within the width of the technological track

$$A(\omega) = \frac{\sigma_y}{\sigma_x} \cdot \left(\frac{S_y(\omega)}{S_x(\omega)} \right)^{\frac{1}{2}} \tag{1}$$

where σ_x, $S_x(\omega)$ – standard deviations and normalized spectral density of the input value; σ_y,, $S_y(\omega)$ – standard deviations and normalized spectral density of the original value; ω – frequency of oscillations of the control effect, s^{-1}.

The theoretical spectral density of oscillations of the initial parameter was found from the expression [23, 24]:

$$S_y(\omega) = \frac{A(\omega)^2 \cdot S_x(\omega) \cdot \sigma_x}{\sigma_y} \tag{2}$$

where $S_x(\omega)$ – normalized spectral density of oscillations of the input quantity.

3 Results and Discussion

Consider the scheme of planning the engineering zone of the field in the track system of agriculture for the operation of the bridge tractor on the coordinate-transport principle of their movement on the land of rectangular correct configuration (Fig. 3).

Changing the direction of movement of the bridge tractor on the turning lane should be done in a circular turn (Fig. 4). The center of rotation of the bridge tractor must be in the area of the transport technological track of one of its sides (left or right, depending on the direction of rotation). Only in this case you can get the desired minimum turning radius and width of the turning lane. At the same time, with this method of reversal, the bridge tractor is moved to the next race in the minimum period of time, which increases productivity. Technically, this can be realized through swivel wheels. Otherwise, when the wheels of the bridge tractor are uncontrollable, its turn on the spot can be realized by lifting one of the sides, where, leaning on the propulsion of the other side as they move, you can rotate around the center of rotation.

Loss of field area under the engineering zone (w_i) in Fig. 3 will be evaluated by a relative indicator, numerically equal:

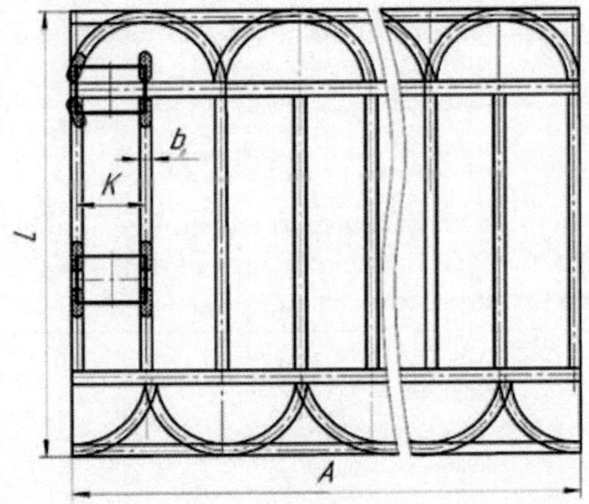

Fig. 3 Scheme of field map planning for the track system of agriculture for the operation of the bridge tractor

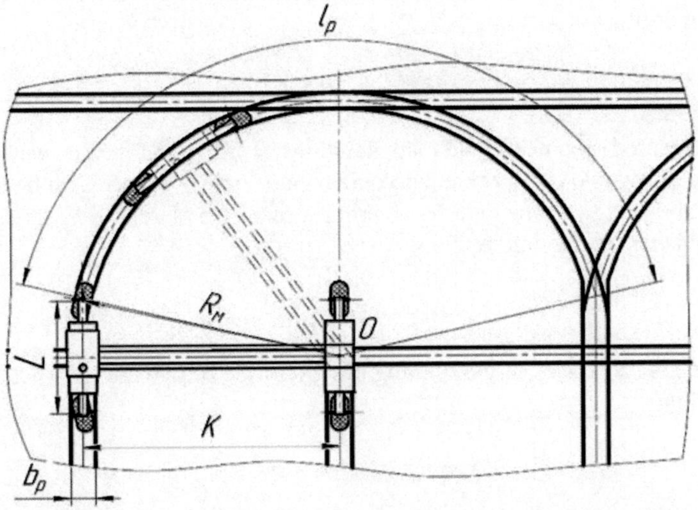

Fig. 4 Diagram of the rotation of the bridge tractor by its circular rotation relative to the center of rotation (point O)

$$w_i = \frac{S_i}{A \cdot L}, \quad (3)$$

where S_i – the area of the engineering zone of the field; L and A – length and width of the field.

The land use ratio (k_s) will be equal to:

$$k_S = 1 - w_i = 1 - \frac{S_i}{A \cdot L}. \quad (4)$$

The area of the engineering zone in Fig. 3 can be determined by the sum of three components:

$$S_i = S_o + S_{ot} + S_{tt}, \tag{5}$$

where S_o, S_{ot} – the area of traces of transport technological tracks on the main field and turning lanes, respectively; S_{tt} – the total area of traces of the running system of the bridge tractor when making turns.

Area S_o for Fig. 3 will be equal to:

$$S_o = b_p \cdot [L - 2(K + b_p)] \cdot \left[\frac{A - b_p}{K} + 1\right], \tag{6}$$

where b_p – width of the transport technological track; K – track width of the tractor.

The width of the transport technological track b_r is represented as the sum of the track width b_s from the tires of the bridge tractor and some width of the technological tolerance c, due, in particular, to the amplitudes of its transverse deviations from rectilinear motion:

$$b_p = b_c + c. \tag{7}$$

In the presented scheme of field map planning according to Fig. 3 on each turning lane there are only two transport technological paths, which is sufficient both for the passage of the bridge tractor and for making turns. Therefore, the S_{ot} area for the two turning lanes on the field will be:

$$S_{\text{ot}} = 4 \cdot b_p \cdot A. \tag{8}$$

The total area S_{tt} traces of the running system of the bridge tractor when making it turns folding:

$$S_{\text{tt}} = \pi \cdot b_p \cdot (A - b_p). \tag{9}$$

After substitution of Eqs. (5–9) in (3) the coefficient of losses of the field area under the engineering zone w_i will be equal to:

$$w_i = \frac{b_c + c}{L \cdot A} \cdot \left([L - 2(K + b_c + c)] \cdot \left[\frac{A - b_c - c}{K} + 1\right] + 4 \cdot A + \pi[A - b_c - c]\right). \tag{10}$$

Analysis of expression (10) confirms the previously mentioned inversely proportional nature of the dependence of w_i on the track width K of the bridge tractor. In this case, increasing the width of the tires b_c of its propulsion and technological tolerance on the contrary increases the loss of field area under the engineering zone. Equation (10) does not have an extremum of the function w_i from the argument K, but allows to establish the point of rational optimum of this dependence, as well

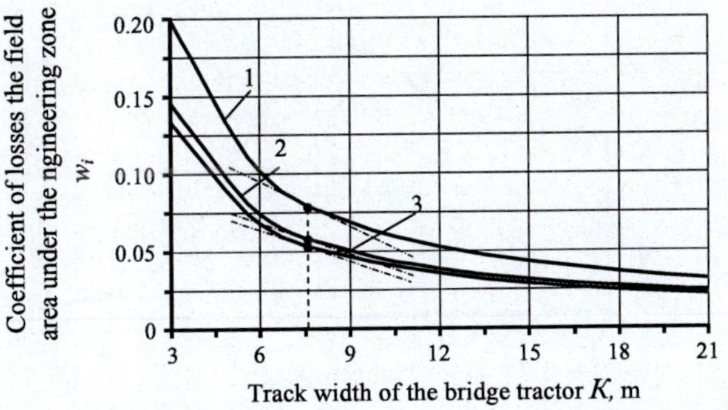

Fig. 5 Dependence of the coefficient of losses of the field area under the engineering zone w_i on the track width K of the bridge tractor depending on the width of the tires of its engines: 1 – $b_{c1} = 0.586$; 2 – $b_{c2} = 0.429$; 3 – $b_{c3} = 0.394$ m

as to estimate the influence of parameters of the bridge tractor running system and technological tolerance from the standpoint of efficient land use in track agriculture. To do this, in the Mathcad environment, three curves of dependence w_i and K were constructed for three variants of tractor tires of the bridge tractor: 1 – $b_{c1} = 0.586$; 2 – $b_{c2} = 0.429$; 3 – $b_{c3} = 0.394$ m (Fig. 5).

In Fig. 5 the area of change of argument K can be conditionally divided into two intervals. In each of the intervals, the dependence w_i on K is close to linear. Moreover, in the range of initial values of K, the function w_i decreases intensively, and for higher values of K – slowly. Suppose that the rational value of K corresponds to the optimum point of the specified two-band curve, which divides the latter into two parts with significantly different properties. Using the least squares method to determine the parameters of the rational function [25], the points of the rational optimum of two-band curves in Fig. 5.

For the considered variants of parameters of tires of engines of the bridge tractor the rational size of track width of the last is necessary for $K = 7.5$ m. The obtained value of the track width when using tires with a width of 0.429 and 0.393 m allows to have the number of losses of the field area under the engineering zone not more than 6%. And when increasing the track width to 9 m, the amount of area loss is at the level of 5%.

The calculations on the influence of the length of field A on the loss of area under the engineering zone w_i and found that the quantitative values of this effect can be considered insignificant. Since, for example, for the path of a bridge tractor $K = 3$ m with tire wheels 15.5R38 diameter which is 0.394 m when reducing the length of the field from 1000 to 100 m loss of field area under the engineering zone w_i changes from 0.151 to 0.133, and at $K = 30$ m – from 0.033 to 0.015 – also changes less than 2.0%. In general, in the range of change A from 100 to 1000 m the value of w_i varies by no more than 2%.

Consider the degree of influence of technological tolerance c on the nature of the increase in field area losses under the engineering zone w_i.

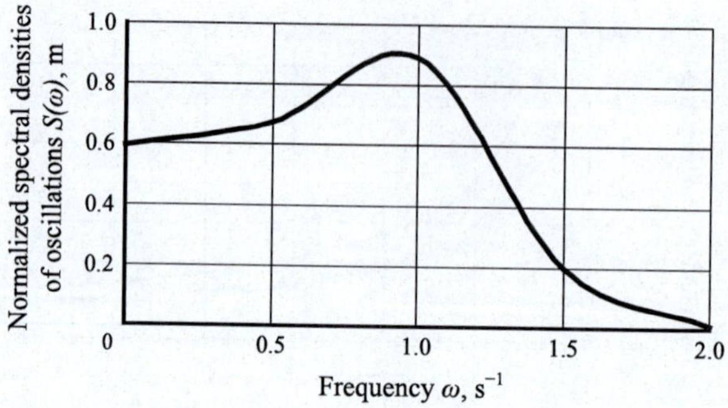

Fig. 6 Normalized spectral densities of oscillations linear transverse displacement of the bridge tractor

The oscillations of the transverse displacement of bridge tractor in the harrowing unit during its working movement are also low-frequency (Fig. 6). The main spectrum of variances is concentrated in the frequency range 0…2.0 s^{-1}. The standard of oscillations of this parameter was ± 0.05 m.

The calculation of the experimental amplitude-frequency characteristic according to expression (1) and its comparison with the theoretical one showed (Fig. 7) that in the operating frequency range (0…2 s^{-1}) of the input signal oscillations the largest difference between theoretical and field data does not exceed 15%.

Analysis of the dependence of w_i and c (Fig. 8) shows that the width of the technological tolerance c significantly affects the loss of field area under the engineering zone.

Thus, for the considered variants of tractor tires, the engines of the bridge tractor with an increase from 0.3 m to the loss of the field area under the engineering zone are increased by 1.5–1.75 times. Therefore, the use of a bridge tractor in the track system of agriculture requires substantiation of the principles of their automatic driving,

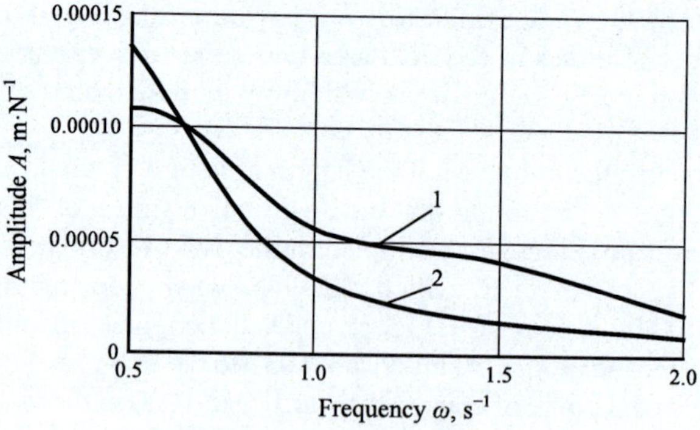

Fig. 7 Theoretical (1) and experimental (2) amplitude-frequency characteristics of oscillations of linear transverse displacement during testing of a bridge tractor as part of a harrowing control unit

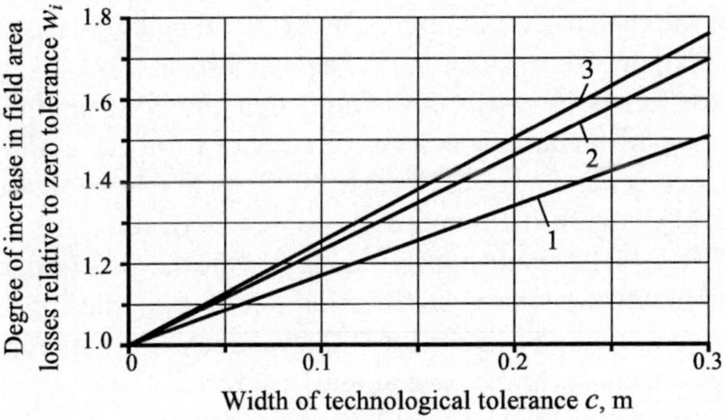

Fig. 8 Influence of technological tolerance width c on the degree of field area losses under the engineering zone relative to zero tolerance depending on the width of the tires of the bridge tractor: 1 – bc1 = 0.586 m; 2 – bc2 = 0.429 m; 3 – bc3 = 0.394 m

which minimizes the amplitude of transverse deviations from a given rectilinear trajectory, and, as a consequence, directly reduce the value of c.

As a result of the conducted researches it is possible to conclude that planning of the field under track system of agriculture with use of the bridge tractor has to consider width of a track of the last and parameters of transport system. The use of modern bridge tractors with a track width of more than 7.5 m allows to achieve the amount of losses of the field area under the engineering zone of not more than 5–6%, which is quite acceptable.

The practical use of the bridge tractor in the track system of agriculture requires substantiation of the principles of their automatic driving, which will minimize the loss of field area under the engineering zone at least 1.5 times.

4 Conclusions

Studies have shown that the loss of field area under the engineering zone significantly depends on the width of the track for the movement of the bridge tractor, the value of which is directly determined by the width of the tires of their wheels. Calculations have established that according to the criterion of the minimum coefficient of losses of the field area under the engineering zone, the rational value of the track width of the latter is $K = 7.5$ m. In practice, this means that when using the tires of a bridge tractor with a width of 0.393…0.429 m, allows you to have the amount of losses of the field area under the engineering zone not more than 6%. At the same time, when increasing the width of the track tractor to 9 m, which is typical for foreign models of so-called "bridge" tractors, the amount of area loss is reduced to 5%.

The loss of the field area under the engineering zone is practically independent of its length. Proof of this is the fact that when reducing the length of the field from 1000 to 100 m, this value changes by no more than 2% in the range of variations in the track width of the road tractor to 30 m. Therefore, in practice, the size of the field when it is arranged under the track system of agriculture does not significantly affect the value of the coefficient of losses of the field area under the engineering zone.

With large transverse deviations of the bridge tractor from the rectilinear trajectory of its movement, increasing the width of the technological tolerance of the permanent technological track by only 0.3 m increases the loss of field area under the engineering zone by 1.5…1.75 times. In practice, this means that the use of a bridge tractor in the track system of agriculture requires justification of the principles of their automatic driving, which, by improving the stability of their movement, will minimize the loss of field area under the engineering zone at least 1.5 times).

References

1. Chamen T (2015) Controlled traffic farming–from worldwide research to adoption in Europe and its future prospects. Acta Technol Agric 18:64–73. https://doi.org/10.1515/ata-2015-0014
2. Gasso V, Sørensen CAG, Oudshoorn FW (2013) Controlled traffic farming: a review of the environmental impacts. Eur J Agron 48:66–73
3. Tullberg JN, Yule DF, McGarry D (2007) Controlled traffic farming–from research to adoption in Australia. Soil Tillage Res 97(2):272–281
4. Onal I (2012) Controlled traffic farming and wide span tractors. Agric Mach Sci 8(4):353–364
5. Lamers JG, Perdok UD, Lumkes LH (1986) Controlled traflic farming syslems in the Netherlands. Soil Tillage Res 8:65–76
6. Bulgakov V, Adamchuk V, Kuvachov V et al (2018) Study of effectiveness of controlled traffic farming system and wide span self-propelled gantry-type machine. Res Agric Eng 64(1):1–7
7. Bulgakov V, Pascuzzi S, Nadykto V, Ivanovs S, Adamchuk V (2021) Experimental study of the implement-and-tractor aggregate used for laying tracks of permanent traffic lanes inside controlled traffic farming systems. Soil Tillage Res 208:104895. https://doi.org/10.1016/j.still.2020.104895
8. Bulgakov V et al (2021) Study of the steering of a wide span vehicle controlled by a local positioning system. J Agric Eng LII:1144. https://doi.org/10.4081/jae.2021.1144
9. Chamen WCT, Dowler D, Leede PR (1994) Design, operation and performance of a gantry system: experience in arable cropping. J Agric Eng Res 59:45–60
10. Pascuzzi S, Anifantis AS, Santoro F (2020) The concept of a compact profile agricultural tractor suitable for use on specialised tree crops. Agriculture 10:123. https://doi.org/10.3390/agriculture10040123
11. Bulgakov V, Pascuzzi S, Ivanovs S, Nadykto V, Nowak J (2020) Kinematic discrepancy between driving wheels evaluated for a modular traction device. Biosyst Eng 196:88–96. https://doi.org/10.1016/j.biosystemseng.2020.05.017
12. Bulgakov V, Pascuzzi S, Adamchuk V, Ivanovs S, Pylypaka S (2019) A theoretical study of the limit path of the movement of a layer of soil along the plough mouldboard. Soil Tillage Res 195:104406. https://doi.org/10.1016/j.still.2019.104406
13. Guerrieri AS, Anifantis AS, Santoro F, Pascuzzi S (2019) Study of a large square baler with innovative technological systems that optimize the baling effectiveness. Agriculture 9(5):86
14. Bulgakov V, Pascuzzi S, Adamchuk V, Kuvachov V, Nozdrovicky L (2019) Theoretical study of transverse offsets of wide span tractor working implements and their influence on damage to row crops. Agriculture 9:144
15. Pascuzzi S, Santoro F (2017) Analysis of the almond harvesting and hulling mechanization process. A case study. Agriculture (Switzerland) 7(12):100

16. Manetto G, Cerruto E, Pascuzzi S, Santoro F (2017) Improvements in citrus packing lines to reduce the mechanical damage to fruit. Chem Eng Trans 58:391–396
17. Baldoin C, Balsari P, Cerruto E, Pascuzzi S, Raffaelli M (2008) Improvement in pesticide application on greenhouse crops: results of a survey about greenhouse structures in Italy. Acta Hortic 801:609–614. https://doi.org/10.17660/ActaHortic.2008.801.69
18. Bulgakov V, Pascuzzi S, Anifantis AS, Santoro F (2019) Oscillations analysis of front-mounted beet topper machine for biomass harvesting. Energies 12:2774
19. Pascuzzi S, Blanco I, Anifantis AS, Scarascia Mugnozza G (2016) Hazards assessment and technical actions due to the production of pressured hydrogen within a pilot photovoltaic-electrolyzer-fuel cell power system for agricultural equipment. J Agric Eng XLVII:507:89–93
20. Bulgakov V, Pascuzzi S, Beloev H, Ivanovs S (2019) Theoretical investigations of the headland turning agility of a trailed asymmetric implement-and-tractor aggregate. Agriculture 9:224. https://doi.org/10.3390/agriculture9100224
21. Bulgakov V, Kuvachov V, Nozdrovický L et al (2018) The study of movement of the wide span tractor–based field machine unit with power method of its control. Acta Technol Agric 4:163–168
22. Bulgakov V, et al (2017) The study of movement wide span tractor (vehicles) with kinematic (turn of wheels) method of its control. Mech Agric Conserv Resour 63(5):188–193
23. Priestley MB (1982) Spectral Analysis and Time Series. Academic Press, London, New York. ISBN 978–0125649018
24. Fisher RA (1992) Statistical methods for research workers. In: Kotz S, Johnson NL (eds) Breakthroughs in Statistics. Springer Series in Statistics (Perspectives in Statistics). Springer, New York, NY. https://doi.org/10.1007/978-1-4612-4380-9_6
25. Nadykto V, Velichko O (2014) Determining the optimum point of the curve and ways to determine it. Mach Technol Agro-Ind Complex 2(53):16–18

Research into Power and Load Parameters of Flexible Screw Conveyors for Transportation of Agricultural Materials

Volodymyr Bulgakov, Simone Pascuzzi, Valerii Adamchuk, Jüri Olt, Zinovii Ruzhylo, Oleksandra Trokhaniak, Francesco Santoro, Margus Arak, Janusz Nowak, and Hristo Beloev

Abstract In order to carry out experimental research and determine the performance criteria for the transportation of agricultural materials, a test unit of the combined screw conveyor has been designed by the authors. The axial speed and capacity of bulk material transportation on a curved route have been determined for the articulated-section operating device of the screw conveyor. The calculation has proved that the angular velocities of adjacent sections are almost the same. As a result of the research into the power and load parameters of flexible screw conveyors, graphic relations between the values of the torque T and power N, on the one hand, and the rate of rotation of the operating device's drive shaft, on the other hand, have been plotted for the process of transportation of bulk materials in screw conveyors, when the process pipeline is fully filled with grain material.

Keywords Agricultural material · Screw conveyor · Transportation

---

V. Bulgakov · Z. Ruzhylo · O. Trokhaniak
National University of Life and Environmental Sciences of Ukraine, Kyiv, Ukraine

S. Pascuzzi (✉) · F. Santoro
Department of Agricultural and Environmental Science, University of Bari Aldo Moro, Bari, Italy
e-mail: simone.pascuzzi@uniba.it

V. Adamchuk
Institute of Mechanics and Automation of Agricultural Production of the National Academy of Agrarian Sciences of Ukraine, Glevakha, Vasylkiv Dist., Kyiv Region, Ukraine

J. Olt · M. Arak
Institute of Forestry and Engineering, Estonian University of Life Sciences, 56 Kreutzwaldi Str., 51006 Tartu, EE, Estonia

J. Nowak
Department of Operation of Machines and Management of Production Processes, University of Life Sciences in Lublin, 13, Akademicka Str., 20-950 Lublin, Poland

H. Beloev
Department of Agricultural Machinery, University of Ruse Angel Kanchev, BG 7017 Ruse, Bulgaria

S. Pascuzzi and F. Santoro (eds.), *Farm Machinery and Processes Management in Sustainable Agriculture*, Lecture Notes in Civil Engineering 289,
https://doi.org/10.1007/978-3-031-13090-8_6

1 Introduction

Flexible screw conveyors are widely used in the transportation of agricultural materials in various production processes and efficiently perform the functions required from them [1]. However, the existing screw-type operating devices cannot fully meet the performance requirements to such types of conveyors [2]. Continuous auger flights quickly deteriorate due to the action of alternating cyclic loads, while the combined and sectioned operating devices are material-intensive parts. Their operation results in the increased consumption of power and heavy wear of the inner surfaces of the flexible casings [3, 4].

The completed theoretical and experimental investigations on the conditions in the zones of the loading and transhipment of bulk agricultural materials in case of single- and double-line screw conveyors as well as the intake of the materials by active loading spouts have provided for selecting the rational design, kinematic and dynamic parameters of the operating devices as well as their operating conditions [5–7].

In the papers [4, 8], the parameters and operating conditions are determined for vertical and inclined screw conveyors transporting agricultural materials and the rational parameters are established for their operating devices. The results of the research into the motion of materials in screw conveyors with rotary casings are presented in the paper [9].

The development and analysis of screw conveyors as well as their contribution to the damaging of materials are discussed in the papers [10].

In view of the above-said, it is necessary to improve the existing and develop new designs of auger operating devices for screw conveyors, determine their optimum design and kinematic parameters that would provide for raising their performance in the process of transporting agricultural materials.

The application of swivel-joint screw operating devices can ensure high efficiency in the performance of loading/discharge processes in the transportation of bulk agricultural materials on curved routes.

The aim of the research was to raise the performance efficiency of flexible screw conveyors in the transportation of agricultural materials by way of developing new auger-type operating devices.

2 Materials and Methods

In order to improve the performance of screw conveyors, the authors have developed the design of a sectioned shaft with a radial ball connection between the sections

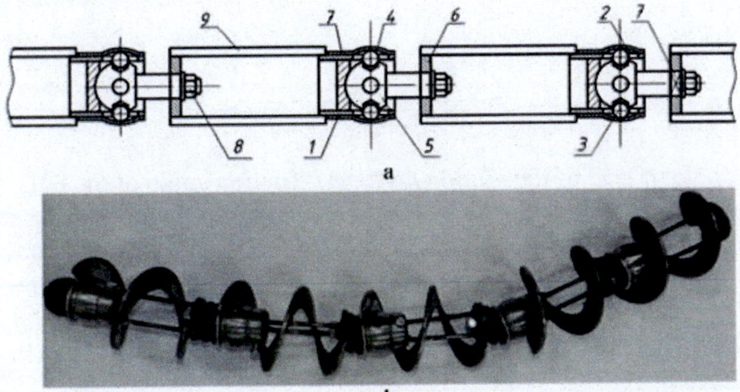

a

b

Fig. 1 Structural layout of articulated shaft **a** and picture of articulated-section screw operating device **b**

(Fig. 1). At the right end of each section, the cylindrical bushing 1 is fixed. The latter has a system of parallel slots 2, which are placed with equal spacing on the circumference and interface the rolling elements 3. On the other side, the rolling elements 3 are contained in the inner spherical surface of the bushing 4, which enables the swivelling rotation of the spherical pivot 5.

On the opposite side of each section, inside its left end, the connecting bushing 6 equipped with the square or profiled hole 7 is rigidly installed normally to the centreline of the section. The respective end of the spherical pivot 5 from the adjacent section is inserted into the hole 7 and rigidly fixed in it with the nut 8. On the outer circumference of the connecting bushing 6, the linking rods 9 are rigidly fixed with equal spacing. The opposite ends of the linking rods 9 are connected to the outer surface of the bushing 1. Onto each cylindrical section, a screw section is attached (Fig. 1).

When a section of the auger is rotated, the rotational motion is transferred via the rolling elements 3 to the spherical pivot 5 and to the adjacent screw sections.

In order to analyse the process of the bulk material transportation by the screw conveyor with an articulated-section screw operating device, the axial velocity of the bulk material in a horizontal high-speed screw conveyor is determined with the use of the following relation [11–13]:

$$V_{ax} = \frac{T_s(\omega - \omega_k)}{2\pi} \tag{1}$$

where T_s – lead of helix; ω – angular velocity of the bulk material; ω_k – angular velocity of the circumferential component of motion of the bulk material. The relation (1) can be represented in the following form:

$$V_{ax} = \frac{T_s\omega}{2\pi} \cdot k_v$$

where

$$k_v = \frac{k_s}{tan\,\theta\, tan(\theta + \varphi) + 1}, \tag{2}$$

where: θ – angle of lead of the section helix; φ – friction angle of the bulk material with respect to the surface of the helix, where $\varphi = arctan\,\mu_1$; μ_1- friction factor; k_s – coefficient that represents the design solution of the sectioned screw conveyor, k_s = 0.98–0.99.

In case the bulk material is transported at an angle of α to the horizontal surface, its axial velocity is reduced, which is taken into account with the coefficient k_α:

$$V_{ax\alpha} = V_{ax} \cdot k_\alpha. \tag{3}$$

In case the sectioned screw conveyor is positioned vertically [14–16]:

$$k_\alpha = 1 - S_c = 1 - \frac{\omega_k}{\omega} = 1 - \sqrt{\frac{tan(\theta + \varphi)}{\mu_2 \rho}} \tag{4}$$

where S_c – coefficient of dynamic similarity; ρ – power-speed coefficient, $\rho = \frac{\omega^2 R_c}{g}$; R_c – external radius of helix, g – acceleration of gravity, $g = 9.81$ m s^{-2}.

Accordingly, the axial transportation velocity of the bulk material in the vertical part of the flexible screw conveyor is equal to:

$$V_{ax\,B} = \frac{T_s \omega}{2\pi} \cdot \frac{k_s \left[1 - \sqrt{\frac{g\, tan(\theta+\varphi)}{\omega^2 \mu_2 R_c}}\right]}{tan\,\theta\, tan(\theta + \varphi) + 1}. \tag{5}$$

In case of a flexible conveyor, in which the route turns in space, but the loading is done in the horizontal direction, under the condition of maintaining a constant capacity of $Q = const$ along the route, the coefficient of charge ψ at the loading point has to take into account its changing as a result of the changes in the axial velocity of transportation [17]:

$$\psi_o V_o = \psi_\alpha V_\alpha. \tag{6}$$

Therefore, the amount of the change in V_α on the transportation route is extremely important. The change in the axial velocity of transportation can be approximated by the parabola of the following form:

$$V_\alpha = V_h - (V_h - V_w)\left(\frac{2\alpha}{\pi}\right)^\lambda, \tag{7}$$

where V_h and V_w – horizontal and vertical components of the transportation velocity, respectively. for $\alpha = 90°$ we can write $V_w = V_h \cdot k_\alpha$.

Taking into account the expression (4),

$$V_\alpha = V_h\left[1 - S_c\left(\frac{2\alpha}{\pi}\right)^\lambda\right]; \tag{8}$$

or

$$V_\alpha = V_h\left[1 - \frac{1}{\omega}\sqrt{\frac{g\,tan(\theta+\varphi)}{\mu_2 R_c}}\left(\frac{2\alpha}{\pi}\right)^\lambda\right] \tag{9}$$

where the parameter λ depends on the rheological properties of the bulk material and to a first approximation can be assumed to be equal to $\lambda = 2$. Therefore, the curve presented in expression (9) can be considered a parabola and the dependence of the velocity V_α on the angle α can, in the first approximation, be considered parabolic.

In the case of a freely suspended screw conveyor, the centreline of the flexible auger has the form of a catenary line, which varies according to the following relation:

$$y = a\left(ch\frac{x}{a} - 1\right), \tag{10}$$

where the parameter a depends on the elevation of the discharge end opening and can be determined from the following relation:

$$\frac{h}{a} + 1 = ch\left(\frac{l_x}{a}\right), \tag{11}$$

where h – elevation; l_x – horizontal projection of the conveyor length.

The current angle of pipeline inclination is determined by the following relation:

$$tan\,\alpha = \frac{dy}{dx} = sh\frac{x}{a}, \tag{12}$$

whence

$$\alpha = arctan\left(sh\frac{x}{a}\right). \tag{13}$$

Hence, in case of a freely suspended flexible (sectioned) screw conveyor, the axial velocity of the bulk material transportation is equal to:

$$V_{ax\alpha} = V_h\left[1 - \frac{1}{\omega}\sqrt{\frac{g\,tan(\theta+\varphi)}{\mu_2 R_c}}\left(\frac{2\,arctan\left(sh\frac{x}{a}\right)}{\pi}\right)^\lambda\right]. \tag{14}$$

The coefficient of charge ψ_α in the most unfavourable cross-section may not exceed $\psi_\alpha \leq 0.7$. Accordingly, in the loading area, the rational coefficient of charge has, in accordance with (6), to be equal to:

$$\psi_o = \frac{\psi_\alpha V_\alpha}{V_o} = 0.7 \cdot k_\alpha. \quad (15)$$

The screw conveyor capacity is determined by the following relation:

$$Q = \pi \psi_o \cdot V_{ax\alpha}\left(R_k^2 - R_c^2\right), \quad (16)$$

where R_k – curvature radius of the pipeline.

The capacity of the screw conveyor with a changing transportation route is determined on the basis of the following relation:

$$N = Q \cdot L \cdot w, \quad (17)$$

where w – specific power input for the transportation:

$$w = \frac{\mu_2 \cdot \rho \cdot R_c^2 \cdot \omega_f^2 \cdot \omega \cdot cos\,\beta}{V_f}, \quad (18)$$

where L – length of transportation; ω_f – angular velocity of the section; V_f – axial speed of material transportation by a hinged-sectional working body of a screw conveyor along a curved track; β – is the angle of inclination of the material transportation trajectory.

The axial velocity and capacity of the bulk material transportation performed by the articulated-section operating device of the screw conveyor on the curved route have been determined.

For a flexible (sectioned) screw conveyor with a changing route, the elementary power is input for the transportation of the bulk material on an interval dl and is determined by the following relation [18]:

$$dN = Q \cdot w(l)dl, \quad (19)$$

where $w(l) = \frac{\mu_2 \rho R_c^2 \omega \omega_f^2(l)\, cos[\beta(l)]}{V_f(l)}$.

In accordance with (1), the angular velocity of the flow is found by the following formula:

$$\omega_{f\alpha} = \omega - \frac{2\pi V_{ax\alpha}}{T_s}. \quad (20)$$

The inclination angle β of the bulk material transportation trajectory is determined by the formula:

$$tan\,\beta = \frac{V_{ax\alpha}}{\omega_{f\alpha} R_c} = \frac{V_{ax\alpha}}{\left[\omega - \frac{2\pi V_{ax\alpha}}{T_s}\right] R_c}. \quad (21)$$

Axial speed of material transportation by a hinged-sectional working body of a screw conveyor with a curved track V_f take equal $V_{ax\alpha}$.

Hence, the bulk material transportation capacity of the flexible (sectioned) screw conveyor with a changing transportation route is determined by the following relation:

$$N = Q \int_0^L w(l)dl = Q\mu_2 \rho R_c^2 \omega \int_0^L \frac{\left(\omega - \frac{2\pi V_{ax\alpha}}{T}\right)^2 cos\,\beta}{V_{ax\alpha}} dl. \quad (22)$$

It should be noted that $V_{ax\alpha}$, depends on the length of the transport, since x is a horizontal projection L, i.e. length of transportation.

From the obtained analytical dependence (23), it can be seen that with an increase in the angle β of inclination of the material transportation trajectory, the values of the energy-power parameters of this screw conveyor increase, and an increase in the pitch of the screw turns in the direction of material movement leads to a decrease in the energy-power parameters of transportation.

In the ball joint drive, the same as in case of the universal-joint drive, the swivel joint transfers the rotary motion with certain non-uniformity, which is caused by the shift of the sections with respect to each other due to the motion of the balls.

The basic fixed coordinate system $Oxyz$ is chosen as follows. Its origin O is placed at the centre of the swivel joint, the axis Ox is aligned with the centreline of the first section, the axis Oz is aligned with the axis of rotation with respect to the first section (Fig. 2). Accordingly, the rotated (by an angle of α) coordinate system has its axis $O'x'$ aligned with the centreline of the second section, its axis $O'z'$ is aligned with the axis Oz.

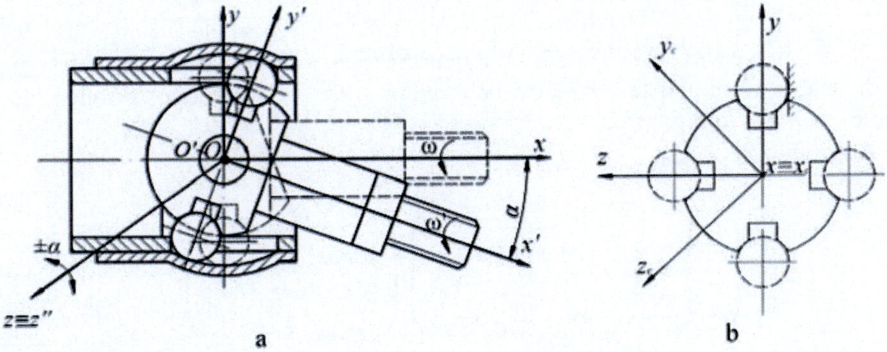

Fig. 2 Schematic model of swivel-joint drive: **a** – side view; **b** – front view

Additionally, for each section its own couple of coordinate systems is assigned – $O_c x_c y_c z_c$ and $O_c' x_c' y_c' z_c'$, which are rigidly connected with the respective sections and at the initial instance $t = 0$ coincide with the above-mentioned basic and rotated coordinate systems.

For the transition from one coordinate system to the other, the homogeneous coordinate systems $Oxyz1$, $O'x'y'z'1$, $O_c x_c y_c z_c 1$, $O_c' x_c' y_c' z_c' 1$ are used.

Hence, the rotation of one section ($O_c' x_c' y_c' z_c'$) in the basic coordinate system $Oxyz$ about the axis Ox can be represented in the matrix form as follows:

$$\begin{pmatrix} x \\ y \\ z \\ 1 \end{pmatrix} = \begin{pmatrix} cos(\omega t) & -sin(\omega t) & 0 & 0 \\ sin(\omega t) & cos(\omega t) & 0 & 0 \\ 0 & 0 & 1 & 0 \\ 0 & 0 & 0 & 1 \end{pmatrix} \begin{pmatrix} x_c \\ y_c \\ z_c \\ 1 \end{pmatrix} \tag{23}$$

In the coordinate system $O_c x_c y_c z_c$, the equation of the working surface of the upper slot is as follows:

$$-z_c - r = 0, \tag{24}$$

where r – radius of the ball (half of the slot width).

In the basic coordinate system, the equation of the rotating surface of the slot is as follows:

$$y\,sin(\omega t + \varphi_i) - z\,cos(\omega t + \varphi_i) - r = 0, \tag{25}$$

where $\varphi_1 = 0$ for the upper slot, $\varphi_i = \frac{\pi}{2}(i - 1)$ for the slot of the following i-th ball.

The rotation of the coordinate system $O_c' x_c' y_c' z_c'$ relative to the rotated coordinate system $O_c x_c y_c z_c$ is described by the similar relation (23).

Accordingly, the i-th ball will have the following current coordinates of its centre in the coordinate system $O'x'y'z'$:

$$C_i'\left\{x_{ci}' = 0;\ y_{ci}' = R\,cos(\omega' t + \varphi');\ z_{ci}' = sin(\omega' t + \varphi')\right\}.$$

The relation between the system $Oxyz$ and the rotated coordinate system $O'x'y'z'$ in the matrix form is determined by the rotation of the second section through the angle (-α) (acc. to Fig. 2).

Accordingly:

$$\begin{pmatrix} x \\ y \\ z \\ 1 \end{pmatrix} = \begin{pmatrix} cos\,\alpha & sin\,\alpha & 0 & 0 \\ -sin\,\alpha & cos\,\alpha & 0 & 0 \\ 0 & 0 & 1 & 0 \\ 0 & 0 & 0 & 1 \end{pmatrix} \begin{pmatrix} 0 \\ R\,cos(\omega' t + \varphi') \\ R\,sin(\omega' t + \varphi') \\ 1 \end{pmatrix} \tag{26}$$

In the fixed coordinate system $O'x'y'z'$, the current coordinates of the balls rolling together with the second section are:

$$\begin{aligned} x_{ci} &= R\,sin\,\alpha\,sin(\omega' t + \varphi_i^{'}); \\ y_{ci} &= R\,cos\,\alpha\,cos(\omega' t + \varphi_i^{'}); \\ z_{ci} &= R\,sin\,\alpha\,cos(\omega' t + \varphi_i^{'}). \end{aligned} \tag{27}$$

The next step is to break provisionally the kinematic link between the sections and allow each of them to move independently with the same angular velocity: $\omega' = \omega$.

In that case, the coordinates of the ball centres will be described by the relations (27) at $\omega' = \omega$, $\varphi' = \varphi$, their distances from each of the flat surfaces of the slots are determined by means of substituting their coordinates into the Eq. (25):

$$l = |R\,cos\,\alpha\,cos(\omega t + \varphi_i)sin(\omega t + \varphi_i) - R\,sin\,\alpha\,cos(\omega t + \varphi_i) - r|. \tag{28}$$

It is obvious that at $t = 0$, $l = r$, that is, ball 1 touches the surface of the slot. Also, due to the smallness of the angle α, we can take $sin\,\alpha \approx sin(\omega t + \varphi_i)$.

The expression (28) can be transformed as follows:

$$\begin{aligned} l = &|R(cos\,\alpha - 1 + 1)cos(\omega t + \varphi_i)sin(\omega t + \varphi_i) - R\,sin(\omega t + \varphi_i) \\ &cos(\omega t + \varphi_i) - r| \\ &= |R(cos\,\alpha - 1)cos(\omega t + \varphi_i)sin(\omega t + \varphi_i) - r| \\ &= \left|\frac{R\,sin[2(\omega t + \varphi_i)](1 - cos\,\alpha)}{2} + r\right|. \end{aligned} \tag{29}$$

At the current instance t, the ball in the broken kinematic link lags behind the surface of the slot by the following amount:

$$\Delta l = l - r = \frac{R\,sin[2(\omega t + \varphi_i)](1 - cos\,\alpha)}{2} \tag{30}$$

The analysis of the relation (25) indicates that at the instance, when $\varphi_1 = 0$, $\varphi_2 = \frac{\pi}{2}$, $\varphi_3 = \pi$, $\varphi_4 = \frac{3\pi}{2}$, there is no clearance between the ball and the surface of the slot. However, the clearance in the intermediate states results in the following angular deviation $\Delta\varphi$ of the ball:

$$\Delta\varphi = arcsin\{(1 - cos\,\alpha)sin[2(\omega t + \varphi_i)]\}.$$

Respectively, when the kinematic link is closed, the second section will be displaced relative to the first one through the following angle of rotation:

$$\omega' t + \varphi_i^{'} = \omega t + \varphi_i - \Delta\varphi.$$

Thus, the angular velocity of the next section is determined by the following relation:

$$\omega' = \omega - \frac{\Delta\varphi}{t} = \omega - \frac{arcsin\{(1 - cos\,\alpha)sin[2(\omega t + \varphi_i)]\}}{t}. \quad (31)$$

The analysis of the relation proves that the angular velocities of the adjacent sections are almost equal to each other.

In order to carry out experimental research into the performance of the screw conveyors in the transportation of agricultural materials, a test unit of the combined screw conveyor has been developed. The unit comprises the loading pipeline 1 and the discharge pipeline 2, as shown in Fig. 3a. They are made in the form of closed casings 3 and 4 with round cross-sections, in which the helical spirals 5 and 6 are installed. The helical spirals are connected to the drive shafts 7 and 8 of the motors 9 and 10.

In the area of the drive shafts, the pipelines are connected with each other via the two Sects. 11 and 12 of the transfer pipe. Each section is connected at one end to the opening in the closed casing 3 or 4, respectively, their other ends are connected with each other. The helical spirals are installed so that their centrelines coincide with the axes of the motor drive shafts. The pipelines in the area of their connection by the transfer pipe sections can be positioned as in the horizontal plane (Fig. 3b), so in the vertical one (Fig. 3c).

In the course of operation, the bulk material is fed into the intake area of the loading pipeline, then the spiral 5 in the casing transports it towards the transfer

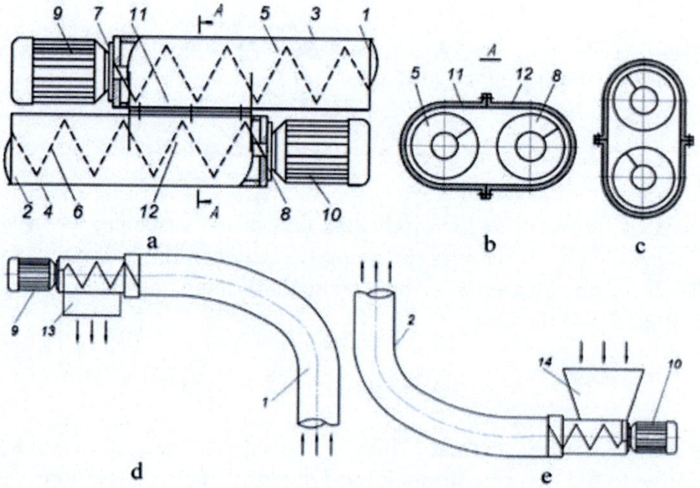

Fig. 3 Schematic model of flexible screw conveyor: **a** – general view; **b** – horizontal arrangement of pipelines; **c** – vertical arrangement of pipelines; **d** – single pipeline alternative of loading conveyor; **e** – single pipeline alternative of discharge conveyor

pipe. Thereafter, the material from the transfer pipe is transferred onto the spiral 6 and transported to the discharge area.

The conveyor can also be operated as a single pipeline. In that case, the transfer pipe sections are equipped with respective special devices: discharge window 13 is installed on the loading pipeline, loading hopper 14 – on the discharge pipeline.

In the first case, the single pipeline conveyor operates in the intake or loading mode (Fig. 3d), in the second case – in the feeding or discharge mode (Fig. 3e).

During the experiments, the bulk agricultural material is fed from the hopper into the loading pipeline, the operating device transports it to the transfer pipe of the test unit, then the operating device in the discharge pipeline carries it to the discharge area.

For the analysis of the obtained results, the graphic relations have been plotted between the torque T and the motor capacity N, on the one hand, and the operating device rotation frequency n, on the other hand, at different values of the bulk material lifting height h and the process pipeline curvature radius R_k. The peak (maximum) values of the data obtained in the experiments have been used for plotting the diagrams.

In the experiments, the values of torque and motor capacity were recorded as percentage of the nominal ratings. The motor capacity was determined as the product of the motor power rating (2.2 kW) and the value for the selected mode of operation. The same was applied to the torque.

The experimental investigations were carried out with the use of the 3.5 m long material transportation pipeline. Wheat grain, peas and commercial salt were used as the bulk material.

3 Results and Discussion

Basing on the results of the completed experiments, the graphic relations have been plotted that show how the values of torque T and capacity N change with the operating device drive shaft rotation frequency, in case the process pipeline is fully filled with bulk agricultural material (Fig. 4).

The analysis of the graphic relations has proved that a rise in the operating device rotation frequency n from 300 to 600 rpm results in the reduction of the torque T, the decrease is $\Delta T = 7.4\%$ for wheat grain, for commercial salt $\Delta T = 5\%$.

As regards the capacity N, the relation shows a rather clearly marked trend of the linear growth of N with the increase of the operating device rotation frequency. For example, in case of wheat grain $\Delta N = 54\%$, in case of commercial salt $\Delta N = 59\%$.

Also, experimental research has been carried out with a purpose of determining the effect of changes in the material lifting height h and process pipeline curvature radius R_k on the values of the torque T and capacity N of the conveyor drive, when transporting commercial salt, wheat and peas.

In view of the fact that in the previous research the effect of changes in the operating device rotation frequency on the value of the torque was analysed in detail,

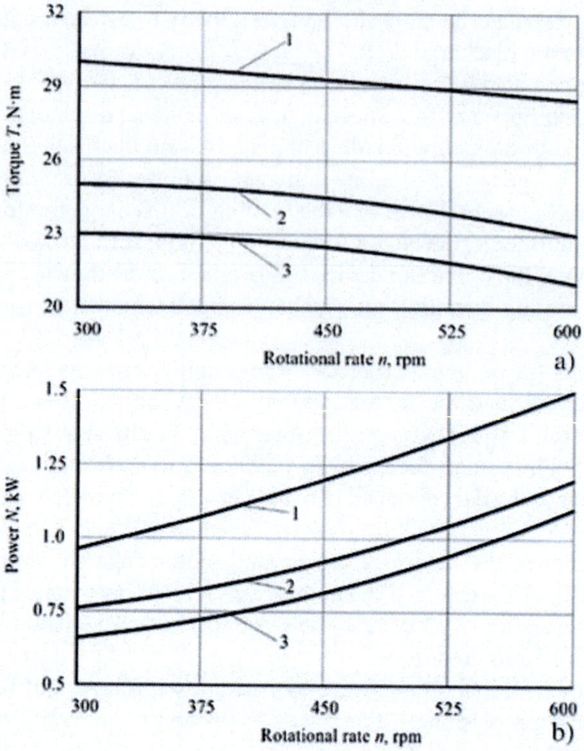

Fig. 4 Graphic relations between torque T **a** and power N **b** and operating device rotation frequency n: 1 – commercial salt; 2 – peas; 3 – wheat

in the described experiments the operating device rotation frequency was maintained constant, its value was equal to 450 rpm.

Using the results of the experimental research, the graphic relations have been plotted between the value of the torque output by the drive of the articulated-section screw operating device, on the one hand, and the height h of the material motion path (Fig. 5a) and the curvature radius R_k of the process pipeline (Fig. 5b), on the other hand.

The analysis of the diagrams (Fig. 5a) indicates that an increase in the process pipeline curvature radius R_k from 0.6 to 1.8 m results in the reduction of the torque T, the decrease ΔT is equal to 14% for wheat grain, for peas $\Delta T = 13\%$, for commercial salt $\Delta T = 10\%$.

The analysis of the graphic relations (Fig. 5b) has proved that an increase in the height of the bulk material motion path from 0.3 to 1.7 m brings about a growth in the torque T, where the increase ΔT is equal to 25% for wheat grain, for peas ΔT $= 30\%$, for commercial salt $\Delta T = 29\%$.

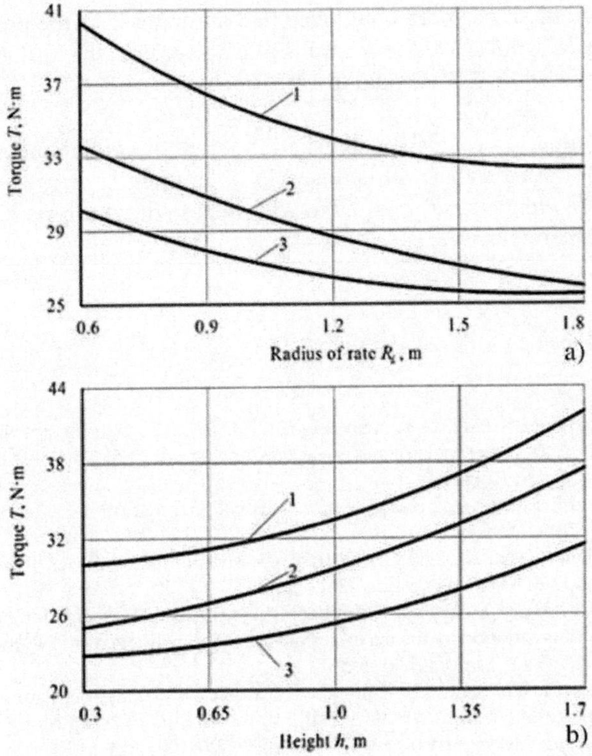

Fig. 5 Relations between torque T imparted to operating device and process pipeline curvature radius R_k **a** and material motion path height h **b** at $n = 450$ rpm: 1 – commercial salt; 2 – wheat; 3 – peas

4 Conclusions

On the basis of the completed research into the power and load parameters of flexible screw conveyors, the graphic relations have been plotted that show, how the values of the torque T and capacity N change during the transportation of bulk material by a screw conveyor, when the conveyor operating device drive shaft rotation frequency changes, provided that the process pipeline is fully filled with grain material.

It has been established that an increase in the operating device rotation frequency n from 300 to 600 rpm results in a decrease in the torque T, where the decrease ΔT is equal to 7.1% for wheat grain, 7.4% for peas, for commercial salt $\Delta T = 5\%$.

As regards the value of the capacity N, the obtained graphic relations show a rather clearly marked trend of its linear growth, when the operating device rotation frequency is increased. For example, in case of wheat grain $\Delta N = 50\%$, in case of peas $\Delta N = 54\%$, for commercial salt $\Delta N = 59\%$.

On the basis of the results of the completed research into the effect of changes in the curvature radius of the process pipeline with an articulated-section operating device on the torque, it has been concluded that an increase in the process pipeline curvature radius R_k from 0.6 to 1.8 м results in a decrease in the torque T. In case of wheat grain the decrease ΔT is equal to 14%, in case of peas $\Delta T = 13\%$, in case of commercial salt $\Delta T = 10\%$.

At the same time, it has been established that the torque T increases, when the height, to which the bulk material is raised in its motion, is increased from 0.3 to 1.7 m. The torque increase ΔT is equal to 25% in case of wheat grain, for peas $\Delta T = 30\%$, for commercial salt $\Delta T = 29\%$.

References

1. Hevko R, Vitrovyi A, Klendii O, Liubezna I (2017) Design engineering and substantiation of the parameters of sectional tools of flexible screw conveyers. Bull Transilvania Univ Brasov 10(59):39–46. Brasov, Romania
2. Lech M (2001) Mass flow rate measurement in vertical pneumatic conveying of solid. Powder Technol 114(1–3):55–58
3. Hevko RB, Klendiy MB, Klendiy OM (2016) Investigation of a transfer branch of a flexible screw conveyer. INMATEH Agric Eng 48(1):29–34
4. Rogatynska O, Liashuk O, Peleshok T, Liubachivskyi R (2015) Investigation of the process of loose material transportation by means of inclined screw conveyers. Bull I Pyliui Ternopil Natl Tech Univ 79:137–143. Ternopil/Ukraine
5. Bulgakov V, Pascuzzi S, Ivanovs S, Kaletnik G, Yanovich V (2018) Angular oscillation model to predict the performance of a vibratory ball mill for the fine grinding of grain. Biosyst Eng 171:155–164. https://doi.org/10.1016/j.biosystemseng.2018.04.021
6. Manetto G, Cerruto E, Pascuzzi S, Santoro F (2017) Improvements in citrus packing lines to reduce the mechanical damage to fruit. Chem Eng Trans 58:391–396. https://doi.org/10.3303/CET1758066
7. Bulgakov V, Pascuzzi S, Adamchuk V, Kuvachov V, Nozdrovicky L (2019) Theoretical study of transverse offsets of wide span tractor working implements and their influence on damage to row crops. Agriculture 9:144. https://doi.org/10.3390/agriculture9070144
8. Rohatynskyi RM, Diachun AI, Varian AR (2016) Investigation of kinematics of grain material in a screw conveyor with a rotating casing. Bull Kharkiv Petro Vasylenko Natl Tech Univ Agric 168:24–31. Kharkiv/Ukraine
9. Lyashuk OL, Rogatynska OR, Serilko DL (2015) Modeling of the vertical screw conveyer loading. INMATEH Agric Eng 45(1):87–94
10. Hevko RB, Rozum RI, Klendiy OM (2016) Development of design and investigation of operation processes of loading pipes of screw conveyors. INMATEH Agric Eng 50(3):89–94
11. Bulgakov V et al (2021) Study of the steering of a wide span vehicle controlled by a local positioning system. J Agric Eng LII:1144. https://doi.org/10.4081/jae.2021.1144
12. Bulgakov V, Pascuzzi S, Nadykto V, Ivanovs S, Adamchuk V (2021) Experimental study of the implement-and-tractor aggregate used for laying tracks of permanent traffic lanes inside controlled traffic farming systems. Soil Tillage Res 208:104895. https://doi.org/10.1016/j.still.2020.104895
13. Bulgakov V, Pascuzzi S, Ivanovs S, Nadykto V, Nowak J (2020) Kinematic discrepancy between driving wheels evaluated for a modular traction device. Biosyst Eng 196:88–96. https://doi.org/10.1016/j.biosystemseng.2020.05.017
14. Bulgakov V, Pascuzzi S, Adamchuk V, Ivanovs S, Pylypaka S (2019) A theoretical study of the limit path of the movement of a layer of soil along the plough mouldboard. Soil Tillage Res 195:104406. https://doi.org/10.1016/j.still.2019.104406

15. Bulgakov V, Pascuzzi S, Beloev H, Ivanovs S (2019) Theoretical investigations of the headland turning agility of a trailed asymmetric implement-and-tractor aggregate. Agriculture 9:224. https://doi.org/10.3390/agriculture9100224
16. Pascuzzi S, Anifantis AS, Santoro F (2020) The concept of a compact profile agricultural tractor suitable for use on specialised tree crops. Agriculture 10:123. https://doi.org/10.3390/agriculture10040123
17. Guerrieri AS, Anifantis AS, Santoro F, Pascuzzi S (2019) Study of a Large Square Baler with Innovative Technological Systems that Optimize the Baling Effectiveness. Agriculture 9(5):86. https://doi.org/10.3390/agriculture9050086
18. Bulgakov V, Pascuzzi S, Anifantis AS, Santoro F (2019) Oscillations analysis of front-mounted beet topper machine for biomass harvesting. Energies 12:2774. https://doi.org/10.3390/en12142774

Assessment of a Big Square Baler Able to Reduce Soil Impurities During Baling Process. First Evaluations

Arrigo Salvatore Guerrieri, Francesco Santoro, Francesco Vicino, and Simone Pascuzzi

Abstract In animal husbandry, results optimization is largely dependent on how the animals are fed so fodder-crops have a fundamental role in this sector. Dried fodder production has relevance in Southern Italy, even if in this territory the agronomic and meteorological conditions generate physical and micro-biological problems in the production of fodder. Some of the fundamental complications are connected to the amount of soil inside the fodder, which often is the vehicle of a large number of CFUs (colony forming units) of Clostridium bacteria, that could have dangerous effects on the welfare of the cows.

The aim of this paper is the evaluation of the performance of a large square baler, manufactured in Southern Italy, equipped with innovative devices which should allow to make prismatic bales compact and with less impurities if compared with the bales produced by similar square presses, built by other manufacturers. The presence and the amount of soil and impurities was assessed through first comparative tests carried out, considering also other two similar square presses. The obtained results, even if have to be validated through other in-depth tests, seem to confirm the benefits of the adopted technical solutions of the analysed large square baler.

Keywords Big square baler · Straw · Hay · Zootechny

1 Introduction

Hay bales are fundamentally used within the zootechny for the feeding of the animals and in this connection the quality of the fodder-crops is basic in the Euro-pean agricultural and economic sector [1]. Several mechanical treatments are involved in the manufacturing of big bales in order to obtain hygiene nutritive forage, even if the

A. S. Guerrieri
Department of Agriculture, Forests, Nature and Energy, University of Tuscia, Viterbo, Italy

F. Santoro (✉) · F. Vicino · S. Pascuzzi
Department of Agricultural and Environmental Science, University of Bari Aldo Moro, Bari, Italy
e-mail: francesco.santoro@uniba.it

S. Pascuzzi and F. Santoro (eds.), *Farm Machinery and Processes Management in Sustainable Agriculture*, Lecture Notes in Civil Engineering 289,
https://doi.org/10.1007/978-3-031-13090-8_7

drying of the forage is accompanied by a competing process of deterioration which can lead up to the loss of nutrients and toxic metabolic products [2]. Unfortunately, the production of fodder in Central and Southern Italy is affected by agronomic and meteorological conditions which generate physical and microbiological problems. Some of the fundamental complications are: (i) the high content of clostridia; (ii) the presence of terricolous anaerobic bacteria; (iii) the dried fodder bales high humidity rate [3]. The number of Clostridium genus bacteria colony forming units (CFU) is largely affected by the presence of soil residual inside the fodder; this event could have effects both for cow wellness and for the quality of milk and of its products [4, 5].

It's well know that balers pick up loose straw or hay and then the compress the picked product into same size and weight bales tying with twine [6–10]. Obviously, these machine have to be designed and set up in agreement with the safety laws in force, in order to carry out all the operation in compliance with the operators' safety [11–17]. The efficiency of a large square baler, produced by Southern Italy manufacturer, having the characteristic of an high cleanliness of the dried fodder is the final goal of this paper.

2 Cicoria HD 1270 Baler

The machine under test is the large square baler, Cicoria HD 1270 (hereinafter "Cicoria's machine") (Fig. 1).

This machine is equipped with a system that feeds from the top the pressing chamber, so allowing the baling to be optimized, making bales compact and with less impurities [18–21]. The feeding from the top greatly improves the cleaning of the product without damaging it, and without changing the nutritional level, thus obtaining a high-quality hay suitable for zootechnical feeding. The separation of soil and stone residues occurs by force of gravity without energy consumption during the quite long way carried out from the product before entering the compression chamber [22–25]. The thrust mechanism of the piston can produce a force of 18 ×

Fig. 1 Cicoria HD 1270 baler under test

10^3 kg in the pressing chamber and the compression of the product does not take place with a rectilinear sliding of the piston in the chamber, but through a semi-circular movement caused by the action of a system of rocker levers [24–27]. This compression system, absolutely different from the solutions commonly adopted on the market, allows a considerable reduction in the energy absorption of the machine. The piston can perform up to 48 strokes/min, with a higher stroke of 76 cm and a lower stroke of 36 cm requiring a power of 50–60 kW at the PTO (1000 rpm) of the tractor, considerably lower than other machines on the market with the same operational capacity. The supply chamber volume ranges from 0.63 to 0.92 m^3, and the baling capacity in terms of bales density varies according to the material: a) density of 180 kg/m^3 for straws; b) density of 260 kg/m^3 for fodder [26–32].

3 Materials and Methods

The presence and the quantity of soil and impurities inside the straw bales set up by the large square baler Cicoria HD 1270 was assessed by experimental tests in which different samples of straw were collected from prismatic bales produced both by the Cicoria machine and by the other two producers' similar square presses that, in the following, are named "producer 1" and "producer 2". The samples were gathered according to a randomized block experimental scheme in order to statistically evaluate impurities and soil quantity. Therefore, three theses were considering, corresponding to the samples obtained from the bales obtained respectively by: the Cicoria machine (thesis 1), the machine "producer 1" (thesis 2) and the machine "producer 2" (thesis 3). For each thesis three repetitions were considered. The collected samples were then processed in Dept. of Agricultural and Environmental Sciences – University of Bari Aldo Moro laboratory, in agreement with the following phases: a) the samples were subjected to ventilated drying at 65 °C for 3 h in a forced air oven ARGOLAB model TCF 200 (Argolab-XS Instruments srl, Carpi – Italy), in order to remove the humidity and stand-ardize the initial conditions for all the samples object of the case study; b) the sample were ground by means a RETSCH model 5657 grinder (Retsch Italia-Verder Scientific srl Industria, Torre Boldone, Bergamo district, Italy), to have 2 mm fineness; c) with ABT model 220-5DM (KERN & SOHN GmbH Industry – Balingen – Germany) electronic analytical balance, has been used a quantity of 2 g from each ground sample and placed into an furnace heat-resistance cup. A number of nine cups (considering the theses and the repetitions) were prepared. The filled cups were put, for 24 h at a temperature 550 °C, inside a Zetalab ZA (Zetalab s.r.l. Industry – Padova – Italy) muffle furnace for drying purpose. At the end of drying, cups contents were weighed to evaluate the difference between the initial and dried state in order to calculate percentage of the ashes content.

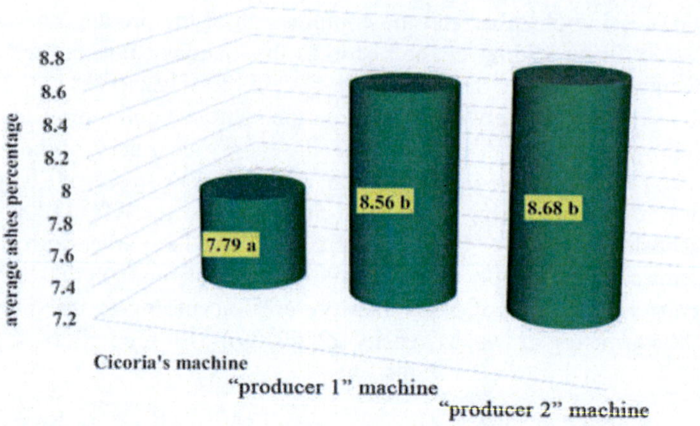

Fig. 2 Statistical analysis results of straw from the three compared balers data

4 Results and Discussion

To statistically analyze the data related to the three theses straw samples was carried out a variance analysis and significance level of 0.05% "Duncan" tests were adopted to compare obtained average values [33, 34]. The results emphasize that the Cicoria's machine produced straw bales with an average value of 7.79% as content of impurities and soil. This value is about 10% lower than machine by "producer 1" average value and 11.4% lower than the machine by "producer 2". Furthermore, it is significant the difference among the ashes percentages for the samples obtained from, respectively, the Cicoria's machine (thesis 1) and the "producer 1" (thesis 2) and "producer 2" (thesis 3) machines. Finally, it is not significant the difference between the ash percentages concerning the "producer 1" (thesis 2) and "producer 2" (thesis 3) machines (Fig. 2).

5 Conclusions

First experimental evaluation of the performance of the "Cicoria Square Bales" HD 1270 large square baler were carried out. This machine is equipped with innovative devices, which improves the effectiveness in producing square bales using less energy despite other machines. The obtained results, even if have to be validated through other in-depth tests, seem to confirm the benefits of the adopted technical solutions. Further tests are in progress to better evaluate square bales content of impurities and soil, considering different operative conditions such as the type of swath, the harvesting height from soil and the soil texture and humidity.

References

1. Guerrieri AS, Anifantis AS, Santoro F, Pascuzzi S (2019) Study of a large square baler with innovative technological systems that optimize the baling effectiveness. Agriculture 9(5):86. https://doi.org/10.3390/agriculture9050086
2. Cazzato E, Laudadio V, Tufarell V (2012) Effects of harvest period, nitrogen fertilization and mycorrhizal fungus inoculation on triticale (xTriticosecale Witt-mack) forage yield and quality. Renew Agric Food Syst 27:278–286
3. Cazzato E et al (2013) Forage yield and quality of emmer (Triticum dicoccum Schübler) and spelt (Triticum spelta L.) as affected by harvest period and nitrogen fertilization. Soil Plant Sci 63:571–578
4. Wilkinson JM, Davies DR (2013) The aerobic stability of silage: key findings and recent developments. Grass Forage Sci 68:1–19
5. Shinners KJ, Huenink BM, Muck RE, Albrecht KA (2009) Storage characteristics of large round and square alfalfa: low moisture wrapped bales. Trans ASABE 52:401–407
6. Maguire SM, Godwin RJ, O'Dogherty MJ, Blackburn K (2007) A dynamic weighing system for determining individual square bale weights during harvesting. Biosyst Eng 98:138–145
7. Manetto G, Cerruto E, Pascuzzi S, Santoro F (2017) Improvements in citrus packing lines to reduce the mechanical damage to fruit. Chem Eng Trans 58:391–396. https://doi.org/10.3303/CET1758066
8. Cerruto E, Manetto G, Santoro F, Pascuzzi S (2018) Operator dermal exposure to pesticides in tomato and strawberry greenhouses from hand-held sprayers. Sustainability 10:2273. https://doi.org/10.3390/su10072273
9. Pascuzzi S, Santoro F (2017) Evaluation of farmers' OSH hazard in operation nearby mobile telephone radio base stations. In: 16th International Scientific Conference Engineering for rural development Proceedings, vol 16. Jelgava, Latvia, 24–26 May 2017, pp 748–755. ISSN: 1691–5976. https://doi.org/10.22616/ERDev2017.16.N151
10. Pascuzzi S, Santoro F (2017) Analysis of possible noise reduction arrangements inside oive-oil mills: a case study. Agriculture (Switzerland) 7(10):88. https://doi.org/10.3390/agriculture7100088
11. Pascuzzi S, Santoro F (2017) Analysis of the almond harvesting and hulling mechanization process. A case study. Agriculture (Switzerland) 7(12):100. https://doi.org/10.3390/agriculture7120100
12. Pascuzzi S, Anifantis AS, Santoro F (2020) The concept of a compact profile agricultural tractor suitable for use on specialised tree crops. Agriculture 10:123. https://doi.org/10.3390/agriculture10040123
13. Santoro F, Anifantis AS, Ruggiero G, Zavadsky V, Pascuzzi S (2019) Lightning protection systems suitable for stables: a case study. Agriculture 9:72. https://doi.org/10.3390/agriculture9040072
14. Pascuzzi S (2016) Outcomes on the spray profiles produced by the feasible adjustments of commonly used sprayers in "Tendone" vineyards of Apulia (Southern Italy). Sustainability 8(12):1307. https://doi.org/10.3390/su8121307
15. Pascuzzi S, Blanco I, Anifantis AS, Scarascia Mugnozza G (2016) Hazards assessment and technical actions due to the production of pressured hydrogen within a pilot photovoltaic-electrolyzer-fuel cell power system for agricultural equipment. J Agric Eng XLVII:507:89–93. https://doi.org/10.4081/jae.2016.507
16. Sun Y et al (2010) An improved penetrometer technique for determining bale density. Biosyst Eng 105:273–277
17. Bulgakov V et al (2021) Study of the steering of a wide span vehicle controlled by a local positioning system. J Agric Eng LII:1144. https://doi.org/10.4081/jae.2021.1144
18. Bulgakov V, Pascuzzi S, Nadykto V, Ivanovs S, Adamchuk V (2021) Experimental study of the implement-and-tractor aggregate used for laying tracks of permanent traffic lanes inside controlled traffic farming systems. Soil Tillage Res 208:104895. https://doi.org/10.1016/j.still.2020.104895

19. Bulgakov V, Pascuzzi S, Ivanovs S, Nadykto V, Nowak J (2020) Kinematic discrepancy between driving wheels evaluated for a modular traction device. Biosyst Eng 196:88–96. https://doi.org/10.1016/j.biosystemseng.2020.05.017
20. Bulgakov V, Pascuzzi S, Adamchuk V, Ivanovs S, Pylypaka S (2019) A theoretical study of the limit path of the movement of a layer of soil along the plough mouldboard. Soil Tillage Res 195:104406. https://doi.org/10.1016/j.still.2019.104406
21. Bulgakov V, Pascuzzi S, Beloev H, Ivanovs S (2019) Theoretical investigations of the headland turning agility of a trailed asymmetric implement-and-tractor aggregate. Agriculture 9:224. https://doi.org/10.3390/agriculture9100224
22. Bulgakov V, Pascuzzi S, Adamchuk V, Kuvachov V, Nozdrovicky L (2019) Theoretical study of transverse offsets of wide span tractor working implements and their influence on damage to row crops. Agriculture 9:144. https://doi.org/10.3390/agriculture9070144
23. Bulgakov V, Pascuzzi S, Anifantis AS, Santoro F (2019) Oscillations analysis of front-mounted beet topper machine for biomass harvesting. Energies 12:2774. https://doi.org/10.3390/en12142774
24. Pascuzzi S, Santoro F, Manetto G, Cerruto E (2018) Study of the correlation between foliar and patternator deposits in a "Tendone" vineyard. Agric Eng Int CIGR J 20(3):97–107
25. Bulgakov V, Pascuzzi S, Ivanovs S, Kaletnik G, Yanovich V (2018) Angular oscillation model to predict the performance of a vibratory ball mill for the fine grinding of grain. Biosyst Eng 171:155–164. https://doi.org/10.1016/j.biosystemseng.2018.04.021
26. Pascuzzi S, Cerruto E (2015) An innovative pneumatic electrostatic sprayer useful for tendone vineyards. J Agric Eng XLVI:458:123–127. https://doi.org/10.4081/jae.2015.458
27. Pascuzzi S (2015) A multibody approach applied to the study of driver injures due to a narrow-track wheeled tractor rollover. J Agric Eng 46:105–114. https://doi.org/10.4081/jae.2015.466
28. Pascuzzi S, Santoro F (2015) Exposure of farm workers to electromagnetic radiation from cellular network radio base stations situated on rural agricultural land. Int J Occup Saf Ergon 21(3):351–358. https://doi.org/10.1080/10803548.2015.1081774
29. Blanco I, Pascuzzi S, Anifantis AS, Scarascia Mugnozza G (2014) Study of a pilot photovoltaic-electrolyzer-fuel cell power system for a geothermal heat pump heated greenhouse and evaluation of the electrolyzer efficiency and operational mode. J Agric Eng XLV:238:111–118. (eISSN 2239-6268). https://doi.org/10.4081/jae.2014.238
30. Blanco I, Sotirios Anifantis A, Pascuzzi S, Scarascia Mugnozza G (2013) Hydrogen and renewable energy sources integrated system for greenhouse heating. J Agric Eng 44,e45; 226–230. https://doi.org/10.4081/jae.2013.(s1):e45
31. Baldoin C, Balsari P, Cerruto E, Pascuzzi S, Raffaelli M (2008) Improvement in pesticide application on greenhouse crops: results of a survey about greenhouse structures in Italy. Acta Hortic.801:609–614. https://doi.org/10.17660/ActaHortic.2008.801.69
32. Gomez KA, Gomez AA (1976) Statistical Procedures for Agricultural Research, Second edn. Wiley-Interscience, Hoboken. ISBN 978–0471870920
33. Román FD, Hensel O (2014) Numerical simulations and experimental measurements on the distribution of air and drying of round hay bales. Biosyst Eng 122:1–15 (2014)
34. Bianchi B, Tamborrino A, Santoro F (2013) Assessment of the energy and separation efficiency of the decanter centrifuge with regulation capability of oil water ring in the industrial process line using a continuous method. J Agric Eng XLIV, 2S:278–282J. https://doi.org/10.4081/jae.2013.298

The Methane Potential of Corn Cob Cores

Artur Pawłowski, Dawid Wojcieszak, and Jacek Przybył

Abstract Constant climate change and political changes are leading to the search for better solutions in the field of energy materials. An important milestone in the near future will also be the introduction of carbon footprint measurements by the EU. Many companies will have to think about reducing the carbon footprint of their products in order to still be able to compete with western companies. In order to do so, they may need to invest in biogas installations. These installations neutralize biodegradable waste and at the same time provide the possibility to produce electricity and heat. It should be emphasized that methane fermentation is a complicated process of biological decomposition of organic matter carried out under anaerobic conditions. Chemical properties of corn cob cores were determined by elemental analysis, and energy properties were evaluated by heat of combustion and heating value. The ash and volatile content of corn cobs. This is important because carbon (C) supporting various microbial groups affects anaerobic fermentation. Nitrogen (N) is the building material for the amino acids that are involved in protein synthesis and is an important element for microbial growth. The relation of the two elements has a very important effect on methane fermentation. In this paper, the biogas potential is presented based on crop residues of Ambrosini variety (FAO 220). It is one of the most popular early varieties grown in Poland on weaker and mosaic soils both for grain and for high energy silage.

Keywords Corn cob cores · Methane potential · Anaerobic fermentation

1 Introduction

One of the materials with very high potential is corn. Its high biogas potential is mainly due to the cobs with the grain. However, the use of final products such as grain for energy purposes is always controversial. Therefore, the use of maize straw

A. Pawłowski (✉) · D. Wojcieszak · J. Przybył
Department of Biosystems Engineering, Faculty of Environmental Engineering and Mechanical Engineering, Poznan University of Life Sciences, Poznan, Poland
e-mail: artur.pawlowski@up.poznan.pl

S. Pascuzzi and F. Santoro (eds.), *Farm Machinery and Processes Management in Sustainable Agriculture*, Lecture Notes in Civil Engineering 289,
https://doi.org/10.1007/978-3-031-13090-8_8

for these purposes including cob cores, leaves or stalks as inedible residues from maize production is a much better option as it does not disrupt or compete with food production in any way [2]. Additionally, the balance of carbon dioxide produced during biomass combustion is zero because it is absorbed in the renewable process of this fuel for example during photosynthesis [3]. Therefore, biomass is the most independent energy source.

Until recently, no one was seriously interested in this material. As it turns out, every year Polish fields produce up to 4 million Mg of corn straw [4]. Unfortunately, the residues that stay in the field are a reservoir for many pathogens including the well-known and problematic European corn borer (*Ostrinia nubilalis*). Therefore, the best way would be to collect and remove them from the field. Such a lignin- and cellulose-rich residue base, after earlier preparation, can be a good substrate for the production of new materials, including combustion and processing in biogas installations.

Corn crop residues are non-homogeneous and represent about 50% of the plant dry matter. They mainly consist of four fractions, stalks, leaves, cob cores otherwise known as setts, and cover leaves sometimes named husks [5]. Such parts of maize residues as stalks, cores, and leaves contain fibers composed of lignin, hemicellulose, and cellulose with the highest energy value, which are much more difficu to metabolizelt for methane bacteria. These polymers are difficult to biodegrade and are determined as lignocellulose.

Special attention should be focused on the FAO standard. This is a system of numerical symbols which is an indicator of the earliness of varieties. It informs about the length of the vegetation season needed to accumulate dry matter. The first number means the basic class of earliness. The second number tells about the earliness group within the basic class, and the third number informs about the color of the grain. It is important to say, that varieties with a higher FAO can produce a much higher yield. This is connected with the fact that the variety is a later, the vegetation season is longer. On the other hand, hybrids with a low FAO, for example early ones, yield less than medium-early and late ones, while they are characterized by a significantly higher content of cobs in the yield of green matter and dry matter. In this paper presents the energy potential of maize using the example of Ambrosini variety, which is an early variety.

2 Materials and Methods

2.1 Corn Cobs Cores and Its Characteristics

The research material consisted of corn cob cores the Ambrosini variety KWS company (FAO 220). The cob cores were collected on October 2020 from maize plantations located on a farm in Kiedrowo, Poland, near Wągrowiec [52.85196,

17.3860]. Corn cob cores were harvested by hand from the field. The yield of corn cobs Ambrosini variety was 3,3 Mg/ha of dry matter content 36%.

2.2 Laboratory Investigations

The content of total solids (TS) in the corn cobs corse was measured by drying and weighing at a temperature of 105 ± 1 °C, according to the standard EN 12,880: 2004. The content of organic dry matter (LOI) equivalent to loss on ignition (at 550 °C ± 25 °C) EN 12,879:2004 was measured in the corn cobs cores.

The ash percentage content was calculated from the weight of the combustion residue according to modified ISO 1171. Samples were heated to 710 ± 10 °C and kept at this temperature until a constant weight was reached. For this purpose, a muffle furnace was used.

The carbon, hydrogen, nitrogen and sulphur percentages were determined with a Flash 2000 elemental analyser (Thermo Fisher Scientific, USA) in CHNS/O configuration according to the EN 15,104 standard. The instrument was calibrated with standard Methionine (Thermo Fisher Scientific, USA).

2.3 Methane Yield

The methane yield in the maize stover fractions was measured according to the DIN standard 38 414-S8, in the Ecotechnology Laboratory at the Department of Biosystems Engineering, Poznan University of Life Sciences. Daily biogas production was measured every 24 h, with an accuracy of ± 0.01 dm^3. The qualitative and quantitative composition of the fermentation gases was determined with a Geotech GA5000 gas analyser whenever the gas volume in the tube exceeded 450 ml. The GA5000 gas analyser had the following quality certifications: ATEX II 2G Ex ib IIA T1 Gb (Ta ¼ 10 C to þ50 C), IECEx, CSA and calibration certificate UKAS ISO 17025.

2.4 Energetic Calculations

Energy was calculated in the basis at the results of corn cobs cores methane fermentation. Methane energy yield coefficient was 9.97 kWh/Nm3, the cogeneration unit efficiency was 40% for electrical energy and 45% for heat [1]

$$E_e = V_{CH_4} \cdot R_{aCH_4} \eta_e \text{ [MWh]} \quad (1)$$

where:

Table 1 Physical parameters of corn cobs cores

Substrate	TS [%]	LOI [% TS]	Ash [%]
Cobs cores	94,4 ± 0.12	97.5 ± 0.11	2.55 ± 0.11

± standard deviation.

Table 2 Chemical composition of corn cobs cores

Substrate	C [%]	H [%]	N [%]	S [%]
Cobs cores	1,0[a] ± 0,02	46,9[a] ± 0,25	5,9[a] ± 0,11	0,07[a] ± 0,00

[a] a considerable difference was not found between average values in columns according to the HSD Tukey test (ANOVA) for the investigated factors.

E_e – amount of energy generated [MWh];
V_{CH4} – amount of methane generated [Nm^3];
R_{aCH4} – methane energy yield coefficient [9.97 kWh/m^3];
η_e – electrical efficiency of the cogeneration unit [0.40 for electrical energy and 0.45 for heat].

3 Results

3.1 *Physical Properties and Chemical Composition of Corn Cob Cores*

The following results were obtained from the study physical parameters of corn cobs cores (Table 1). The content of total solids (TS) in the corn cobs cores was 94,4% and the content of organic dry matter was 97,5%. Such a value is a good indication of the material's potential for methane fermentation, because biogas is produced only from organic matter. In contrast, the percentage of organic dry matter for corn silage often used for methane fermentation was 95% [6]. The ash percentage content in this experiment was with result of 2.5%.

ANOVA analysis showed that the elemental content in the cob cores of maize cultivar Ambrosini with FAO 220 was at a high level (Table 2).

3.2 *Biogas and Methane Yield*

The static methane fermentation graph indicates that the process was without any problems. For maize cob cores with an earliness standard (FAO 220), fermentation lasted 33 days (Fig. 1).

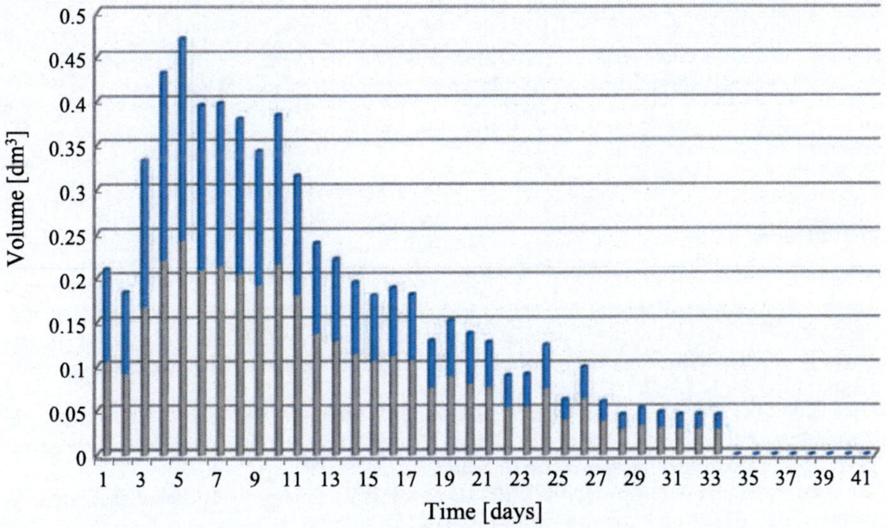

Fig. 1 Daily biogas production during methane fermentation of the corn cobs cores

Table 3 Methane yield in biogas process of the corn cobs cores

Substrate	Biogas [m^3/Mg TS]	CH_4 [%]	CH_4 [m^3/Mg TS]	HRT
Cobs cores	443,6 ± 11,9	53,2	248,1 ± 4,67	33

Statistical analysis showed that the daily biogas yield from cob cores of maize variety with FAO 220 was 443,6 m^3/Mg TS and methane yield from cob cores was 248,1 m^3/Mg TS (Table 3).

3.3 *Energetic Calculations*

Based on the obtained results, presented the energy value of the analyzed material (Table 4). With methane yield of 248,1 m^3/Mg TS, total electricity was 989,4 kWh/Mg TS, but total heat was 1038,0 kWh/Mg TS.

Thanks to the research provided very interesting information on the energy potential of one hectare of maize plantation (Table 5).

Table 4 Energetic value of corn cobs cores

Substrate	Methane yield [m^3/Mg TS]	Total electricity calculated [kWh/Mg TS]	Total heat calculated [kWh/Mg TS]
Cobs cores	248,1	989,4	1038,9

Table 5 Energy potential of corn cob cores per hectare in anaerobic digestion

Substrate	Cobs cores yield [Mg TS/ha]	Total electricity calculated [MWh/ha]	Total heat calculated [kWh/ha]	Total energy value [kWh/ha]
Cobs cores	2,11	2,09	2,19	4,28

References

1. Lalak J, Kasprzycka A, Murat A, Paprota EM, Tys J (2014) Pretreatment methods of lignocelulosic biomass to improve methane fermentation process (Obróbka wstępna biomasy bogatej w lignocelulozę w celu zwiększenia wydajności fermentacji metanowej). Agrophysics (Agrofizyka) 21(1):52–56
2. Greenhalf MKC, Ouadi M, Jahangiri H, Hornung A, Briens C, Berruti F (2020) Effect of torrefaction pretreatment on pyrolysis of corn cobs (Wpływ obróbki wstępnej toryfikacji na pirolizę kolb kukurydzy). Engineering (Inżynieria), tom 7
3. Grzybek A, Gradziuk P, Kowalczyk K (2001) Straw as an Energy Fuel (Słoma energetyczne paliwo), The Village of Tomorrow (Wieś Jutra), p 15
4. Wojcieszak D, Przybył J, Mioduszewska N, Zaborowicz M, Dach J, Kowalik I (2015) Energy use of maize straw (Energetyczne zagospodarowanie słomy kukurydzianej). W: Current Issues in Bioengineering systems (Aktualne Problemy Inżynierii Biosystemów). W: Red. M. Lipiński, J. Przybył. Wyd. Poznań University of Life Sciences (Uniwersytetu Przyrodniczego w Poznaniu), pp 293–301
5. Woźniak M et al (2021) Chemical and structural characterization of maize stover fractions in aspect of its possible applications. Materiały 14:1527. https://doi.org/10.3390/ma14061527
6. Kozłowski K et al (2018) Laboratory simulation of an agricultural biogas plant start-up. Chem Eng Technol 41(4):711–712. https://doi.org/10.1002/ceat.201700390

Energy Comparison of Corn Stover Fraction

Dawid Wojcieszak, Artur Pawłowski, and Jacek Przybył

Abstract The extensive use of fossil fuels for energy production has raised concerns about their harmful effects on the environment. Findings from British Petroleum (BP), fossil fuels still account for 84% of global primary energy consumption. Over the last decade, research efforts have increasingly focused on employing biomass as a raw material for producing biofuels used for thermochemical con-version processes. The use of biomass on a large scale may lead to controversial competition for arable land, water and consequently, food. Therefore, only waste materials and agricultural by-products and residues should be used for the pro-duction of biofuels. One of the primary by-products of agricultural production is corn stover. Therefore, this study aimed to determine the of chemical and energetic proper-ties of corn stover fraction. The corn stover fractions are characterised by different average heat of combustion values. The highest combustion heat value, i.e. 18.2 MJ · kg^{-1}, is found in the casing leaves. The lowest calorific value, i.e. 17.7 MJ · kg^{-1}, is found in corn stalks and cob cores.

Keywords Corn cob cores · Methane potential · Anaerobic digestion

1 Introduction

1.1 Challenges for Energy

Energy is one of the key sources of modern life and a major factor in economic growth. It is therefore recognized that energy determines the economic and social development of the world [1]. The extensive use of fossil fuels for energy production has raised concerns about their harmful effects on the environment [2–4]. As a result, interest in developing alternative, environmentally-friendly energy sources continues

D. Wojcieszak (✉) · A. Pawłowski · J. Przybył
Department of Biosystems Engineering, Faculty of Environmental Engineering and Mechanical Engineering, Poznan University of Life Sciences, Poznan, Poland
e-mail: dawid.wojcieszak@up.poznan.pl

S. Pascuzzi and F. Santoro (eds.), *Farm Machinery and Processes Management in Sustainable Agriculture*, Lecture Notes in Civil Engineering 289,
https://doi.org/10.1007/978-3-031-13090-8_9

to grow [5]. Although renewable energy is the world's fastest growing energy sector, fossil fuels supply the vast majority of the world's energy needs [6]. Findings from British Petroleum (BP), fossil fuels still account for 84% of global primary energy consumption [7]. British Petroleum estimates that oil, gas and coal reserves will be completely consumed by 2052, 2060 and 2090 respectively [8]. On this basis, it can be concluded that currently engineering sciences should focus their activities on the search for new, renewable energy sources (RES).

1.2 Agriculture Biomass

Over the last decade, research efforts have increasingly focused on employing biomass as a raw material for producing biofuels used for thermochemical conversion processes. The use of biomass on a large scale may lead to controversial competition for arable land, water and consequently, food. Therefore, only waste materials and agricultural by-products and residues should be used for the production of biofuels.

The last decade has seen a steady increase in global maize grain production. According to the Food and Agriculture Organization of the United Nations (FAO), the production of maize grain in the last fifteen years has risen from 714.2 million tonnes in 2005 to 1 162.3 million tonnes in 2020, an increase of 39% (Fig. 1).

One of the primary by-products of agricultural production is corn stover [9, 10]. These harvest corn straw residues contain four key fractions, i.e., husks, leaves, cobs, and stalks, which are structurally and morphologically distinct. According to Shinners et al. [11], 1 kg of maize grain (dry weight) includes 15% of cob cores, 22% of leaves, 14% of cover leaves, and 50% of stalks.

Therefore, this study aimed to determine the of chemical and energetic properties of corn stover fraction.

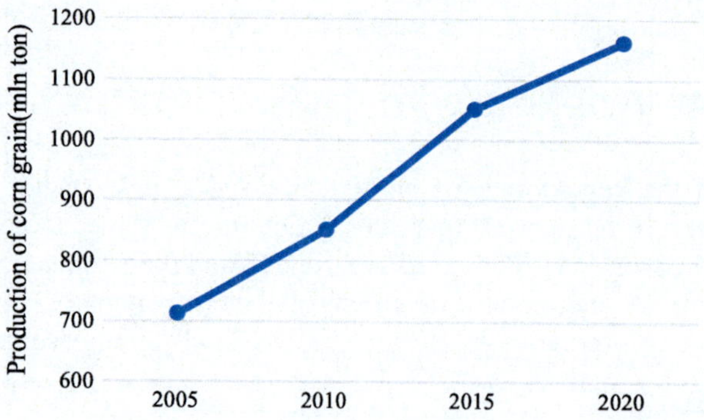

Fig. 1 World corn grain production 2005–2020 (own elaboration based on FAO Statistic, 2021)

2 Materials and Methods

2.1 *Corn Stover Fraction and Its Characteristics*

Research material consisted of corn stover fraction after corn grain harvest Podium cultivar KWS, FAO 200, from a plantation located in the northern part of Wielkopolska. The final plant density was 74.8 thou/ha. Maize harvesting was carried out manually. Harvested plants were transported to the laboratory of the Department of Biosystems Engineering of the University of Life Sciences in Poznañ and divided into fractions: cob cores, cover leaves, stem leaves, stems and grain.

2.2 *Laboratory Investigations*

The content of total solids (TS) in the corn cobs corse was measured by drying and weighing at a temperature of 105 ± 1 °C, according to the standard EN 12,880: 2004.

The ash percentage content was calculated from the weight of the combustion residue according to modified ISO 1171. Samples were heated to 710 ± 10 °C and kept at this temperature until a constant weight was reached. For this purpose, a muffle furnace was used.

The carbon, hydrogen, nitrogen and sulphur percentages were determined with a Flash 2000 elemental analyser (Thermo Fisher Scientific, USA) in CHNS/O configuration according to the EN 15,104 standard. The instrument was calibrated with standard Methionine (Thermo Fisher Scientific, USA) [2].

2.3 *Combustion Heat and Calorific Value*

The combustion heat tests were carried out in an automatic calorimeter (LECO AC600) calibrated with benzoic acid (LECO, New York, NY, USA) in accordance with the PN-EN ISO 18125: 2017–07 standard [2]. To determine the heat of combustion, samples of corn stover fraction from varieties with FAO 200 standards were used, and the moisture content was determined separately for immediately after harvesting and seasoned samples (average moisture content of 6%).

The calorific value of the corn cob cores of the studied cultivars was calculated using Eq. (1):

$$LHV = HHV + P(W_a + 8.94(H_a)), \tag{1}$$

where LHV (lower heating value) is the calorific value in the analytical state (J/g), HHV is higher heating value (J/g), P is the heat ofwater evaporation at 25 °C,

1%content = 24.42 (J/g), W_a is the moisture content of the analytical sample (%), and Ha is the hydrogen content of the analytical sample (%).

2.4 *Statistical Analysis*

STATISTICA 13.3 software (TIBCO Software Inc., Palo Alto, CA, USA) software was used for statistical analysis of the results. The calculations included an analysis of variance (ANOVA) for the one-factor system, followed by Tukey's HSD test (honest significant difference test) for each variable at = 0.05.

3 Results

3.1 *Physical and Chemical Properties of Corn Stover Fractions*

The heat of combustion indicates the energy value of biomass [12], and the value of the heat of combustion of biomass depends on its moisture, chemical composition, and ash.

The moisture of corn stover fraction ranged from 33.9 to 79.7%. The moisture content of the stalks was the highest. The highest yield was stalks (13.5 $Mg \cdot ha^{-1}$) and the husks was the lowest yield (1,6 $Mg \cdot ha^{-1}$). Yield of cobs was 3.3 $Mg \cdot ha^{-1}$ (Table 1).

Ash content is a crucial parameter directly affecting the heat of biomass combustion [12, 13]. Noteworthy is the lowest ash content amounted 2.2% was corn cobs. Statistical analysis showed significant differences in the ash content of the corn Stover fraction (Fig. 2).

Elemental analysis is commonly used to evaluate energetic biomass [14]. ANOVA analysis indicated that elemental contents were significantly different in the corn stover fractions. Only the nitrogen content was not significantly different in the corn stover fractions (Table 2).

Table 1 Yield and moisture content of corn stover fraction

Fraction	Yield [$Mg \cdot ha^{-1}$]	Moisture content [%]
Cobs	3.3 ± 0.7	64.9
Husk	1.6 ± 0.8	55.8
Leaves	2.5 ± 0.5	33.9
Stalks	13.5 ± 3.9	79.7

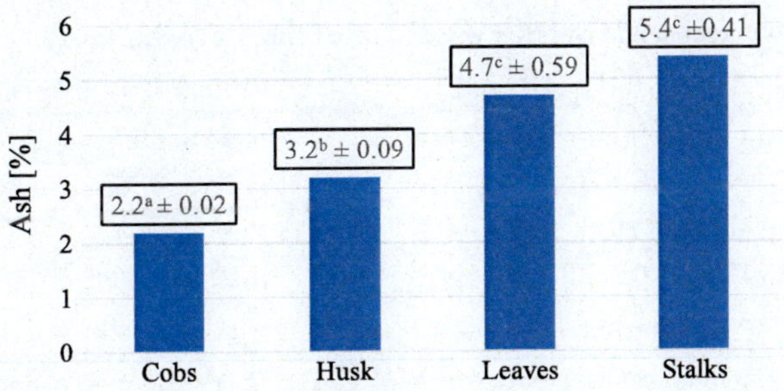

Fig. 2 Ash content of corn stover fractions

Table 2 Chemical composition of elements (C, N, H, O) corn stover fraction

Fraction	C [%]	H [%]	N [%]	O [%]
Cobs	$44.8^{c} \pm 0.04$	$0.5^{b} \pm 0.04$	$5.9^{a} \pm 0.04$	$45.0^{b} \pm 1.00$
Husk	$43.8^{b} \pm 0.17$	$0.5^{b} \pm 0.06$	$6.0^{a} \pm 0.03$	$43.1^{ab} \pm 0.92$
Leaves	$43.3^{a} \pm 0.06$	$1.0^{c} \pm 0.05$	$5.7^{a} \pm 0.04$	$41.9^{ab} \pm 0.49$
Stalks	$43.7^{ab} \pm 0.11$	$0.3^{a} \pm 0.00$	$5.6^{a} \pm 0.07$	$41.3^{a} \pm 0.78$
n	2	2	2	2

Mean value ($n = 2$) ± standard deviation; identical superscripts (a, b, c, d) denote no significant difference ($p < 0.05$) between mean values according to post-hoc Tukey's HSD test.

3.2 *Heat of Combustion and Calorific Value*

The value of biomass during the combustion process can be assessed using two indicators. The first indicator is the heat of combustion (HHV) and the second indicator is the calorific value (LHV). ANOVA analysis indicated that heat of combustion corn stover fraction was significantly different (Table 3). This certainly has to do with the different moisture content of the fraction.

Table 3 Heat of combustion and calorific value of fresh corn stover fraction

Fraction	Average HHV [$MJ \cdot kg^{-1}$]	Average LHV [$MJ \cdot kg^{-1}$]
Cobs	$6.2^{c} \pm 0.06$	$5.0^{c} \pm 0.05$
Husk	$8.0^{b} \pm 0.07$	$6.8^{b} \pm 0.07$
Leaves	$11.9^{a} \pm 0.03$	$10.7^{a} \pm 0.03$
Stalks	$3.6^{d} \pm 0.01$	$2.5^{d} \pm 0.01$
N	3	3

Mean value ($n = 3$) ± standard deviation; identical superscripts (a, b, c, d) denote no significant difference ($p < 0.05$) between mean values according to post-hoc Tukey's HSD test.

Table 4 Heat of combustion and calorific value of dried corn stover fraction

Fraction	Average HHV $[MJ \cdot kg^{-1}]$	Average LHV $[MJ \cdot kg^{-1}]$
Cobs	$17.8^{a} \pm 0.17$	$16.5^{a} \pm 0.17$
Husk	$18.2^{a} \pm 0.15$	$17.0^{a} \pm 0.15$
Leaves	$18.0^{a} \pm 0.04$	$16.8^{a} \pm 0.04$
Stalks	$17.7^{a} \pm 0.06$	$16.6^{a} \pm 0.06$
n	3	3

Mean value ($n = 3$) ± standard deviation; identical superscripts (a, b, c, d) denote no significant difference ($p < 0.05$) between mean values according to post-hoc Tukey's HSD test.

The heat of combustion and calorific value of dried to 6% corn stover fractions was same. ANOVA analysis confirmed the absence of significant differences (Table 4),

References

1. Cergibozan R (2022) Renewable energy sources as a solution for energy security risk: empirical evidence from OECD countries. Renew Energy 183:616–626
2. Wojcieszak D, Przybył J, Czajkowski Ł, Majka J, Pawłowski A (2022) Effects of harvest maturity on the chemical and energetic properties of corn stover biomass combustion. Materials 15:2831
3. Aghbashlo M, Tabatabaei M, Hosseini SS, Dashti BB, Mojarab SM (2018) Performance assessment of a wind power plant using standard exergy and extended exergy accounting (EEA) approaches. J Clean Prod 171:127–136
4. Aghbashlo M, Hosseinpour S, Tabatabaei M, Dadak A (2017) Fuzzy modeling and optimization of the synthesis of biodiesel from waste cooking oil (WCO) by a low power, high frequency piezo-ultrasonic reactor. Energy 132:65–78
5. Khalife E, Tabatabaei M, Demirbas A, Aghbashlo M (2017) Impacts of additives on performance and emission characteristics of diesel engines during steady state operation. Prog Energy Combust Sci 59:32–78
6. EIA (2019) World Energy Outlook 2019 with Projections to 2050. U.S. Energy Information Administration, Office of Energy Analysis, U.S. Department of Energy, Washington, DC, p 20585
7. BP (2020) BP Statistical Review of World Energy
8. BP (2011) BP Statistical Review of World Energy June 2011: What's inside?
9. Wojcieszak D et al (2020) Chemical composition of maize stover fraction versus methane yield and energy value in fermentation process. Energy 198:117258
10. Woźniak M et al (2021) Chemical and structural characterization of maize stover fractions in aspect of its possible applications. Materials 14:1527
11. Shinners KJ, Binversie BN, Muck RE, Weimer PJ (2007) Comparison of wet and dry corn stover harvest and storage. Biomass Bioenergy 31:211–221

12. Galhano dos Santos R, Bordado JC, Mateus MM (2018) Estimation of HHV of lignocellulosic biomass towards hierarchical cluster analysis by Euclidean's distance method. Fuel 221:72–77
13. Demirbas A (2002) Relationships between heating value and lignin, moisture, ash and extractive contents of biomass fuels. Energy Explor Exploit 20:105–111
14. Kantorek M, Jesionek K, Polesek-Karczewska S, Ziółkowski P, Badur J (2019) Thermal utilization of meat and bone meals. Performance analysis in terms of drying process, pyrolysis and kinetics of volatiles combustion. Fuel 254:115548

Economic Effects of Food Industry Waste Management in the Context of Sustainable Development

Agnieszka Starek-Wójcicka, Monika Stoma, Emilia Osmólska, Leszek Rydzak, and Paweł Sobczak

Abstract As a result of food production around the world, millions of tons of by-products are generated, mainly in fruit and vegetable processing plants and in the production of alcoholic and non-alcoholic beverages. Unfortunately, they are often difficult to manage, causing environmental, social and economic impacts. The scale and type of waste management problems are related to the specificity of individual sectors of the food industry, the dispersion of production sources and the overwhelming number of small and medium-sized food processing enterprises. In waste management, there is a need to reduce and prevent waste generation through the analysis of technological processes, in particular the economic practices of handling this type of production raw materials. Waste has measurable values—material and energy. Rational management of depleted resources requires treating waste as a valuable raw material that can be reused, processed or, as a last resort, recovered for energy. The issue of Food Waste is currently progressing, bringing together all divisions of waste management practice. Therefore, the article presents the proper management of waste, which will allow for its management while reducing the fixed costs of running a business. By-products containing significant amounts of antioxidant compounds, which have a positive effect on the quality of the finished product, were also characterized.

Keywords Food industry · Waste management · By-product · Sustainability · Economic effects

A. Starek-Wójcicka · L. Rydzak
Department of Biological Bases of Food and Feed Technologies, Faculty of Production Engineering, University of Life Sciences in Lublin, 20-612 Lublin, Poland

M. Stoma · E. Osmólska (✉)
Department of Power Engineering and Transportation, Subdepartment of Logistic and Business Management, Faculty of Production Engineering, University of Life Sciences in Lublin, 20-612 Lublin, Poland
e-mail: emilia.osmolska@up.lublin.pl

P. Sobczak
Department of Food Engineering and Machines, Faculty of Production Engineering, University of Life Sciences in Lublin, 20-612 Lublin, Poland

S. Pascuzzi and F. Santoro (eds.), *Farm Machinery and Processes Management in Sustainable Agriculture*, Lecture Notes in Civil Engineering 289,
https://doi.org/10.1007/978-3-031-13090-8_10

1 Introduction

Sustainable development is understood as the integration of social and economic development and broadly understood progress while taking into account their positive impact on the natural environment and striving to ensure a balance between profits and costs of this development. This issue has recently become one of the most important goals of the European Union. An example of activities undertaken in this area is, among others, maximizing the use of resources of biological origin in order to protect the natural environment and reduce production costs.

The growing social awareness of the impact of food on health with the simultaneously changing standard of living of the population in the economic sphere, although this process is often described as being too slow, causes that more and more attention is focused on the technologies of obtaining new products or enriching existing ones. Properly carried out technological processes allow for the management of post-production waste and use them as basic raw materials and/or additives to many food products. Some of them may show pro-health properties, extremely valuable, others—they may be functional additives replacing not only synthetic compounds, but also supporting the functioning of the human body (dietary fiber, polyphenolic compounds). Their use is a challenge of time, stimulates the development of the economy and is the engine of scientific progress [1–4].

The article describes the main sources and possibilities of full and economic utilization of raw materials and by-products produced during production processes.

2 Types of Waste and their Economic Aspects of Use

The waste material generated in the production of wine is grape seeds (Fig. 1), characterized by strong antioxidant properties. The polyphenols they contain (mainly proanthocyanidins and catechins) have the ability to protect cells against free radicals, chelate metal ions, and reduce the amount of primary and secondary products formed in the oxidation process [5, 6].

The extract of these organic antioxidants can be used in food production, especially in animal products. For example, Libera and Stasiak [8] added this valuable waste material to raw maturing meat. During the two-month refrigerated storage, the authors of the study found that the addition of grape seed extract effectively inhibited the processes of fat hydrolysis occurring in raw-ripening gammon (reducing the content of free fatty acids by 33%). Grape seed extract also counteracted adverse oxidative changes. The color of the product after the cut was redder compared to the control sample (without added gammon) and with the addition of sodium ascorbate. No effect of the additives used on the value of active acidity of the gammons was found. Slight differences were noted only during the two-month storage of the gammons, where the pH value decreased by about 5%.

Fig. 1 The extract of grape seeds [7]

The high antioxidant potential of grape seed extract was used by Reddy et al. [9] by adding it to lamb vacuum-packed and stored in a refrigerator at 4 °C. They showed that the product enriched with grape seed extract had significantly lower TBARS values compared to the control. Adding grape seed extract to lamb significantly reduced the total number of microorganisms and coliform bacteria. Interestingly, the by-products of wine production, in this case grape seed flour, made it possible to reduce the oxidation level of frankfurters. In addition, an increase in the content of protein (from 11.84 to 13.26%) and fiber (from 0.00- to 0.229%) was observed in the products. The color values (L *, a * and b *) of the product tended to decrease in relation to the amount of the additive used, which was probably related to the darker color of the flour [10].

The fruit and vegetable industry allows to enjoy the benefits of fruit and vegetables throughout the year, but at the same time generates significant amounts of plant residues (Fig. 2), which have a negative impact on the environment.

The pomace has the largest share in the generated waste. They are characterized by a high water content, therefore they are considered unstable and microbiologically unstable material. The resulting large mass of pomace in a relatively short period

Fig. 2 Apple pomace during the production of juice [11]

of time is an extremely important problem for processing plants [12–15]. Hence, waste from the fruit and vegetable industry is very often used as animal feed due to the high content of organic acids, sugars, nitrogen-free compounds, fatty substances and vitamins [16]. They are often used in plant cultivation as fertilizer or compost [17]. Another way to manage the pomace is the process of anaerobic processing, producing energy in the form of biogas [18, 19]. The various directions of waste management are presented in Fig. 3.

Recently, there is a need for more innovative ways to use by-products, such as functional food additives to enrich cereal, dairy, fish and meat products [21–23]. Scientific research by Jung et al. [2015] proved that properly prepared apple pomace (forced air-dried or freeze-dried) can be a valuable addition to chicken pate or dried beef "Beef jerky". Meat products enriched with this valuable waste had much higher dietary fiber content and radical scavenging activity compared to the control sample. The addition of apple pomace resulted in a slight reduction in the hardness, elasticity, cohesiveness and chewiness of these meat products compared to the sample without their participation. The changes in textural properties presumably resulted from the presence of fiber in the additive used.

An interesting way to use watermelon skins is to make instant noodles from them (Fig. 4). Scientists took up this topic by following the statistics of the amount of instant noodle consumption in the world, which, according to the WINA data,

Fig. 3 Global food waste management an implementation [20]

Fig. 4 Ready waterline—Watermelon skin noodles [26]

is increasing year by year [24]. Unfortunately, instant noodles contain chemicals like TBHQ (Tert-Butyl Hydroquinone) that can cause hyperactivity in adolescents, asthma, rhinitis and dermatitis. Therefore, they are willingly added by producers due to the fact that they significantly extend the shelf life or delay the rancidity processes [25]. According to the latest research, pasta made from watermelon peels can be a healthy replacement for these traditionally consumed products.

The plant identification results showed that watermelon peel and seeds contain many biologically active substances: Alkaloids, triterpenoids, flavonoids, tannins, etc. The antioxidant ingredients in instant noodles (Waterline) can therefore provide measurable health benefits [27].

This method of waste processing is also economically most justified. According to the data of the Economic Research Service in the USA, in 2020 over 100,000 ha of watermelons were cultivated, which gave 1 7,233,000 kg. In terms of production, watermelon is one of the three most cultivated crops in the US, next to onions and head lettuce. Such a large production of this fruit generates a huge amount of waste in the form of skins. Had it not been for an attempt to develop them, they would have posed a serious environmental problem. However, the costs of disposal of this waste can be included/transformed/invested in its proper treatment as a valuable raw material. It is enough to perceive them not as a problem, but as a source of a natural, pro-health and innovative additive or intermediate. The more that phytocomponents, usually lost in traditional methods of treating fruit or vegetable waste, can be obtained using extraction with inexpensive solvents approved for use in the food industry or using modern techniques. For example, ultrasonically assisted extraction enables the extraction of bioactive ingredients in a very shorter time, at low temperature, with less energy and solvent requirements [28–31].

Another trend that has been observed in the food market for several years is the growing interest of producers and consumers in various types of flours, especially those made from fruit and vegetable by-products. Such products can be used as thickeners, gelling agents, fillers and water-retaining agents, and even for the production of edible films [32]. The literature includes research on the incorporation of FVBF (Fruit and Vegetable By-products' Flour) into several products. Due to these texturing properties, adding (even in small doses) FVBF to liquid and semi-solid food products increases their viscosity. For example, the introduction of passion fruit flour increased the apparent viscosity of the yoghurt because the casein curd was denser. On the other hand, the use of carrot pomace flour for cakes resulted in a change in sensory characteristics. Compared to apple pomace flour, these characteristics were adversely affected by the carrot product, which consequently contributed to a very low overall consumer acceptance score (mainly due to texture and taste) [33]. Therefore, Gomez and Martinez [34] suggest that FVBF should be skillfully selected to obtain the optimal composition, and thus the appropriate quality of the final product. In general, FVBF can positively affect the color of the final product due to the presence of carotenoids (yellow, orange and red) or anthocyanins (red, purple, blue) [35, 36]. The health benefits of consuming FVBF products are related to the high content of fiber and many bioactive compounds with antioxidant properties, improving the intestinal microflora and increasing the feeling of fullness. Therefore, such a diet reduces the risk of chronic diseases—type diabetes 2, obesity, cardiovascular disease and cancer [37–39].

The bio-raw material with a high recycling potential is the avocado seed, which has as much as 70% more antioxidants than the flesh of the fruit itself (Fig. 5).

Antioxidants counteract the formation of free radicals, and also participate in their neutralization. Free radicals are involved in the destruction of dangerous bacteria, but they also attack other molecules, such as proteins and DNA of cells. As a result, the excess of radicals can contribute to the development of many diseases and accelerate the aging process of the body. This is why antioxidants should be included in our diet if we want to support the immune system [41–43].

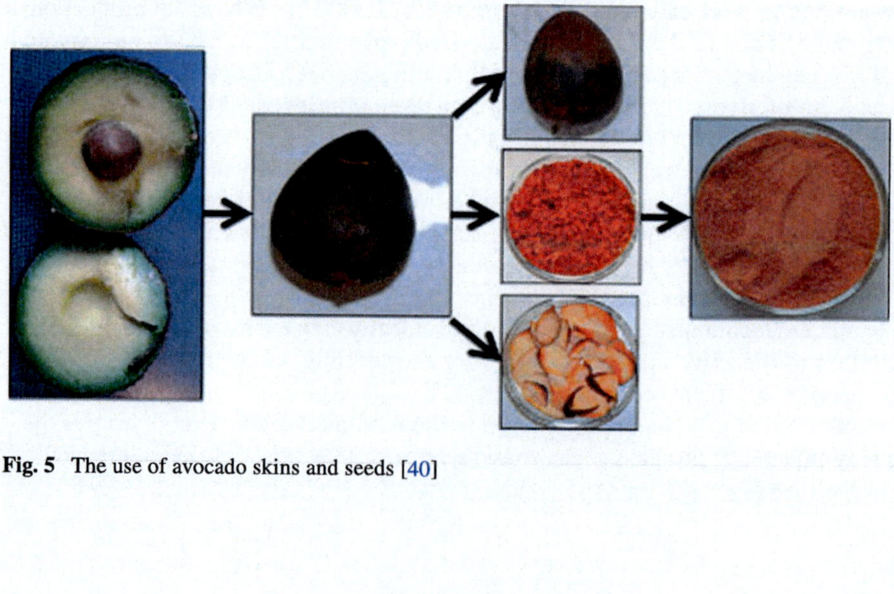

Fig. 5 The use of avocado skins and seeds [40]

Confirmation of this information is the research of Wang et al. [44], who showed that the seeds of many varieties (Slimcado, Booth 7, Booth 8, Choquette, Loretta, Simmonds and Tonnage, and Hass) had a much higher antioxidant capacity than the flesh. Other scientists have proved that avocado seeds contain polyphenolic compounds (protocatechic acid, chlorogenic acid, syringic acid and rutin), which are very powerful antioxidants [45]. As reported by Kosińska et al. [46] the beef burger samples containing 0.1% freeze-dried extract and 0.5% fresh powdered avocado seed showed no significant differences from BHA (0.05%), but differed from the control in terms of TBARS value. During 8 days of storage, the burger meat enriched with these additives showed no signs of oxidation.

Interestingly, recently it has been found that properly processed avocado peel is also a rich source of antioxidants [47]. Products with its participation can therefore strengthen the immune system and fight free radicals. Rotta et al. [48] by adding dried avocado peel to tea, showed that such a drink has good antioxidant activity, which results from the high content of phenolic, flavonoid compounds present in this part of the fruit. Moreover, avocado peel tea showed antioxidant activity and a flavonoid content higher than that of apple peel tea. Microbiological tests have proven that proper preparation (drying) of the skins, packaging and subsequent storage allows to obtain a final product with a long shelf life. The tea enriched with avocado peel was positively assessed by the panel of tasters in the sensory analysis.

3 Conclusions

Waste causes economic losses and are a burden for all countries of the world. Not using "leftovers" means wasting labor and other inputs (land, energy, etc.) needed to produce, distribute and consume. It is true that waste management (setting up infrastructure for collection, sorting and treatment) is costly. However, when it is established, it is capable of generating income and creating jobs.

Prospects for the use of food industry by-products containing health-promoting substances, especially phenolic compounds with antioxidant and bactericidal properties are relatively large, but so far little known. In many cases, seeds (pips) and skins are waste from technological processes, therefore their reuse as a source of antioxidants can bring measurable economic benefits and contribute to the reduction of pollution introduced by the food industry to the environment.

Therefore, it is necessary to take into account both environmental aspects (waste management, protection against pollution) and economic (profitability of processing). If the agri-food industry plants wisely organize the handling of waste produced in a company or institution, they can achieve many financial and image benefits without negative impact on the environment. However, in order to be able to manage waste effectively, it is necessary to establish an in-depth cooperation between entrepreneurs and research units. Each of the parties mentioned has an influence on the others and only in the course of agreements, exchange of views and detailed arrangements is it possible to achieve both common and own interests.

References

1. Thi NBD, Kumar G, Lin CY (2015) An overview of food waste management in developing countries: current status and future perspective. J Environ Manage 157:220–229
2. Mak TM, Xiong X, Tsang DC, Iris KM, Poon CS (2020) Sustainable food waste management towards circular bioeconomy: policy review, limitations and opportunities. Biores Technol 297:122497
3. Kavitha S, Yukesh Kannah R, Gopalakrishnan Kumar M, Gunasekaran J, Rajesh Banu L (2020) Food Waste to Valuable Resources, 1st edn. Academic Press, Cambridge, pp 1–13
4. Goh BHH et al (2020) Progress in utilisation of waste cooking oil for sustainable biodiesel and biojet fuel production. Energy Convers Manage 223:113296
5. Mielnik MB, Olsen E, Vogt G, Adeline D, Skrede G (2006) Grape seed extract as antioxidant in cooked, cold stored turkey meat. LWT Food Sci Technol 39(3):191–198
6. Wereńska M (2013) Naturalne antyutleniacze stosowane do mięsa. Nauki Inżynierskie i Technologie 1(8):79–90
7. Viterna (2021) What is OPC grape seed extract? https://www.viterna.at/What-is-OPC-grape-seed-extract. Accessed 27 Apr 2021
8. Libera J, Stasiak DM (2015) Wpływ ekstraktu z pestek Vitis vinifera na barwę i stabilność oksydacyjną dojrzewającego baleronu. Kształtowanie jakości żywności. Wydawnictwo Naukowe PTTŻ, pp 161–170
9. Reddy GB, Sen AR, Nair PN, Reddy KS, Reddy KK, Kondaiah N (2013) Effects of grape seed extract on the oxidative and microbial stability of restructured mutton slices. Meat Sci 95(2):288–294
10. Özvural EB, Vural H (2011) Grape seed flour is a viable ingredient to improve the nutritional profile and reduce lipid oxidation of frankfurters. Meat Sci 88(1):179–183
11. Sadownictwo (2017) Koszty produkcji soku zagęszczonego. https://www.sadownictwo.com.pl/koszty-produkcji-soku-zageszczonego. Accessed 29 July 2017
12. Shalini R, Gupta DK (2010) Utilization of pomace from apple processing industries: a review. J Food Sci Technol 47(4):365–371
13. Kruczek M, Drygaś B, Habryka C (2016) Pomace in fruit industry and their contemporary potential application. World Sci News 48:259–265
14. Majerska J, Michalska A, Figiel A (2019) A review of new directions in managing fruit and vegetable processing by-products. Trends Food Sci Technol 88:207–219
15. Esparza I, Jimenez-Moreno N, Bimbela F, Ancín-Azpilicueta C, Gandía LM (2020) Fruit and vegetable waste management: conventional and emerging approaches. J Environ Manage 265:110510
16. Wadhwa M, Bakshi MP, Makkar HP (2015) Waste to worth: fruit wastes and by-products as animal feed. CAB Rev 10(31):1–26
17. Burg P, Zemánek P, Michálek M (2011) Evaluating of selected parameters of composting process by composting of grape pomace. Acta Universitatis Agriculturae et Silviculturae Mendelianae Brunensis 59(6):75–80
18. Valenti F et al (2018) Anaerobic co-digestion of multiple agricultural residues to enhance biogas production in southern Italy. Waste Manage 78:151–157
19. Zapalowska A, Bashutska U (2019) The use of agricultural waste for the renewable energy production. Наукові праці Лісівничої академії наук України 18:138–144
20. C40 Knowledge (2018) Global Food Waste Management: An implementation guide for cities. https://www.c40knowledgehub.org/s/article/Global-Food-Waste-Management-An-implementation-guide-for-cities?language=en_US. Accessed 01 June 2018
21. Lyu F, Luiz SF, Azeredo DRP, Cruz AG, Ajlouni S, Ranadheera CS (2020) Apple pomace as a functional and healthy ingredient in food products: a review. Processes 8(3):319
22. Iqbal A, Schulz P, Rizvi SS (2021) Valorization of bioactive compounds in fruit pomace from agro-fruit industries: present Insights and future challenges. Food Biosci 44:101384
23. Jung J, Cavender G, Zhao Y (2015) Impingement drying for preparing dried apple pomace flour and its fortification in bakery and meat products. J Food Sci Technol 52(9):5568–5578

24. Instant Noodles, Demand rankings. https://instantnoodles.org/en/noodles/demand/table/. Accessed 11 May 2021
25. Sikander M, Malik A, Khan MSG, Qurrat-ul-ain, KR (2017) Instant noodles: are they really good for health. A review. Electron J Biol Chem 13(3)
26. Ipitex (2021) Waterline (Watermelon Skin Noodles) as an alternative antioxidant instant noodle products. https://ipitex.nrct.go.th/index.php/international/project_detail/0303000241. Accessed 11 Feb 2021
27. Rahmita NA (2021) Waterline (watermelon skin noodles) as an alternative antioxidant instant noodle products. Doctoral dissertation, Universitas Islam Negeri Maulana Malik Ibrahim
28. AgMRC (2021) Watermelon. https://www.agmrc.org/commodities-products/vegetables/watermelon. Accessed 01 Aug 2021
29. Seabra IJ, Braga ME, Batista MT, de Sousa HC (2010) Fractioned high pressure extraction of anthocyanins from elderberry (Sambucus nigra L.) pomace. Food Bioprocess Technol 3(5):674–683
30. Kua YL, Gan S, Morris A, Ng HK (2016) Ethyl lactate as a potential green solvent to extract hydrophilic (polar) and lipophilic (non-polar) phytonutrients simultaneously from fruit and vegetable by-products. Sustain Chem Pharm 4:21–31
31. Kumar K, Srivastav S, Sharanagat VS (2021) Ultrasound assisted extraction (UAE) of bioactive compounds from fruit and vegetable processing by-products: a review. Ultrason Sonochem 70:105325
32. Foste M, Verheyen C, Jekle M, Becker T (2020) Fibres of milling and fruit processing by-products in gluten-free bread making: a review of hydration properties, dough formation and quality-improving strategies. Food Chem 306:125451
33. Espírito-Santo AP, Lagazzo A, Sousa ALOP, Perego P, Converti A, Oliveira MN (2013) Rheology, spontaneous whey separation, microstructure and sensorial characteristics of probiotic yoghurts enriched with passion fruit fiber. Food Res Int 50(1):224–231
34. Gomez M, Martinez MM (2018) Fruit and vegetable by-products as novel ingredients to improve the nutritional quality of baked goods. Crit Rev Food Sci Nutr 58(13):2119–2135
35. Velasco-Arango VA, Hleap-Zapata JI, Ordóñez-Santos LE (2021) Effect of adding guava (Psidium guajava) epicarp extract flour on the physicochemical, textural, colour and sensory properties of Frankfurters. Food Technol Biotechnol 59(2):185–193
36. Santos D, da Silva JAL, Pintado M (2022) Fruit and vegetable by-products' flours as ingredients: a review on production process, health benefits and technological functionalities. LWT 154:112707
37. Fardet A (2010) New hypotheses for the health-protective mechanisms of whole-grain cereals: what is beyond fibre? Nutr Res Rev 23(1):65–134
38. Acosta-Estrada BA, Gutierrez-Uribe JA, Serna-Saldívar SO (2014) Bound phenolics in foods, a review. Food Chem 152:46–55
39. Padayachee A, Day L, Howell K, Gidley MJ (2017) Complexity and health functionality of plant cell wall fibers from fruits and vegetables. Crit Rev Food Sci Nutr 57(1):59–81
40. Science direct (2021) Temperature and pretreatment effects on the drying of Hass avocado seeds. https://www.sciencedirect.com/science/article/abs/pii/S0961953415301392. Accessed 13 Dec 2021
41. Alkhalf MI, Alansari WS, Ibrahim EA, ELhalwagy ME (2019) Anti-oxidant, anti-inflammatory and anti-cancer activities of avocado (Persea americana) fruit and seed extract. J King Saud Univ Sci 31(4):1358–1362
42. Dabas D, Elias RJ, Ziegler GR, Lambert JD (2019) In vitro antioxidant and cancer inhibitory activity of a colored Avocado seed extract. Int J Food Sci
43. Vo TS, Le PU (2019) Free radical scavenging and anti-proliferative activities of avocado (Persea americana Mill.) Seed extract. Asian Pac J Trop Biomed 9(3):91
44. Wang W, Bostic TR, Gu L (2010) Antioxidant capacities, procyanidins and pigments in avocados of different strains and cultivars. Food Chem 122(4):1193–1198
45. Pahua-Ramos ME et al (2012) Hypolipidemic effect of avocado (Persea americana Mill) seed in a hypercholesterolemic mouse model. Plant Foods Hum Nutr 67(1):10–16

46. Kosińska A, Karamać M, Estrella I, Hernández T, Bartolomé B, Dykes GA (2012) Phenolic compound profiles and antioxidant capacity of Persea americana Mill. Peels and seeds of two varieties. J Agric Food Chem 60(18):4613–4619
47. Salazar-López NJ et al (2020) Avocado fruit and by-products as potential sources of bioactive compounds. Food Res Int 138:109774
48. Rotta EM, de Morais DR, Biondo PBF, dos Santos VJ, Matsushita M, Visentainer JV (2016) Use of avocado peel (Persea americana) in tea formulation: a functional product containing phenolic compounds with antioxidant activity. Acta Scientiarum Technol 38(1):23–29

Ranking of Plant Biomass by Fuel Value Index

Artur Kraszkiewicz, Artur Przywara, Milan Koszel, and Stanisław Parafiniuk

Abstract Using the fuel value index (FVI), the suitability for energy purposes of selected types of wood and herbaceous biomass frequently used in Poland was assessed. These raw materials were compared in lump wood, briquettes, and pellets assortments. The heat of combustion, density, moisture, and ash content was determined for the tests' biofuels. The differentiation of the obtained results for the performed determinations resulted in significant changes of the FVI, which had values from 0.18 to 2.05. The obtained results show that the compaction of wood and herbaceous biomass significantly increases its value. A significant dominance of pellets and briquettes made of wood raw materials was noticed in the performed classification. It is mainly due to the high density of such products. On the other hand, pellets and briquettes from wheat and rye straw were often higher than the indications for lump wood, referring to the values of the FVI. The low FVI values for oat straw and chamomile biomass observed in the test conditions were primarily related to the high content of the mineral fraction in these fuels (possibly resulting from soil contamination), resulting in more ash after combustion.

Keywords Biomass properties · Solid biofuels · Sustainable agriculture

1 Introduction

Plant biomass is a significant source of energy globally, and the demand for this raw material will continue. In a system and prosumer energy, the use of wood raw materials and herbaceous biomass commonly found in agriculture, such as cereal straw, is widespread. At the same time, applying the principles of the sustainable development concept in the power industry creates conditions for increasing the use of standard plant materials, commonly found in the agricultural environment and

A. Kraszkiewicz (✉) · A. Przywara · M. Koszel · S. Parafiniuk
Department of Machinery Exploitation and Management of Production Processes, University of Life Sciences in Lublin, 20-612 Lublin, Poland
e-mail: artur.kraszkiewicz@up.lublin.pl

S. Pascuzzi and F. Santoro (eds.), *Farm Machinery and Processes Management in Sustainable Agriculture*, Lecture Notes in Civil Engineering 289,
https://doi.org/10.1007/978-3-031-13090-8_11

those that are by-products from the agri-food sector. At the same time, the productization of waste materials enables rational management of forest resources. And the use of biomass fuels, which by definition are neutral regarding CO_2 emissions, can significantly reduce this greenhouse gas in the atmosphere [1–5].

In terms of using plant biomass for energy purposes, specific chemical and physical properties of biofuels made from it are essential. Compared to wood biomass, herbaceous biomass, in particular, less popular than wheat or rye straw, has a lower energy value and density and higher ash content and moisture. Additionally, there is a higher content of nitrogen and sulfur in such biomass. These features lower the value of such biomass and indicate wood biomass as better for combustion. An unquestionable solution to improving the density and moisture of biomass is its agglomeration to form briquettes or pellets [1, 5–10].

The different physical and chemical properties of solid biofuels produced from such raw materials often do not allow for a meaningful assessment. The literature on the topic describes solutions that summarize the basic properties of such fuels and allow for a quantitative assessment of their quality. A frequently used solution is the fuel value index (FVI), which depends on the calorific value, density, moisture, and ash content of the fuel and is an essential tool for assessing the rank of firewood species [11–16].

In this study, using the fuel value index (FVI), the obtained values were calculated and compared for wood species frequently used in Poland and the herbal biomass used. These raw materials were compared in lump wood, briquettes, and pellets assortments.

2 Material and Methods

The raw material for the research was obtained from the area of the Lublin voivodeship, which is located in the east of Poland (Fig. 1).

Its area is characterized by low forest cover, with the dominant agricultural production on the predominantly fertile soils made of loess and "loess-like" silt deposits [17].

Plant species, characteristic for this area, were selected for the research, the biomass of which is and can be used for energy purposes. In the group of wood raw materials, birch, pine, and black locust wood were selected. In the group of herbaceous materials, wheat, rye, oat, and hemp (canabis) straw were used and by-products resulting from the cleaning of ryegrass seeds and chamomile flowers.

The collected raw materials were crushed with a hammer mill on sieves with 10 mm perforation, and then pellets were made on a flat die with 8 mm holes and 28 mm thick. In contrast, briquettes were made using a hydraulic briquetting machine with a thickening sleeve diameter of 50 mm and a compaction pressure of 100 bar. At the same time, due to the problematic shredding of the locust wood and hemp straw, and thus the resulting too large and uneven dimensions of the grains, pellets (hemp straw and locust wood) and briquettes (locust wood) were not made of them. In

Fig. 1 Map showing location of study site

addition, the wood was dried in the air circulating dryers to a moisture content of 16% as the value of this parameter after its seasoning and acceptable during combustion [4, 7].

The basic energy and physical properties of the produced fuels were determined using the following methods:

- moisture content (MC)—weighing method according to EN ISO 18134-3: 2015 [18],
- higher heating value (HHV)—following EN ISO 18125: 2017 [19], for three repetitions;
- ash content (AC)—following EN ISO 18122: 2016 [20], for three repetitions.
- diameter (D) and length (L)—direct measurement with a caliper, according to EN ISO 16127: 2012 [21], for three repetitions of randomly selected 100 ± 1 g fuel portions (pellets) or ten briquettes or wood chips;
- density (BD)—calculated based on geometric features in randomly selected samples and the weight of the biofuel batch according to the formula (1):

$$BD = \frac{4 \cdot m}{\pi \cdot D^2 \cdot L} \tag{1}$$

where:

BD—density of briquettes or pellets (kg m^{-3}),
m—weight of briquette or pellet portion (kg),
D—briquette or pellet diameter (m),
L—length of the briquette or pellet (m).

The fuel value index (FVI) was calculated to classify these biofuels, which is used by many researchers [12, 14, 16, 22]. The heat of combustion and the fuel's density is considered positive features, and the ash and moisture content that reduce its value is negative. However, the moisture content of the biomass must vary depending on the variant of its processing. Therefore, following the example of such researchers as Bhaat and Todaria [23] or Deka et al. [24], the ratio without this parameter was also calculated. The formulas of the indicators used are presented in Eqs. 2 and 3.

$$\mathrm{FVI} = \frac{HHV \cdot BD}{AC \cdot MC} \tag{2}$$

or

$$\mathrm{FVI}_{\mathrm{without\,MC}} = \frac{HHV \cdot BD}{AC} \tag{3}$$

where:

FVI—fuel value index (–),
HHV—higher heating value of fuel ($\mathrm{MJ \cdot kg^{-1}}$),
BD—density of fuel ($\mathrm{g \cdot cm^{-3}}$),
AC—ash content of fuel (%),
MC—moisture content of fuel (%).

3 Results

The determined chemical and physical parameters of the analyzed biofuels are presented in Table 1.

Concerning biomass energy use for energy purposes, the essential thing is the amount of energy it contains, expressed in the heat of combustion. This parameter was diversified for the analyzed biofuels due to the different assortments of raw materials used. The highest heat of combustion of about 19 $\mathrm{MJ \cdot kg^{-1}}$, was found in ryegrass waste, black locust wood, and pine wood. The remaining raw materials were characterized by the heat of combustion from less than 17 $\mathrm{MJ \cdot kg^{-1}}$ (oat straw) to slightly over 18 $\mathrm{MJ \cdot kg^{-1}}$ (hemp straw) (Table 1). Ash content for wood raw materials was close to 1%, slightly higher values were observed for herbaceous biomass (2–4%), and for chamomile waste, it was close to 10%—most probably this raw material was contaminated with soil (Table 1). The following parameters considered—the moisture content and the density—resulted from biomass processing, forming lump wood, pellets, and briquettes. The highest moisture in the range of 16.13–16.24% characterized fresh wood. The processing of biomass, wood, and other herbaceous biomass, its seasoning, shredding, and compaction caused the value of this parameter to decrease and stabilize, ranging from less than 10 to slightly over 11%.

Another important parameter in terms of the assessment of biofuels is their density. For energy purposes, tree species and more dense formed biofuels are preferred.

Table 1 Properties of biofuels

Fuel assortment	HHV ($MJ{\cdot}kg^{-1}$)	AC (%)	MC (%)	BD ($g{\cdot}cm^{-3}$)
Birch wood	17.83	1.29	16.23	0.61
Pine wood	18.63	0.97	16.24	0.49
Black locust wood	19.00	0.95	16.13	0.72
Birch sawdust pellets	17.83	1.29	9.55	0.92
Pine sawdust pellets	18.63	0.97	9.69	0.97
Birch sawdust briquettes	17.83	1.29	9.55	0.99
Pine sawdust briquettes	18.63	0.97	9.69	1.04
Wheat straw pellets	17.78	2.31	11.42	1.13
Rye staw pellets	17.78	3.40	10.82	1.03
Oat straw pellets	16.89	6.45	10.83	1.02
Raygras waste pellets	19.13	3.71	9.63	1.15
Camomile waste pellets	17.68	9.70	9.66	1.10
Wheat straw briquettes	17.78	2.31	11.42	0.95
Rye straw briquettes	17.78	3.40	10.82	0.86
Oat straw briquettes	16.89	6.45	10.83	1.00
Raygras waste briquettes	19.13	3.71	9.91	0.98
Camomile waste briquettes	17.68	9.70	9.99	1.00
Canabis straw briquettes	18.09	2.51	10.98	0.83

The reason for this is a higher energy content per unit volume and a more stable course of the combustion process [11, 24]. The most significant differentiation related to biomass processing was observed during the analysis of fuel density. The least processed raw material in the form of lump wood had a density from 0.49 (pine wood) to 0.72 $g{\cdot}cm^{-3}$ (locust wood). The compaction of raw materials into the form of pellets and briquettes increased the fuel density, which ranged from 0.83 (hemp straw briquettes) to 1.15 $g{\cdot}cm^{-3}$ (ryegrass pellets).

The values of the calculated FVI values, respectively, in the variants, considering the moisture content in the fuel, are shown in Figs. 2 and 3.

A significant differentiation of the FVI values was observed during the research, ranging from 0.18 to 2.05 (Fig. 2). At the same time, considering these values within the scope of wood raw materials, a 2–threefold increase in the value for pellets and briquettes was observed. It is mainly due to a reduction in moisture condensation and an increase in density. At the same time, in the group of raw materials of herbaceous biomass, the values of this indicator did not exceed the value of 1, and the differences in values resulted mainly from the type of raw material assessed and not the method of compaction. It is also crucial that the values for pellets and briquettes from herbaceous biomass were comparable and often higher than lump wood (Fig. 2).

After disregarding the moisture content in the fuel, the index values increased significantly, and the distribution of values was similar. Furthermore, a higher value was observed for wood pellets and briquettes than for lump wood, with the discrepancy being slightly smaller. On the other hand, for herbaceous biomass, the lowest

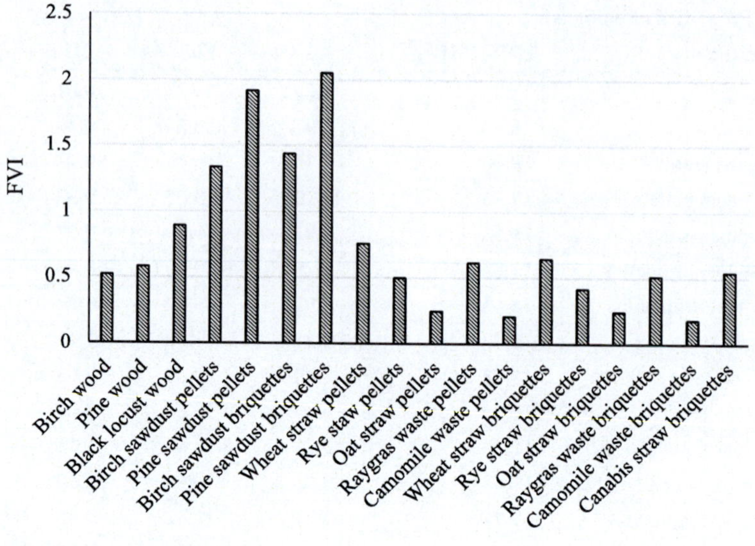

Fig. 2 Ranking on the basis of FVI

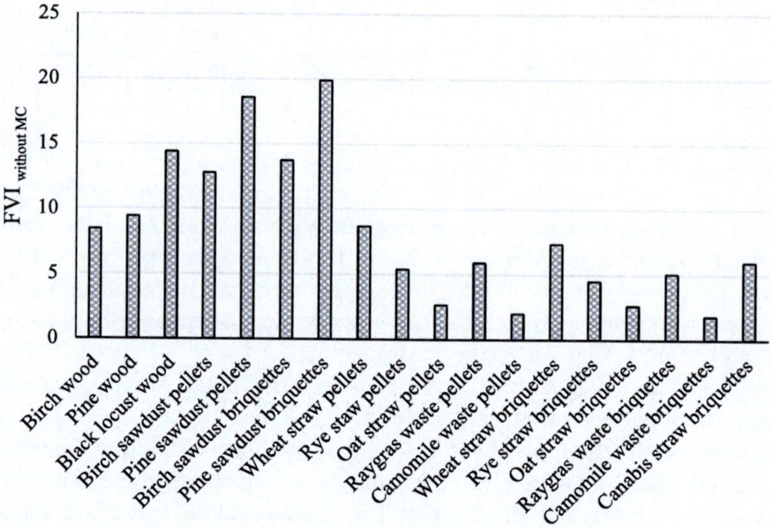

Fig. 3 Ranking on the basis of FVI without moisture content (MC)

values were also observed for biofuels made of oat straw and chamomile biomass. In this case, despite the lower density relating to fuels formed from herbaceous biomass, an increase in the value of lump wood is noticeable (Fig. 2).

Comparing the obtained values with the results of other researchers in the respective assortments, similar values of the FVI for wood and herbaceous biomass were noticed. However, when referring to the research, which did not consider moisture, only the index values for pellets and wood briquettes were comparable. The others were much smaller [25–28].

4 Conclusions

The use of plant biomass for energy purposes is an essential element of Poland's energy policy, which is part of the world's energy strategy. Using different plant biomass for energy purposes, which is processed into various geometric forms, the qualitative assessment of biofuels due to their different physical and chemical characteristics is complicated. The classification of fuels in terms of their energy content is insufficient to select the best fuel. Regarding the course of combustion processes, other parameters are also critical, such as fuel density, moisture, and ash content in the fuel.

The conducted research allowed to draw the following conclusions:

1. Compaction of wood and herbaceous biomass significantly increases its value in terms of FVI. The obtained values indicate a significant dominance of pellets and briquettes made of raw wood materials. It is due to the high density of such products. Nevertheless, wheat and rye straw pellets and briquettes with FVI values often exceeded the indications for lump wood.
2. Low values of the FVI for oat straw and chamomile biomass, observed in the test conditions, are primarily related to the high content of the mineral fraction in these fuels (possibly due to soil contamination), resulting in combustion with a higher amount of ash.
3. The FVI used during the research in its definition does not consider other chemical features of biofuels, e.g., sulfur and nitrogen content or the chemical composition of ash, which directly affect the combustion process and emissions to the atmosphere. Hence, it would be advisable to develop such a tool to consider other biofuels features in its formula.

References

1. Mc Kendry P (2002) Energy production from biomass (part 1): overview of biomass. Biores Technol 83:37–46
2. Fournel S, Marcos B, Godbout S, Heitz M (2015) Predicting gaseous emissions from small-scale combustion of agricultural biomass fuels. Biores Technol 179:165–172
3. Obernberger I, Brunner T, Bärnthaler G (2006) Chemical properties of solid biofuels – significance and impact. Biomass Bioenerg 30:973–982
4. Olsson M, Kjällstrand J (2002) Emissions from burning of softwood pellets. In: Proceedings of the first world conference on pellets. pp 111–114. Stockholm, Sweden
5. Van Loo S, Koppejan J (2007) Handbook of biomass combustion and co-firing. IEA Bioenerg Task 32:266–272
6. Demirbas A (2004) Combustion characteristics of different biomass fuels. Prog Energy Combust Sci 30(2):219–230

7. Demirbas A (2005) Potential applications of renewable energy sources, biomass combustion problems in boiler power systems and combustion related environmental issues. Prog Energy Combust Sci 31:171–192
8. Kraszkiewicz A, Kachel-Jakubowska M, Niedziółka I (2015) Analysis of selected physical and chemical properties of plant biomass of agricultural origin in terms of its energy use. Bulg J Agric Sci 21:1295–1299
9. Krzyżaniak M, Stolarski MJ, Szczukowski S, Tworkowski J (2014) Thermophysical and chemical properties of biomass obtained from willow coppice cultivated in one- and three-year rotation cycles. J Elem 1:161–175
10. Wang ZJ, Pei D (2011) Physical properties of pellets made from sorghum stalk, corn stover, wheat straw, and big bluestem. Ind Crops Prod 33:325–332
11. Goel VL, Behl HN (1996) Fuel wood quality of promising tree species for alkaline soil sites in relation to tree age. Biomass Bioenerg 10:57–61
12. Bhatt BP, Todaria NP (1992) Woodfuel characteristics of some mountain species. Forest Ecol Manag 47:363–366
13. Abbot PG, Lowore JD (1999) Characteristics and management potential of some indigenous firewood species in Malawi. For Ecol Manage 11:111–121
14. Kataki R, Konwer D (2001) Fuelwood characteristics of some indigenous tree species of north-east India. Biomass Bioenerg 20:17–23
15. Kataki R, Konwer D (2002) Fuelwood characteristics of indigenous tree species of north-east India. Biomass Bioenerg 22:433–437
16. Ojelel S, Otiti T, Mugisha S (2015) Fuel value indices of selected woodfuel species used in Masindi and Nebbi districts of Uganda Energy. Sustain Soc 5:14
17. Zielony R, Kliczkowska A (2012) Regionalizacja przyrodniczo-leśna Polski 2010. CILP, Warszawa
18. EN ISO 18134-3:2015 (2015) Solid biofuels - Determination of moisture content - Oven dry method - Part 3: Moisture in general analysis sample. Polish Committee for Standardization: Warsaw, Poland
19. EN ISO 18125:2017 (2017) Solid biofuels. Determination of calorific value. Polish Committee for Standardization: Warsaw, Poland
20. EN ISO 18122:2016 (2016) Solid biofuels - Determination of ash content. Polish Committee for Standardization: Warsaw, Poland
21. EN ISO 16127:2012 (2012) Solid biofuels - Determination of length and diameter of pellets. Polish Committee for Standardization: Warsaw, Poland
22. Purohit AN, Nautiyal AR (1987) Fuel wood value index of Indian mountain tree species. Int Tree Crops J 4:177–182
23. Bhaat BP, Todaria NP (1990) Fuelwood characteristics of some mountain trees and shrubs. Biomass 21:233–238
24. Kraszkiewicz A (2022) Influence of geometrical features of solid biofuels on the implementation of the combustion process. Combustion Science and Technology. Published online: 03 January 2022
25. Deka D, Saikia P, Konwer D (2007) Ranking of fuelwood species by fuel value index. Energy Sour Part A Recov Utilization Environ Effects 29(16):1499–1506
26. Baqir M, Kothari R, Singh RP (2019) Characterization and ranking of subtropical trees in a rural plantation forest of Uttar Pradesh, India, as fuel wood using fuel wood value index (FVI). Environ Dev Sustain 21:763–776
27. Marquez-Reynoso MI, Ramírez-Marcial N, Cortina-Villar S, Ochoa-Gaona S (2017) Purpose preferences and fuel value index of trees used for firewood in El Ocote biosphere reserve, Chiapas, Mexico. Biomass Bioenergy 100:1–9
28. Bhatt BP, Lemtura M, Changkija S, Sarkara B (2017) Fuelwood characteristics of important trees and shrubs of Eastern Himalaya. Energy Sour Part A Recov Utilization Environ Effects 39(1):47–50

Energetic Potential of Dairy Cow Breeding in Poland

Wiktor Bojarski, Jakub Pulka, Patrycja Pochwatka, Piotr Bresz, Mateusz Nowak, and Jacek Dach

Abstract Due to the high cattle population, Poland occupies a leading position among the European Union countries in terms of the amount of manure produced. However, its use in this country is primarily limited to fertilizing farmland. As a result, its storage in fields in the form of often unbent or improperly compressed piles contributes to the emission of greenhouse gases. Among the gases responsible for global warming, methane and nitrous oxide are emitted from improperly stored manure. These gases have a 25 and 298 times greater impact on the climate, respectively, than carbon dioxide. However, the use of cow feces as a substrate for biogas plants allows, on the one hand, to reduce gaseous emissions during manure storage and, on the other hand, to produce "green" fuel (biomethane) and/or electricity and heat. The paper presents the potential of biomethane production from cow manure, amounting to 1,826 million m^3 or 7.282 TWh of electric energy.

Keywords Agriculture · Anaerobic digestion · Emissions · Manure · Biogas

1 Introduction

1.1 Climate and Agriculture

There has been an inevitable interdependence between agriculture and climate since the dawn of time. It is even assumed that the climate was most likely the factor thanks to which agriculture was invented [1]. Unlike other branches of the economy,

W. Bojarski · J. Pulka · P. Bresz · M. Nowak · J. Dach (✉)
Faculty of Environmental Engineering and Mechanical Engineering, Department of Biosystems Engineering, Poznan University of Life Sciences, Poznań, Poland
e-mail: jacek.dach@up.poznan.pl

P. Pochwatka
Faculty of Production Engineering, Department of Environmental Engineering and Geodesy, University of Life Sciences in Lublin, Lublin, Poland

S. Pascuzzi and F. Santoro (eds.), *Farm Machinery and Processes Management in Sustainable Agriculture*, Lecture Notes in Civil Engineering 289,
https://doi.org/10.1007/978-3-031-13090-8_12

agriculture is extremely sensitive to the impact of the climate and its possible disturbances. Atmospheric phenomena related to the climate, such as insolation of too high or too low temperatures, the intensity of rainfall or their absence, the influence of wind, and other factors directly affect the agricultural economy. It is related to the biological aspect of farming. World agrarian production changes production volume every year, mainly due to weather conditions [2].

All even minimal changes in climate cause serious problems in agriculture. They necessitate the emergence of newer and newer ways of solving the existing problems, thus changing the production method so far. An example can be the periods of cooling down and warming the climate, which altered the range of selected crops, caused the deadlines for some fieldwork to shift significantly, and influenced the change in the way of cultivating crops [3].

1.2 Animal Production

Excessive emission of greenhouse gases to the atmosphere is warming the climate. Gases that have a particularly strong influence on global warming are methane and nitrous oxide (25 and 298 times more powerful than CO_2, respectively) [4]. Agriculture is a strong source of both methane and nitrous oxide. The source of methane in agriculture is primarily animal production and, more precisely, the resulting feces and intestinal fermentation of ruminants [5, 6].

The gases mentioned above are also produced from ruminant excrements. These feces contains a significant amount of organic matter and water. Such residues constitute an ideal medium for the growth of methanogenic bacteria. The multiplication of these bacteria means that the resulting feces also emit methane and non-methane organic compounds. Additionally, improperly stored manure (animal feces and bedding) is, apart from methane, a source of nitrous oxide emissions [7].

According to Eurostat data, it shows that in seventeen countries of the European Union, where cattle production accounts for about 75% of EU cattle, the number of these animals in 2020 exceeded 59.5 million head [8]. Such a huge number of cattle may indicate the enormous methane emission into the atmosphere. Therefore, cows are perceived as the world's largest anthropological source of methane.

1.3 Mechanism of Emission from Manure

In Poland, the main application of manure is the fertilization of arable lands. For this purpose, it is transported from the barns to the fields, where it is formed into piles, and then periodically (2–3 times a year), it is spread over the fields. However, very often, the manure is not compacted during the creation of piles. This situation favors the accumulation of oxygen in the formed pile. This oxygen is then used by the bacteria that start to break down organic matter. As a result of this process, the temperature inside the pile increases, reaching even 40–60 °C in just several hours after the pile formation. As a consequence, the oxygen contained in a heap is consumed by actively growing bacteria. The formed prism begins to collapse, which consequently prevents ventilation. The anaerobic conditions are created. Thus, the process of anaerobic decomposition of organic matter begins. As a result of this process, energy is released in chemical form, i.e., methane emission begins. It replaces the heat emitted so far. These conditions are very similar to those in a biogas reactor. In addition to methane emissions, there are also emissions of nitrous oxide. It is formed as a result of nitrification and denitrification processes of nitrogen present in a heap [9].

1.4 Manure as a Substrate for Biogas Plants

High manure production provides Poland with wide prospects for its use for biogas production. It is currently the best possible method of managing the remains of livestock production, which is environmentally friendly, and at the same time, brings economic benefits. Of the nearly 100 million Mg of manure produced annually, only less than one percent is used for biogas production. This is a small amount considering that the digestate resulting from the work of the biogas plant can be used for fertilizing fields just like manure. This by-product of biogas plant operation is a rich nitrogen, phosphorus, and potassium source. The use of digestate to fertilize arable lands has a beneficial effect on the cultivated plants [10, 11]. Moreover, it has been proven that the digestate is responsible for neutralizing insects and plant pathogens that are dangerous to crops [12].

The study aims to estimate the energy potential (production of biomethane and electricity) of the Polish dairy farming sector using biogas installations. To achieve this work aim, it will be necessary to analyze the number of dairy cows in Poland in individual provinces and then carry out energy calculations in the field of biomethane and electricity production from planned biogas plants.

2 Materials and Methods

2.1 Manure and its Characteristics

Data on the exact number of farms in Poland producing fertilizers in the form of manure (including urine) or slurry will be available from the summer of 2022. For this article, assumptions were made to calculate the amount of manure produced by dairy cows. The Central Statistical Office publishes data on the number of dairy cows in individual provinces [13]. Based on the market analysis, it was assumed that the overwhelming majority of dairy cows in Poland are farmed in a manure system where, in addition to solid fertilizer (manure), liquid leachate (urine) is also produced. According to the official document [14], a typical well-run dairy cow (milk production of over 8,000 L per year) produces 16.4 Mg of manure and 8.2 Mg of urine per year. Hence, it was possible to determine the amount of generated fertilizers after multiplying the number of cows in a given voivodeship and the production of manure and urine.

The energy efficiency of manure and urine as a result of the anaerobic digestion process was taken from the data of the own analysis. Manure and urine samples were taken from typical cattle farm in western Poland and tested for methane efficiency with standard methodology DIN 38 414/S8. On this basis, they will receive the weight of manure and urine converted into biogas potential expressed in cubic meters of biomethane.

2.2 Energetic Calculations

The amount of electricity that could be produced was calculated by multiplying the volume of biomethane, the energy value of methane and the electrical efficiency coefficient of the cogeneration unit, according to the following formula:

$$E_e = V_{CH4} \cdot W_{CH4} \cdot \eta_e [MWh] \tag{1}$$

where:

E_e—amount of electric energy [MWh];
V_{CH4}—biomethane volume [m^3];
W_{CH4}—energetic value of methane [0.009968 MWh/m^3];
η_e—electric efficiency of CHP [0.4].

The electric power that can be obtained in the analyzed voivodeships from biogas plants fed with cow feces should be calculated according to the following formula:

$$P_e = \frac{E_e}{t} [MW] \tag{2}$$

where:

P_e—electric power [MW];
E_e—amount of electric energy [MWh];
t—yearly working time of CHP [8400 h for well working biogas plants].

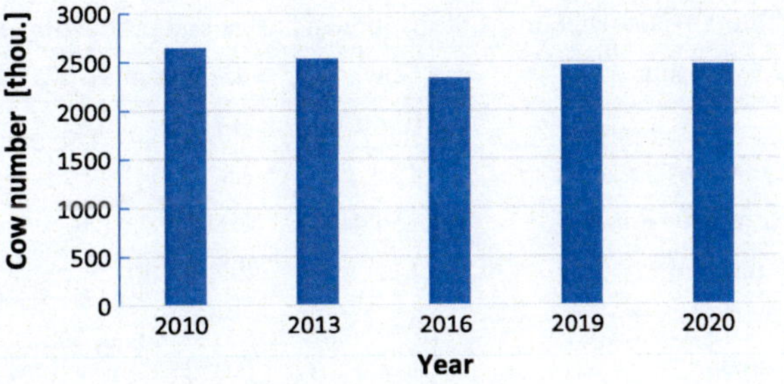

Fig. 1 The number of dairy cows in Poland between 2010–20

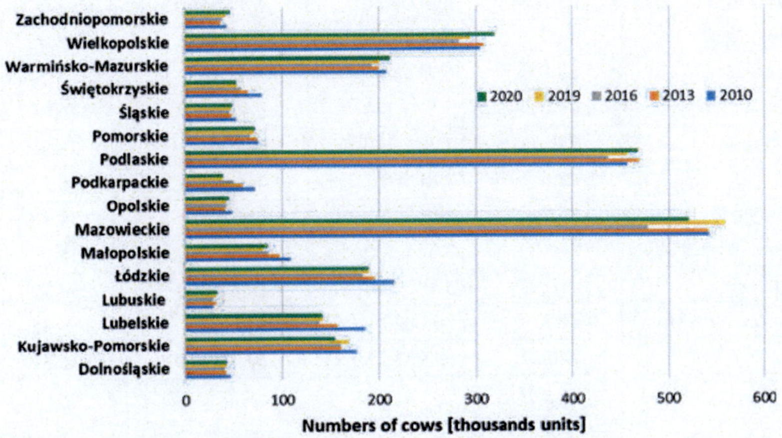

Fig. 2 The number of dairy cows in analyzed voivodeship 2010–20

3 Results

3.1 Dairy Cow Production

The number of dairy cows in Poland has remained at a more or less similar level since 2010 (Fig. 1).

As shown in Fig. 2, there is a huge difference in the production scale between voivodships. The energy potential of cow breeding (related to the use of manure in biogas plants) will therefore vary greatly from region to region.

3.2 Energetic Calculations

The following values were adopted as energy efficiency: Manure (21.56% of dry matter)—45.04 m^3 CH_4/Mg of fresh mass and urine (0.79% of dry matter)—1.23

Table 1 Calculated mass of manure and urine and production of methane

Voivodeship	Cows	Manure mass	Urine mass	CH_4 prod
	[thou.]	[Mg]	[Mg]	[mln m^3]
Dolnośląskie	42.1	682 020	353 640	31.15
Kujawsko-Pomorskie	154.4	2 501 280	1 296 960	114.25
Lubelskie	141	2 284 200	1 184 400	104.34
Lubuskie	33	534 600	277 200	24.42
Łódzkie	188.4	3 052 080	1 582 560	139.41
Małopolskie	85.4	1 383 480	717 360	63.19
Mazowieckie	520.4	8 430 480	4 371 360	385.09
Opolskie	44.7	724 140	375 480	33.08
Podkarpackie	38.9	630 180	326 760	28.79
Podlaskie	467.7	7 576 740	3 928 680	346.09
Pomorskie	72.3	1 171 260	607 320	53.50
Śląskie	48.8	790 560	409 920	36.11
Świętokrzyskie	53.2	861 840	446 880	39.37
Warmińsko-Mazurskie	210.8	3 414 960	1 770 720	155.99
Wielkopolskie	319.9	5 182 380	2 687 160	236.72
Zachodniopomorskie	47.1	763 020	395 640	34.85
Poland	**2468.1**	**39 983 220**	**20 732 040**	**1826.34**

m^3 CH_4/Mg of fresh mass. Manure and urine mass as well as biomethane production is presented in Table 1.

The methane production potential ranges from 24 million m^3 in the Lubuskie voivodship to 385 million m^3 in the Mazowieckie voivodship. For Poland, the potential for methane production in biogas plants fed with cow manure is 1,826 million m^3. In Q1 2022, the average European natural gas benchmark TTF price was 92.61 EUR/MWh (0.923 EUR/m^3). This means that the total value of potentially produced biomethane is 1,686 million EUR based on the natural gas market price. However, as "green" gas from manures, biomethane should be priced significantly higher than natural gas as a fossil fuel. Using biomethane (non-separated from biogas) as fuel in CHP units for electricity production, the amount of electric energy produced is 7.282 TWh with electric energy power of 867 MW.

References

1. Hetherington R (2012) Living in a dangerous climate. Cambridge University Press, Cambridge
2. Arora NK (2019) Impact of climate change on agriculture production and its sustainable solutions. Environ Sustain 2:95–96

3. Raza A, Razzaq A, Mehmood SS, Zou X, Zhang X, Xu J (2019) Impact of climate change on crops adaptation and strategies to tackle its outcome: a review. Plants (Basel) 8(2):34
4. Pochwatka P, Kowalczyk-Juśko A, Sołowiej P, Wawrzyniak A, Dach J (2020) Biogas plant exploitation in a middle-sized dairy farm in Poland: energetic and economic aspects. Energies 13(22):6058
5. Scarlat N, Fahl F, Dallemand JF, Monforti F, Motola V (2018) A spatial analysis of biogas potential from manure in Europe. Renew Sustain Energy Rev 94:915–930
6. Liu S, Proudman J, Mitloehner FM (2021) Rethinking methane from animal agriculture. CABI Agric Biosci 2:22
7. Mielcarek P (2012) Weryfikacja wartości współczynników emisji amoniaku i gazów cieplarnianych z produkcji zwierzęcej. Inżynieria Rolnicza 267–276. (in Polish)
8. Eurostat Homepage. https://ec.europa.eu/eurostat. Accessed 19 Apr 2022
9. Mazurkiewicz J (2022) Energy and economic balance between manure stored and used as a substrate for biogas production. Energies 15:413
10. Kozłowski K, Pietrzykowski M, Czekała W, Dach J, Kowalczyk-Juśko A, Jóźwiakowski K, Brzoski M (2019) Energetic and economic analysis of biogas plant with using the dairy industry waste. Energy 183:1023–1031
11. Marks S, Dach J, Morales FJF, Mazurkiewicz J, Pochwatka P, Gierz L (2020) New trends in substrates and biogas systems in Poland. J Ecol Eng 21:19–25
12. Cao Y, Chang Z, Wang J, Ma Y, Fu G (2013) The fate of antagonistic microorganisms and antimicrobial substances during anaerobic digestion of pig and dairy manure. Biores Technol 136:664–671
13. Central Statistical Office (2021) Statistical Yearbook of Agriculture. Department of Statistical Publishing. (in Polish)
14. Regulation of the Council of Ministers of February 12, 2020 on the adoption of the "Action Program to reduce water pollution with nitrates from agricultural sources and to prevent further pollution." (in Polish)

Effect of Mechanical Treatment on the Efficiency of Methane Fermentation of Biowaste

Piotr Bresz, Jakub Pulka, and Wiktor Bojarski

Abstract In 2016–2020, the share of municipal waste collected from households amounted to approx. 84%, while in 2020, based on data from the Central Statistical Office, the average citizen produced 342 kg. To combat this situation, the principles of the separate collection have been extended, allowing for the reuse of individual materials, such as paper, glass, metal, or even plastic. The way to deal with the growing amount of municipal waste is to implement waste-to-energy processes. In the case of biowaste, the best solution seems to be utilizing the methane fermentation process, however, due to the high level of contamination, this substrate requires purification. One of the solutions seems to be implementing mechanical treatment - separation of the over-sieve fraction >80 mm from the under-sieve fraction of <80 mm from municipal waste and grinding - to obtain a uniform, homogeneous mass. The over-sieve fraction consists mainly of packaging waste, while the under-sieve fraction is composed of biowaste, whose parameters are similar to the substrates used in agricultural biogas plants. In the methane fermentation process, the biowaste can be utilized to obtain electricity, heat, and fuel in the form of biomethane. The substrate's biogas efficiency varies from 107.27 to 115.31 $m^3 \cdot Mg^{-1}$ of fresh mass.

Keywords Biowaste · Mechanical treatment · Methane production

1 State of the Art

1.1 A Subsection Sample

Municipal waste is waste generated in households, excluding end-of-life vehicles and non-hazardous waste [1]. According to the Central Statistical Office, about 13.12 million tons of waste was collected in Poland in 2020, which is almost 365,000 tons

P. Bresz (✉) · J. Pulka · W. Bojarski
Faculty of Environmental Engineering and Mechanical Engineering, Department of Biosystems Engineering, Poznan University of Life Sciences, Poznań, Poland
e-mail: breszpiotr@gmail.com

S. Pascuzzi and F. Santoro (eds.), *Farm Machinery and Processes Management in Sustainable Agriculture*, Lecture Notes in Civil Engineering 289,
https://doi.org/10.1007/978-3-031-13090-8_13

more than in 2019. This value includes selectively collected waste and mixed waste [2].

Over the 5 years, an increase in the mass of selectively collected waste can be observed. In 2016, about 2.94 million tons of waste were selectively collected, which accounted for about 25.2% of the total mass of waste. In 2020, a total of about 4.97 million tons of waste was selectively collected, which accounted for 37.9% of the total waste collected. Based on data from the Central Statistical Office on selectively collected waste (2016–2020), it can be estimated that the average annual increase was about 208 thousand tons.

1.2 *Municipal Waste Collection*

Pursuant to the Act, a waste producer is anyone whose activity or livelihood gives rise to waste, as well as anyone who carries out preliminary treatment, mixing or other operations resulting in a change in the nature or composition of such waste [1]. The ideal is to dispose of waste quickly and hygienically, as regular collection of waste and emptying of containers is necessary to ensure the sanitary condition of the places provided for waste collection [3]. During periods of high air temperatures, waste disposal frequencies should be higher. The organic compounds in the waste are decomposed more quickly, and thus begin to generate odor. Decomposition of the organic fraction can also pose a threat in the form of increase in the number of bacteria dangerous to humans [4].

The basic element in a well-designed municipal waste management system is waste segregation. Due to the increasing use of natural resources, waste segregation and recycling of possible fractions is very important. Recycling saves energy, raw materials and reduces the negative impact on the environment [5].

1.3 *The Biowaste Management*

According to the Waste Act, biodegradable waste is defined as all waste that degrees aerobic or anaerobic decomposition with the participation of microorganisms. These include decomposable garden waste, park waste, food and kitchen waste from households, caterers and retailers, and comparable waste from food processing plants [1]. On the other hand, biowaste does not include agricultural waste, sewage sludge, forestry waste and biodegradable [6]. However, those materials (especially sewage sludge) can be a valuable fertilizers [7].

Renewable energy plays an increasingly important role in the global economy, as well economy, as well as in Poland. One of the forms of energy possible to produce from waste is biogas. It can be obtained from landfills, which, according to the law, should have an installation for its capture, as well as produced from municipal waste [6, 8]. However, the above process belongs to the historical type of waste handling.

Nowadays, thanks to separate collection and separation of green waste from biowaste, the fermentation process of biowaste may give better results than before [9]. This process is also much more economic then composting, which is also effective for treatment some biowaste like sewage sludge [10, 11]. After the introduction of comprehensive biowaste collection, a significant increase in undesirable substances such as plastic or mineral waste in biowaste has been observed, which may amount to 10–15% [12]. The digestion process by separating the redundant fractions from municipal waste and separating the clean biowaste fraction will produce a higher biogas yield than from "normal" municipal waste in the same process. The separated residue fraction represents a maximum of about 20–40% of the initial weight of waste entering the installation and can be used for energy production [13]. The residual fraction, often referred to as pre-RDF (uncleaned) or RDF (cleaned and standardized), is the fuel produced by processing non-recyclable waste, which has a high calorific value (16–18 MJ/kg) and can be used for energy purposes [14]. In fact, in order to increase the efficiency of effective biowaste management, the future technologies of effective separation should based on advanced usage of artificial intelligence (neural networks) and image analysis. Those solutions are already introduced in different areas of environmental protection and agronomy [15].

The aim of the paper was to evaluate the effect of different mechanical treatments (two-shaft shredder, beater shredder and sieve 80 mm) on the efficiency of methane fermentation of municipal biowaste.

2 Materials and Methods

2.1 Materials

The test material consisted of municipal waste from a municipal waste management facility collecting waste from a city of over 200 thousand citizens. To differentiate the results, the experiments were repeated three times at three-month intervals. During the course of the experiments, the raw material was morphologically evaluated (Polish Norm PN-93/Z-15006), crushed and subjected to further analysis. The research methodology consisted in the norms PN-EN 14,774–1:2010 (moisture), PN-G-04584:2001 (ash/organic dry matter), PN-EN ISO 16948:2015–07 (C, H, N, and O) and DIN 38 414/S8 (biogas production efficiency) [16–20].

3 Results

3.1 *Waste Characteristics*

Analysis of the morphological composition (Table 1.) showed that food waste of vegetable origin, i.e. those that should be collected as part of the separate collection of biowaste, accounted for an average of 70%.

As can be observed in Table 2 on average 83% of inorganic contaminants were separated from the waste after using sieves with mesh >80 mm in diameter. This value shows how important different types of screens are in the separation process of biowaste from municipal waste.

In the case of separations using a flail separator, the division into over-sieve and under-sieve fractions was impossible to estimate due to the nature of the separator operation. The tests performed allow us to estimate that about 5–10% of the waste was

Table 1 Morphological composition of waste

Morphological group	Mean (%)
Food waste of vegetable origin	70
Food waste of animal origin	1
Paper and cardboard waste	2
Plastic waste (including tires)	8
Plastic waste (BIO bags)	1
Textile wastes	0
Glass waste	0
Metal wastes	0
Waste org. poz. (biodegradable: grass, leaves, wood)	18
Other mineral wastes (including rubble)	0
Hazardous waste	0
Other (unrecognizable) (multi-material)	0

Table 2 Percentage distribution of inorganic pollutants by fraction

Faction group, [mm]	Mean (%)
>100	65
100–80	18
80–60	8
60–40	7
40–20	2
<20	–
<10	–

Table 3 Percentage of kitchen waste by fraction

Faction group, mm	Mean (%)
>100	19
100–80	4
80–60	6
60–40	18
40–20	35
<20	18

separated at the sieve as a fines fraction. It should be noted, however, that the under-sieve fraction samples showed ground plastic particles, confirming that packaging made of thin film mainly LDPE is not able to be separated (Table 3).

3.2 *Waste Parameters*

The substrates analyzed were characterized by a relatively low mean dry matter (DM) ranging from 25.04 to 25.97% (Table 4). A comparable difference was found for the organic dry matter (ODM) content in the analyzed fractions. It ranged from 78.57 to 81.03% DM. The carbon and nitrogen contents were on a comparable level. The average content of carbon was 40.90% DM., while that of nitrogen was 2.25% DM. It has to be underline that average C/N ratio (18.47–19.2) was a little bit lower comparing to optimal (20–25) for anaerobic digestion or even composting [21]. This could be related with higher content of nitrogen compounds in analyzed waste.

The analysis of the obtained results allows us to conclude that the highest production per dry organic matter was characterized by the fraction after the double-shaft shredder (580 m^3) (Table 5). After analyzing the biogas yield results in relation to the fractional composition of the sample (Ta. 1.), a relationship can be observed between the occurrence of the "food waste" fraction and the "org. pos. waste" fraction and the biogas production yield.

Table 4 Substrate petameters by fractions

Substrate parameters, mean							
Substrate	pH	DM [%]	ODM [%DM]	Ash [%DM]	N [%DM]	C [%DM]	C/N
Two-shaft shredder	5.07	25.04	81.03	18.87	2.34	42.01	18.47
Beater shredder	5.32	25.05	78.79	21.21	2.19	40.29	18.66
Sieve 80 mm	5.17	25.97	78.57	21.43	2.22	40.41	19.20

Table 5 Summary of cumulative biogas/methane yields (mean) from all experiments

		Yield [m^3 Mg^{-1} fresh mass]		Yield [m^3 Mg^{-1} DM]		Yield [m^3 Mg^{-1} ODM]	
Substrate	CH_4 [%]	Methane	Biogas	Methane	Biogas	Methane	Biogas
Two-shaft shredder	53.87	62.16	115.31	255.62	473.60	312.11	580.07
Beater shredder	52.26	56.70	108.54	230.31	439.51	290.95	556.03
Sieve 80 mm	51.78	55.51	107.27	215.52	416.26	273.48	528.35

4 Conclusions

On the base of research made in this paper, the followed conclusions were made:

- In the process of mechanical treatment of municipal biowaste, two substrates with favorable parameters were obtained: biowaste and RDF,
- Biogas production is highly dependent on the morphological composition of the waste and the percentage of each fraction in the substrate,
- The average biogas production from this fraction was 580 m^3 Mg^{-1} ODM.

References

1. Act of December 14, 2012 on waste, Journal of Laws 2012, No. 2013, item 21. In Polish
2. Local Data Bank, Central Statistical Office, Category: Environmental status and protection, Group: Municipal waste. https://bdl.stat.gov.pl, Accessed 13 Feb 2022
3. Latosińska J, Miłek D, Komór A, Kowalik R, Pongrácz E, Ylä-Mella J (2021) Selective collection of municipal waste in a residential district with multi-family buildings-case study from Poland. Resources 10:83
4. Ciszewski M, Czekaj T, Szewczyk EM (2015) New insight into bacterial zoonotic pathogens posing health hazards to humans. Med Pr 65:819–829
5. Bayar Y, Gavriletea MD, Sauer S, Paun D (2021) Impact of municipal waste recycling and renewable energy consumption on CO_2 emissions across the European Union (EU) member countries. Sustain 13:1–12
6. Jouhara H, Czajczyńska D, Ghazal H, Krzyżyńska R, Anguilano L, Reynolds AJ (2017) Municipal waste management systems for domestic use. Energy 139:485–506
7. Boniecki P, Dach J, Nowakowski K, Jakubek A (2009) Neural image analysis of maturity stage during composting of sewage sludge. In: Proceedings - 2009 international conference on digital image processing, ICDIP, pp 200–203
8. Marks S, Dach J, Fernandez Morales FJ, Mazurkiewicz J, Pochwatka P, Gierz Ł (2020) New trends in substrates and biogas systems in Poland. J Ecol Eng 21(4):19–25
9. Kozłowski K, Pietrzykowski M, Czekała W, Dach J, Kowalczyk-Juśko A, Jóźwiakowski K, Brzoski M (2019) Energetic and economic analysis of biogas plant with using the dairy industry waste. Energy 183:1023–1031

10. Wolna-Maruwka A, Dach J (2009) Effect of type and proportion of different structure-creating additions on the inactivation rate of pathogenic bacteria in sewage sludge composting in a cybernetic bioreactor. Arch Environ Prot 35:87–100
11. Wolna-Maruwka A, Schroeter-Zakrzewska A, Dach J (2012) Analysis of the growth and metabolic activity of microorganisms in substrates prepared on the base of sewage sludges and their impact on growth and flowering of garden verbena. Fresenius Environ Bull 2012(21):325–336
12. Friege H, Eger Y (2022) Best practice for bio-waste collection as a prerequisite for high-quality compost. Waste Manag Res 40:104–110
13. Hemidat S, Saidan M, Al-Zu'bi S, Irshidat M, Nelles M (2019) Potential utilization of RDF as an alternative fuel to be used in cement industry in Jordan. Sustainability 11:5819
14. Paszkowski J, Domański M, Caban M, Zarajczyk J, Pristavka M, Findura P (2020) The use of refuse derived fuel (RDF) in the power industry. Agric Eng 24(3):83–90
15. Boniecki P, Koszela K, Piekarska-Boniecka H, Nowakowski K, Przybył J, Zaborowicz M, Raba B, Dach J (2013) Identification of selected apple pests based on selected graphical parameters. In: Fifth international conference on digital image processing (ICDIP 2013), vol 8878, p 88782S
16. Solid Biofuels (2010) Determination of moisture content - drier method - Part 1: Total moisture - Reference method. Polish Standard PN-EN 14774-1:2010E
17. Solid Biofuels (2010) Determination of volatile compounds content. Polish standard PN-EN 15148:2010
18. Solid Biofuels (2015) Total carbon, ammonia and hydrogen determination. Polish Standard PN-EN ISO 16948:2015-07
19. Solid Fuels. Determination of total sulfur and fly ash with automatic analyzers. Polish Standard PN-G-04584:2001P. (2001).
20. Bresz P. (2022) Effect of mechanical treatment on the efficiency of methane fermentation of bio-waste, Engineering thesis, pp 34–35
21. Czekała W, Dach J, Ludwiczak A, Przybylak A, Boniecki P, Koszela K, Zaborowicz M, Przybył K, Wojcieszak D, Witaszek K (2015) The use of image analysis to investigate C:N ratio in the mixture of chicken manure and straw. In: Seventh international conference on digital image processing, ICDIP 2015, vol 9631, p 963117

Effect of Nozzle Condition and Type in the Spray Quality in Wheat Plants

Alaa Subr, Marek Milanowski, and Stanisław Parafiniuk

Abstract The objective of this research was to investigate the effect of using different design of agricultural nozzles on the spray effectiveness in the wheat plants canopy. Moreover, the effect of using worn nozzles on the spray quality was also evaluated. Four types nozzles was used in the tests: TT, TTj60, DG, and XR. They are produced by TeeJet (Spraying Systems Co., Illinois USA). The first three types has a pre-orifice design, while XR was considered as control treatment because it is wide used internationally. Nozzles were subjected to the accelerated wear test according to the ASAE S471 standard. The flow rate after finishing the accelerated wear test was reached to 10% higher than the nominal flow rate, and these nozzles was named "Worn nozzles". These nozzles (new and worn) was evaluated concerning its spray quality in wheat plants in the laboratory conditions, and using water sensitive papers (WSP). Spray coverage (%), stains diameter (μm), and stains density on square centimetre were calculated. Results data showed that highest spray coverage was achieved using the XR worn nozzles (19%), while using TTj60 nozzle (new and worn) resulted in the lowest value of coverage (14% for both). Biggest stain diameter was registered when using worn DG nozzle (538 μm), lowest value of stain diameter was 445 μm when using new XR nozzle. Concerning the spray density, XR nozzle (new and worn) produced highest spray density, while using worn TT nozzle resulted in the lowest number of stains on square centimetre (333 stain/cm^2).

Keywords Nozzle wear · Pesticide application · Pre-orifice design · Spray coverage

A. Subr (✉)
Department of Agricultural Machines and Equipment, College of Agricultural Engineering Sciences, University of Baghdad, Al-Jadiriya, 10071 Baghdad, Iraq
e-mail: alaa.kamel@coagri.uobaghdad.edu.iq

M. Milanowski
Department of Thermal Technology and Food Process Engineering, Faculty of Production Engineering, University of Life Sciences in Lublin, Głęboka 31, 20-612 Lublin, Poland

S. Parafiniuk
Department of Machinery Exploitation and Management of Production Processes, University of Life Sciences in Lublin, 28 Głęboka Street, 20-612 Lublin, Poland

S. Pascuzzi and F. Santoro (eds.), *Farm Machinery and Processes Management in Sustainable Agriculture*, Lecture Notes in Civil Engineering 289,
https://doi.org/10.1007/978-3-031-13090-8_14

1 Introduction

The spray application is important part of the pest control process. Ensuring that the spray application parameters are within the acceptable limits is one of the essential jobs for the sprayer operator. One of the spray application parameters is the nozzles type, and it has impact on the spray quality on the target (plants leaves or insects for example). In the market, there are a wide variety of nozzles design produced by different manufacturers. The nozzle size also has an influence in the spray application process. [11] reported an increase in the spray coverage and density, with a decrease in the stains diameter as the nozzle size is smaller, and spray speed and application rate are higher. [9] indicated that the water sensitive papers (WSP) is a good tool to indicate the spray droplets characteristics. However, it has limitations like the color changing when the distance between adjacent stains is decreased. [7] recommended using the 600 dpi scanning resolution of the WSP; they reported that using the image processing methods in analyzing the WSP resulted a similar output as the manual calculating of stains.

Physical properties of the spray (extensional and shear viscosity, dynamic surface tension, and the presence of emulsion droplets) have an influence in the spray atomization [5]. Moreover, [6] reported that the nozzle type has an impact on the droplet size spectrum. The effect of nozzle type was significant on the spray density as well as the interaction between nozzle and collector type [3]. [4] reported that the nozzle DG11003 produced better spray density than the standard nozzle (XR11003) under very windy conditions. Wolf and Daggupati (2009) study resulted showed that TT and XR nozzles produced the highest penetration rate to the lower canopy. They reported also that the nozzles had a significant effect on the number of stains on the square centimeter (spray density). [1] study showed that using TT nozzles resulted in the highest spray density (33.0 droplets on the square centimeter) comparing with other types nozzles.

Using nozzles over long period will cause the nozzle tip dimensions to be bigger, which will result in higher values of flow rate. This increased in the flow rate of worn nozzles will make them delivering higher application rate values. This problem has an influence in the spray application process. [10] reported that the spray coverage percentage increased by 81% when the application rate increased from 187 to 375 l/ha. While the nozzle type (XR, DG, and TTj60) did not have an influence in the spray coverage or deposition under the canopy. [13] reported that worn nozzles has an influence in the performance of spraying process, and it could make deviation in the measurement of droplet size.

[8] reported that spray deposition and coverage are influenced by nozzle type and setting, spray pressure, droplet size, spraying height, and application rate. Guler et al. (2012) study results showed an increase in the spray coverage due to the increase in the flow rate of nozzles. [2] found a significant relation between the spray deposition ($\mu L/cm^2$) with the spray coverage. This relation will make it possible to estimate the spray deposition depending on the percentage coverage measured by WSP, even when this percentage is high (30–80%). The objective of this research was firstly to

find out whether using different design of agricultural nozzles would affect the spray effectiveness in the canopy of wheat plants. Secondly, to investigate the effect of using worn nozzles on the spray quality (spray coverage percentage, spray density, and stains diameter).

2 Materials and Methods

Four types of nozzles were used in the experiment: TT, TTj60, DG, and XR. They are produced by TeeJet (Spraying Systems Co., Illinois USA). The features of these nozzles are in Table 1. The first three types has a pre-orifice design, while XR was considered as control treatment because it is wide used internationally.

A set of nozzles (six nozzles) of each nozzle type were subjected to the accelerated wear test according to the ASAE S471 (1999) standard. This test was done inside a device built for this purpose. It consisted of 1000-L tank, pump, pipes, pressure regulator and guage, and holders to fix the nozzles on the pipes. The abrasive solution used to wear the nozzles tips was consisted of 18.2 kg Kaolin KOM + 300l water. The accelerated wear test was stopped after 100 h of wear. The flow rate (at 3.0 bar pressure) for XR, TT, TTj60, and DG nozzles after finishing the test were 1.51, 1.27,

Table 1 Features of the nozzles used in the study*

Nozzle type and size	Full name	Nominal flow rate, l min^{-1}	Spray angle, °	Pressure range, bar	Design feature	Application rate, l ha^{-1} (at speed of 4 km h^{-1})
TeeJet TT 03	Turbo TeeJet, Wide angle flat spray	1.18	110	1.0–6.0	Twin chamber, large and rounded internal passage	354
TeeJet TTj60 03	Turbo twinjet, Twin** flat spray	1.18	110	1.5–6.0	Dual outlet produce dual 110° flat fan spray patterns	354
TeeJet DG 03	Drift guard flat spray	1.18	110	2.0–5.0	Removable pre-orifice	354
TeeJet XR 03	Extended range flat spray	1.18	110	1.0–4.0	Standard	354

*From the catalogue of TeeJet products, Spraying Systems Co., Illinois USA. Tabulations are based on spraying water at 70 °F (21 °C) and at 3.0 bar pressure
**The angle between the twin spray fans of TTj60 nozzle is 60°.

1.19, and 1.28 l/min, respectively. These nozzles (after finishing the accelerated wear test) were named "Worn nozzles". Another set of nozzles were not subjected to the accelerated wear test, and these nozzles were named "New nozzles".

These nozzles (new and worn) were evaluated afterword concerning its spray quality in wheat plants in the laboratory conditions, and using water sensitive papers (WSP). These papers were fixed inside the plants canopy in three positions; first paper was fixed five centimetre above the ground. Second paper was fixed 25 cm above the ground. The third paper position was 45 cm above the ground. After spraying (at 3.0 bar pressure), the WSP were collected (after drying) and scanned with 600 dpi resolution. The scanned photos were analyzed with Image-Pro Premier 3D, Media Cybernetics, Inc. software. Calibration of photos with micron units was done using a graded scales which scanned with the WSP's. Spray coverage (%), stains diameter (μm), and stains density on square centimetre were calculated.

3 Results and Discussion

Figure 1 shows the effect of nozzle type on the stains diameter. TT and DG nozzles produced the biggest stains diameter (530 and 526 μm, respectively) comparing with TTj60 and XR nozzles (511 and 466 μm, respectively). Using worn nozzles resulted in an increase in the stains diameter (520 μm) comparing with the new nozzles (497 μm, Fig. 2). The reason for this increase is probably due to the increase in the nozzle tip dimensions after the accelerated wear test. This increase in the tip dimensions would result in a drop in the pressure value inside the nozzle body, which in turn will make the droplets more bigger.

Using XR nozzle resulted the highest coverage percentage (16.7%) comparing with the other nozzles (Fig. 3). While the TTj60 nozzle produced the lowest percentage of coverage (13.6%). Worn nozzles produced more coverage percentage

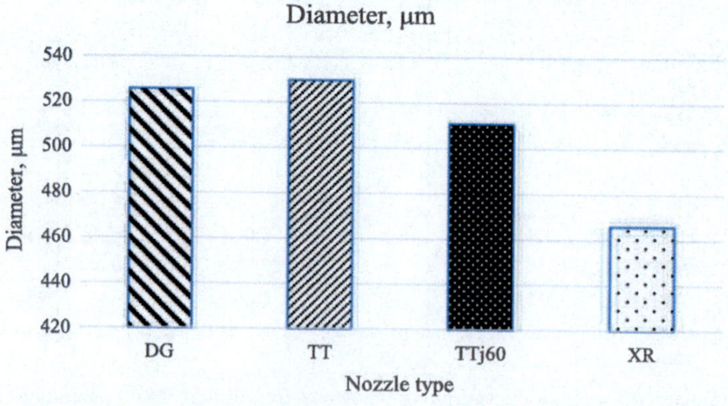

Fig. 1 Effect of nozzle type on the stains diameter

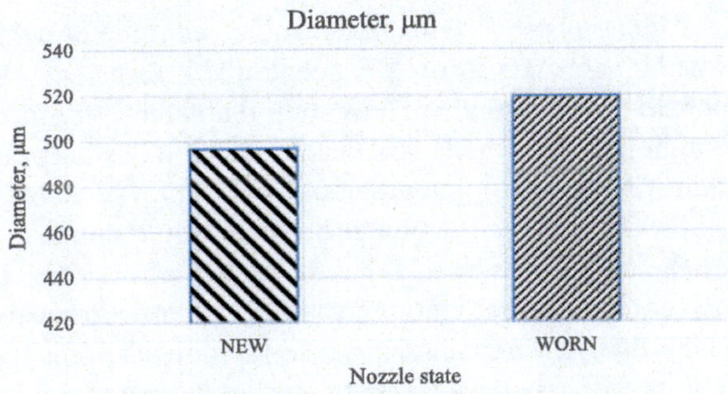

Fig. 2 Effect of nozzle wear on the stains diameter

(15.9%) than the new nozzles (14.3%, Fig. 4), this probably because worn nozzles have higher flow rate than new nozzles, which in the end will result higher application rate. This increase in application rate usually will result in increase in the coverage percentage.

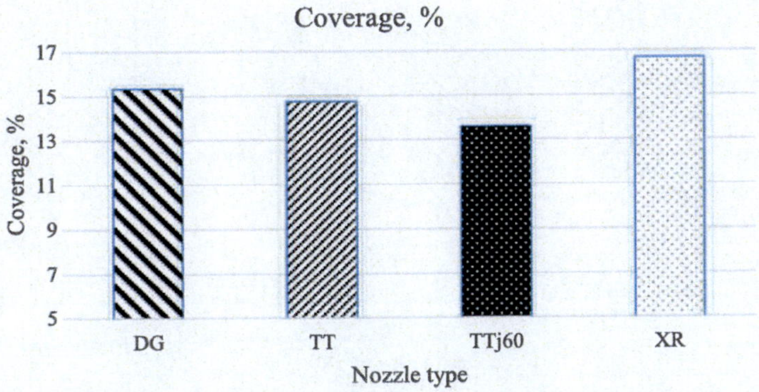

Fig. 3 Effect of nozzle type on the percentage of spray coverage

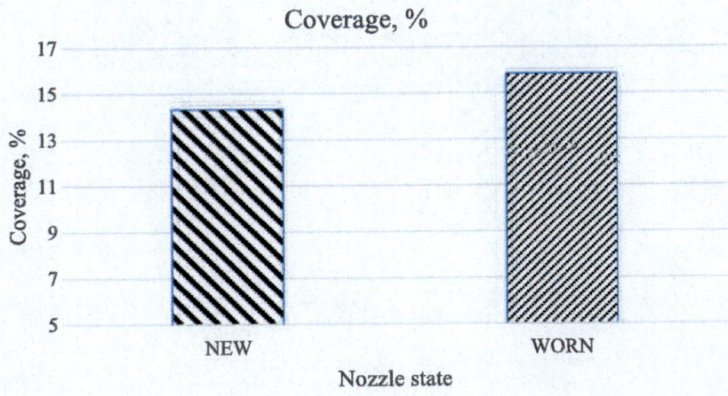

Fig. 4 Effect of nozzle wear on the percentage of spray coverage

The number of stains in the square centimeter, which indicate the spray density, was higher when using XR nozzle (539 stains/cm^2, Fig. 5). Probably because this nozzle type had the smallest stain diameter. New nozzles produced slightly higher number of stains in the square centimeter (421 stains/cm^2) than the worn nozzles (406 stains/cm^2, Fig. 6), the previous explanation probably the reason for this increase.

Producing small diameter droplets would enhance the spray droplets uniformity on the target. However, it has a negative effect since it work on increasing the drift percentage. We can notice from Fig. 2 that worn nozzles produced bigger diameter stains than new nozzles. However, it produced higher percentage of coverage as well. This is probably because worn nozzles deliver more quantity of solution (application rate) than new nozzles.

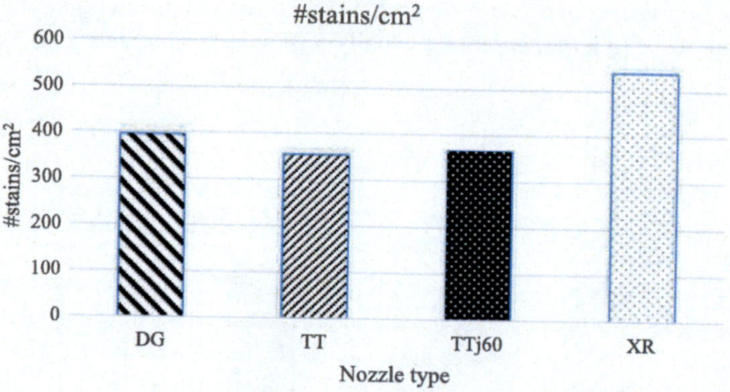

Fig. 5 Effect of nozzle type on the spray density (number of stains in the square centimeter)

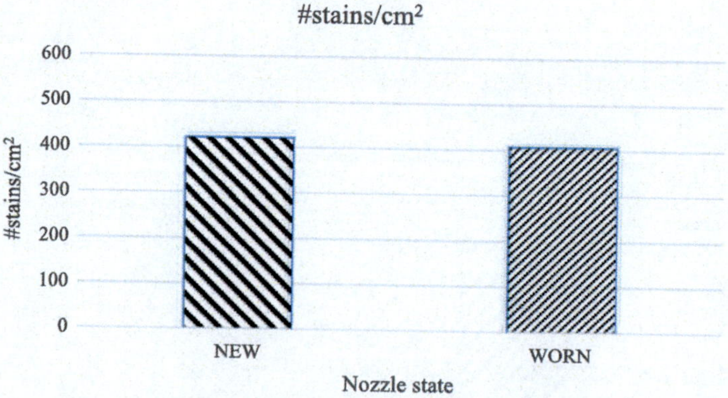

Fig. 6 Effect of nozzle wear on the spray density (number of stains in the square centimeter)

4 Conclusions

From the study results we can concluded the following:

1. Nozzle type has an effect on the stains diameter, percentage coverage, and spray density. XR nozzle produced the highest coverage percentage (16.7%), and highest spray density (539 stains/ cm^2) than other nozzles. However, this nozzle type produced the smallest stains diameter (465 μm).
2. Using worn nozzles produced bigger stains diameter (520 μm) than new nozzles (497 μm). However, it produced in the same time better coverage percentage than new nozzles.

References

1. Aliverdi A, Ahmadvand G (2018) The effect of nozzle type on clodinafop-propargyl potency against winter wild oat. Crop Prot 114:113–119
2. Cerruto E, Manetto G, Longo D, Failla S, Papa R (2019) A model to estimate the spray deposit by simulated water sensitive papers. Crop Prot 124:104861
3. Ferguson JC, Hewitt AJ, O'Donnell CC, Kruger GR (2020) Comparison of water-sensitive paper, Kromekote and Mylar collectors for droplet deposition with a visible fluorescent dye solution. J Plant Prot Res 60:98–105
4. Hassen NS, Sidik NAC (2021) Nozzle type and driving speed effects on spray density of aerial application according to the wind tunnel measurements. J Adv Res Fluid Mech Thermal Sci 84(1):101–110
5. Hilz E, Vermeer AW (2013) Spray drift review: the extent to which a formulation can contribute to spray drift reduction. Crop Prot 44:75–83
6. Krawczuk A, Parafiniuk S, Przywara A, Huyghebaert B, Rabier F, Limbourg Q, Mostade O, Kocira S (2021) Technical parameters of biostimulant spraying a determinant of biometric traits and yield of soybean seeds. Agric Eng 25(1):171–179. https://doi.org/10.2478/agriceng-2021-0014
7. Marçal ARS, Cunha M (2008) Image processing of artificial targets for automatic evaluation of spray quality. Trans ASABE 51(3):811–821
8. Ozkan HE, Paul PA, Derksen RC, Zhu H (2012) Influence of application equipment on deposition of spray droplets in wheat canopy. aspects of applied Biology. Int Adv Pestic Appl 114:317–324
9. Panneton B (2002) Image analysis of water–sensitive cards for spray coverage experiments. Appl Eng Agric 18(2):179
10. Sharpe SM, Boyd NS, Dittmar PJ, MacDonald GE, Darnell RL, Ferrell JA (2018) Spray penetration into a strawberry canopy as affected by canopy structure, nozzle type, and application volume. Weed Technol 32(1):80–84
11. Tadić V, Marković M, Plaščak I, Stošić M, Lukinac-Čačić J, Vujčić B (2014) Impact of technical spraying factors on leaf area coverage in an apple orchard. Tehnički vjesnik 21(5):1117–1124
12. Wolf RE, Daggupati NP (2009) Nozzle type effect on soybean canopy penetration. Appl Eng Agric 25(1):23–30
13. Yao W, Lan Y, Hoffmann WC, Li J, Guo S, Zhang H, Wang J (2020) Droplet size distribution characteristics of aerial nozzles by bell20614 helicopter under medium and low airflow velocity wind tunnel conditions and field verification test. Appl Sci 10(6):2179

Problems with System of Sprayers Testing in Poland Based on 20 Years of Experience

Zbigniew Czaczyk, Jacek Król, Hubert Nagengast, and Daniel Alankiewicz

Abstract It this discussion paper were summarized experiences after twenty years functioning of system for mandatory technical investigations of sprayers (field and orchard). Such mandatory systems should be realized in all EC member states (MS). How it looks today in EC area is difficult to explain. From one side the EC Directives should be introduced follow requirements. This very important issue—specially for food, environment and consumer safety, is not enough supported by the European, and national structures and budgets. Since 2004, after first "SPISE Workshop" the SPISE (Standardised Procedure for the Inspection of the Sprayers in Europe) Working Group was established. There the participants decided, that working should be focused on improving the sustainability of pesticides application unified for MS.

Poland actively taking part in SPISE since beginning. But it is voluntary activity, and not from all MS the national institutions responsible for crop protection are working with.

Conclusions after 20 years of experience, reflections and aspects requiring discussion and optimization on national level in Poland are mentioned—in context of local used PAE (Pesticide Application Equipment) types.

Keywords Sprayer investigations · Pesticide application · Application equipment · Environment pollution · Food safety

Z. Czaczyk (✉)
Poznań University of Life Sciences, Wojska Polskiego 28, 60-637 Poznań, Poland
e-mail: zbigniew.czaczyk@up.poznan.pl

J. Król
Biuro Doradczo-Handlowe, 56-420 Bierutów, Poland

H. Nagengast
Objazdowe Badania Opryskiwaczy PPHU, 11-130 Orneta, Poland

D. Alankiewicz
JAST S.C., 74-110 Banie, Poland

S. Pascuzzi and F. Santoro (eds.), *Farm Machinery and Processes Management in Sustainable Agriculture*, Lecture Notes in Civil Engineering 289,
https://doi.org/10.1007/978-3-031-13090-8_15

1 Introduction

The chemical method of crop protection is used since more than hundred years. The application any active ingredient is conducted with some equipment. It should work safe for the operator, and environment. For large scale of agricultural crop production the manufacturers are developing since decades most modern PAE dedicated to different specific of cultivated crops and also for different field scenarios (weather conditions, BBCH (the phenological development stages of plants), pest). To become good, efficient effect of chemicals application, safe for operator, environment and consumer, it is necessary to use PAE in good technical condition. Also the competences of PAE operator are playing important role. During decades intensification of agricultural production, also intensity and specificity of conducted applications increased. Such situation with significant increase of pesticides use, there was a need to control the technical condition of used PAE. First voluntary procedures of evaluation the technical condition of field sprayers was developed and introduced in The Netherlands and West Germany in 1976 [6]. Then other countries introduced similar actions to improve plant protection safety. First in 1993 was introduced mandatory (field) sprayers investigations in Denmark, Germany and Switzerland. For the air assisted sprayers—mandatory investigations were introduced also in 1993 in Denmark only. In Poland since beginning the technical investigations was introduced as mandatory procedure for field and orchard sprayers in 1999. It was in few years financial supported for cover cost of mandatory test, and also was part of documented modernization (exchange the manometer, spray dose control set, and nozzles. The methods described for sprayers testing evaluated time to time [1, 2], but unfortunately without the opinion of diagnosticians who carry out practically the tests. Over 20 years of experience, the Polish methodology has been largely modified based on the effects of cooperation within SPISE, using ISO [7], EN, Polish Standards PN, and governmental requirements [2–4]. Experiences and problems with mandatory testing of sprayers, to a limited extent, were discussed by specialists during a series of popularizing conferences—Rational Plant Protection Technique (RPPT). Since 2009, after publication of EC Directives: 2009/127/WE, and 2009/128/WE, all MS were obligated to adapt national systems to this requirements [6]. Till end of 2014 should be introduced in all MS the Integrated Plant Production. This requirements obligated also all MS to establish Mandatory systems and methods for technical investigations of PAE. As fare as authors know, this IPP requirements are fulfilled not in all MS. In Poland since 2021 was publish actualized regulation for PAE investigations.

2 Actual Situation

The actualized requirements and investigation procedures evolved a little to be more friendly for diagnosticians/investigators, but in many details we have still problems.

After 11th RPPT conference in 2013, they were not continued. Problems reported by diagnosticians performing PAE practical tests—resulted in the organization of spontaneous discussion workshops coordinated by Professors: Tadeusz Baranowski and Małgorzata Bzowska-Bakalarz, with scientific support and under the patronage of: PTIR (Polish Society of Agricultural Engineering), KOR PAN (Plant Protection Committee of the Polish Academy of Sciences, KTR PAN (Agricultural Technology Committee of the Polish Academy of Sciences and RPT (Agricultural Technical Review) journal. During the first workshop (2014), PAE investigators believed that they could present their problems and proposed changes in a seriously and partner discussion. There was a need to continue the meetings in order to further discuss the possibilities of improving the research system and work comfort. A series of 3 discussion workshops were organized in 2014-16. Representatives of the Ministry of Agriculture and Rural Development, PIORiN (Main Inspectorate of Plant Health and Seed Inspection) [9] were invited to participate, to see existing problems in the practice. The practitioners (investigators/diagnosticians) doing the tests, were the most important participants. The researchers of universities and research institutes taking also actively part with presentations. Many comments were to the existing rules. Many postulates were mentioned- how to improve the methods of practical tests of PAE. Also proposals about efficient conducting, and reporting of test results, and archiving the necessary information were mentioned.

Submitted initiatives were constructive and valuable. Their content can be useful in creating PAE testing systems, mainly where it is not yet fully specified.

The course of the discussion workshops has evidently shown that practical knowledge from investigators is invaluable and should be taken into account in developing and modifying PAE testing methods.

In Poland, various sources (according to different calculation methods) provide significantly divergent information on the numbers of the PAE in use. The differences for tractor powered sprayers oscillate between about 330 thousands from one source, up to 500 thousands from other source.

Additionally, for 20 years the system of recording the examined PAEs has not been improved. Only the number of performed tests is documented and archived. Lack of a coherent system of records of the PAE used.

Another drawback is the poor scope of information on the equipment of the tested PAE, especially on sprayers. During the test, you can easily obtain knowledge about the equipment (nozzles, area serviced in the season, application performance parameters—liquid dose per ha, liquid spraying pressure, driving speed). The obtained information would be the input data for the development of recommendations useful for operators in their working conditions (PAE and field conditions scenarios for the implementation of the application). Until recently, in the accredited letters of SKO only their postal addresses were available. For unknown reasons, the telephone numbers and e-mail addresses could not be made available by PIORiN [9]. Currently, the situation has improved and the above-mentioned data is now available.

Need for classification of sprayers used in Poland on Potential of Drift Reduction were mentioned. The need to classify sprayers used in Poland in terms of drift reduction potential was mentioned. So far, no action has been taken in this regard.

The only material available regarding the drift reduction potential is the German collation from the federal institute with the table heading translated.

Another, rightly perceived problem is the lack of education requirements (technical competences) for candidates for investigators, and the lack of a practical part in the qualification process—an internship in a unit accredited for PAE testing.

Also, the obligation to adjust the spray liquid dose per ha does not take into account the method of selecting nozzles, taking into account the specificity of the application.

Due to the widespread use of automatic dose regulators and other precision farming elements in the equipment for several years, the legislation does not keep pace with the need to define the method of controlling the accuracy of operation of such modern systems.

Some participants reported a problem that PAE owners were not getting their devices ready for the test. Especially in smaller farms, PAE is not washed—as required by the regulation. There is a problem of enforcing this process, respecting the costs and time of both parties.

For such cases, the cost of professional cleaning (inside and outside) of the tested device should be officially specified.

3 Formulated Initiatives and Problems

On the base of ISO 16122 series, Directive 2009/127/WE, and SPISE Manuals, the PAE sprayers testing should be unified in MS, but also adapted adequate to local specifics. The long-standing problem of the correct method of controlling the distribution from the spray boom has been liberalized.

Finally, both methods can be legally used: Patternator as well as the flow rate measuring. This is a compromise as it is known that neither of these methods can be used alone to obtain a complete assessment of the quality of liquid distribution from the boom.

Lack of the obligation to test new sprayers (5 years), and the lack of general records of them, makes it difficult to include all sprayers under inspection. There are no clear rules for changing owners or importing.

Objective instructions should be regularly updated, taking into account the offers of Polish manufacturers (nozzles and sprayers). So far, they have been based mostly on information materials of foreign companies. This applies to a very important aspect—the quality of the atomization—which characterizes the suitability for specific application conditions and the fulfilling of the pesticide label requirements.

Introducing official prices for a compulsory test as for vehicle testing. Tests conducted in the farms of PAE owners should be officially regulated in terms of fees for arrival. No institution offers a demonstration process for carrying out the test with the required scope of activities. Practitioners call for such a model demonstration include a full process of practical and formal activities within approx. 30–45 min. It is related to profitability, which is also important in their commercial functioning.

The current, flexible prices for the test are too low and do not cover the real costs of the test.

4 Conclusions

The knowledge and practical experience delivered to discussion from investigators were very useful. Presented points of view and open discussion about details of test procedure, and problems at this specific work, allowed to create list of postulates addressed to legislators.

Fruitful effects of 3 discussions workshops with possibility to exchange experiences between practitioners, legislators, advisors, engineers, seller of PAE, and researchers confirmed usefulness debates with specialists.

The postulates formulated in the published summaries of the workshops, addressed mainly to legislators, gradually and evolutionarily allow to achieve the effects of changes in line with the expectations of practitioners.

This type of exchange of experiences, information and creating common solutions was used probably the first time and both the effects and the way can be used in other countries developing PAE control systems.

Initiatives to create a national system of registration of sprayers (similar to vehicles) as well as records of diagnosticians and operators with valid authorization to use pesticides are constructive. This will contribute to increasing the safety of agrochemicals application.

Extension of obtaining authorization to test PAE with an internship in an authorized workshop (min. 3 months) is justified.

Sprayers used in IPP should equipped with atomizers having spraying quality technical information from any independent laboratory. This is necessary so that, knowing reliable information about the quality of the atomization, the operator of the sprayer would be able to meet the strict requirements of the instructions for use of the pesticide. The pesticide labels include also required atomization quality, because the operator of the sprayer is personally responsible for any inappropriate side effects of the application. The use of information obtained from practitioners evidently brings constructive conclusions. This allows for efficient and faster optimization of regulations and requirements. This should result in improved comfort and efficiency in implementing mandatory PAE testing.

Acknowledgements Publication was co-financed within the framework of the Polish Ministry of Science and Higher Education's program: "Regional Initiative Excellence" in the years 2019–2022, project no. 005/RID/2018/19, financing amount 12 000 000,00 PLN.

References

1. Czaczyk Z (2021) Luki w przepisach dotyczących badań opryskiwaczy (Gaps in the regulations on sprayer testing). RPT (Agric Tech Rev) 2(264):50–52
2. Czaczyk Z (2016) Review of discussion workshops: how to improve testing procedures and application technique in Poland. In: Book of Abstracts of 6th European Work on SPISE – SPISE 6, pp 88–112. Barcelona
3. Czaczyk Z (2014) Drop-size classification according to requirements of pesticides labels. Prog Plant Prot 54(1):111–120
4. Dz. U (2013) poz. 455: Ustawa z dnia 8 marca 2013 r. o środkach ochrony roślin, p 53
5. Dz. U (1999) nr 20, poz. 175: Rozporządzenie Ministra Rolnictwa i Gospodarki Żywnościowej z dnia 11 lutego 1999 r. w sprawie szczegółowych zasad przeprowadzania badań sprzętu do stosowania środków ochrony roślin, p 5
6. Ganzelmeier H, Wehmann H-J (2007) Second European workshop on standarised procedure for the inspection of sprayers in Europe – SPISE 2 - Mitt. Biol. Bundesanst. Land-Forstwirtsch. 412 Berlin-Dahlem, p 180. Straelen
7. ISO 16122-1 (2015) International Organization for Standardization. Agricultural and forestry machinery — Inspection of sprayers in use — Part 1: General, p 7
8. Regulation (EC) (2009) No. 1107/2009 of the European Parliament and of the Council concerning the placing of plant protection products on the market and repealing Council Directives 79/117/EEC and 91/414/EEC, p 50
9. PIORiN Homepage. http://piorin.gov.pl/. Accessed 21 Mar 2022

Changes in Soil pH and Their Influence on Sustainable Development in the Agriculture

Wioletta Żukiewicz-Sobczak, Paweł Sobczak, Katarzyna Grotkiewicz, Anna Domańska, Jacek Mazur, Patrycja Łusiak, Konrad Żukiewicz, and Małgorzata Bernatek

Abstract The study aimed to evaluate soil pH after combine harvesting, in wheat, rye, and barley crops cultivation according to two agricultural strategies, viz. ecological and conventional. An extensive microbiological analysis was carried out to address the qualitative and quantitative presence of mould fungi in samples of soil taken from wheat crops. The obtained soil pH results varied according to the type of cereal grown. The concentration of mould from the soil samples from organic wheat cultivation was higher as compared with that for the soil samples taken from conventional farming.

Keywords Soil pH · Fungi · Sustainable development · Agriculture

1 Introduction

The condition of the environment and natural resources is very closely related to human impact in various areas of production. Environment, climate, and natural re-sources protection shape our views and actions in the field of sustainable development. The Common Agricultural Policy of the European Union recognizes the need to take measures to protect the environment. Promoting organic farming is one of these measures. Sustainable development in environment protection entails many

W. Żukiewicz-Sobczak · A. Domańska · M. Bernatek
Department of Food and Nutrition, Calisia University, Kalisz, Poland

P. Sobczak (✉) · J. Mazur · P. Łusiak
Department of Food Engineering and Machines, University of Life Sciences in Lublin, Lublin, Poland
e-mail: pawel.sobczak@up.lublin.pl

K. Grotkiewicz
Department of Production Engineering, Logistics and Applied Informatics, University of Agriculture in Krakow, Krakow, Poland

K. Żukiewicz
Department of Power Engineering and Transportation, Faculty of Production Engineering, University of Life Sciences in Lublin, Lublin, Poland

S. Pascuzzi and F. Santoro (eds.), *Farm Machinery and Processes Management in Sustainable Agriculture*, Lecture Notes in Civil Engineering 289,
https://doi.org/10.1007/978-3-031-13090-8_16

positive aspects, not only for soils but also for the entire globe. Sustainability is an issue increasingly more of-ten addressed during scientific symposia around the world. Lack of the natural state of balance leads to various irreversible consequences not only for the environment but also for humans, their quality of life, and health. The conventional strategy uses synthetic fertilisers and chemical plant protection products as well as industrial fodder, which, it was assumed, were to provide higher plant and animal efficiency and, what is important, higher factor productivity, than organic farming. The organic strategy is based on compliance with crop rotation rules, i.e. applying rotation in plant sowing throughout the entire growing season. Furthermore, cultivation in this strategy is based primarily on natural and organic fertilisers and, at the same time, on the limited use of mineral fertilisers and plant protection products. The organic strategy also provides for measures to ensure plant protection and adherence to the standards regulating the available space and cleanliness of livestock housing, including setting up the acceptable animal density on agricultural land [1].

The implementation of environmentally friendly strategies has a decisive impact on the chemical, physical, as well as biological characteristics of the soil. All these aspects have both a direct and indirect impact on one another. An imbalance between these three properties may lead to many adverse consequences for crop cultivation, environment, climate, and human health. Farmers are exposed to a wide range of diverse stimuli, arising from the work performed, that are threatening and harmful to health. These factors include, among others, organic dust of biological origin. Taking into account the place of origin, this type of bio-aerosols may contain a variety of different kinds of crop residues, filamentous fungi, mycotoxins, bacteria, bacterial toxins, mites, as well as fragments of insects and animal excrements with very diverse composition [2].

Natural causes of soil acidification are related to geological and biogeochemical factors as well as climatic conditions. According to Filipek [2013] more than 90% of Polish territory is covered with soil originating from sedimentary rocks, mainly loose clastic ones, transported by glaciers from Scandinavia. The acidity (pH) of the environment is one of the main factors determining the appropriate use of the soil. Excessive lowering of pH of soil used for agriculture significantly reduces the yield. Moreover, it has an adverse impact on the ecosystem and climate in many countries with humid climate [3]. New strategies introduced in agriculture bring with them a number of uncertainties linked to the adaptation of plants to the soil. Currently, in the EU crop market, a gradual replacement is observed of conventional strategy in favour of ecological one. Changing the fertilisation strategy can lead to new soil conditions in crop cultivation. The soil pH and humus content determine the phenomenon of soil fertility as well as the physical, chemical and biological properties of the soil. Rational and effective management of nutrients is related to the assessment of the content of available nutrients in the soil, as well as the agronomic assessment of the soil class. This type of treatment is key in the planning of crops rotation.

Soil pH is a very important soil property. It affects the soil environment as it determines the physical, chemical and biological properties of the soil. Knowledge of the pH value of the soil helps determine its quality and suitability for the cultivation of different crop groups. Proper soil pH, i.e. well-adjusted to plant requirements, has a decisive influence on nutrient use, yield and profitability. Soil pH also affects microbial activity and soil processes.

For growth, development and delivery of adequate yields plants need nutrients taken in appropriate quantities and proportions, depending on the species. Plants absorb nutrients mainly through roots, which is why it is crucial to ensure that appropriate conditions are provided for this process. Soil pH discussed here is one of the main factors influencing the availability of nutrients to plants from soil solution.

Soil pH is one of the most important indicators of its fertility. Most plants thrive predominantly in the case of soil pH ranging from slightly acidic to neutral. The best availability of nutrients for plants occurs when soil pH equals 6.5–7.2. Already under light soil acidification conditions below 6.5 pH, the absorption capacity of certain nutrients, including nitrogen and phosphorus, is significantly reduced [4]. In alkaline substrate, useful micro-organisms develop and the mineral substances are available in a form easily assimilable by plants. In contrast in over acidic soil acidic mould develops, and many minerals form water-insoluble chemical compounds. Acidification of soils is correlated with a decrease in fertility and productivity, and pH values of pHKCl below 4.5 indicate a risk of soil environment degradation [5]. The effects of soil acidification from the point of view of organic and conventional agriculture lead mainly to a reduction in the nutrient absorption by plants, in particular of phosphorus, magnesium and molybdenum, and an increase in the mobility of materials which become dangerous in larger quantities, especially heavy metals and mobile aluminium [6]. Szatnik-Kloc (2004) showed that a high pH of the soil is a significant factor in the increase in toxicity and heavy metals concentration in soils [7].

The yield of the crop depends on the soil condition. High and good quality yields can only be guaranteed by soils with high microbiological activity and proper physical, chemical and biological characteristics, which are the main indicators of soil quality, which, together with soil structure, form the basic foundation of soil fertility [8, 9].

In 2015 Taczyński in his study on the chemical properties of soil in an organic production system in which, among others, rye and oat were grown showed favourable changes in soil pH, measured in KCl, over the eight years of analysis. Between the first (2007) and the last (2014) year of the study, the soil acidity was reduced by one unit, i.e. from 5.1 to 6.1, and thus the pH changed from acid to slightly acid, which, according to the author, was probably the outcome of systematic soil liming and using manure [10].

The aim of the study was to evaluate soil pH after combine harvesting, in wheat, rye, and barley crops cultivation done according to the two agricultural strategies, viz. ecological and conventional. The direct effect of soil pH on soil microbial activity in terms of the quantitative and species composition of mould in soil samples from conventional and organic wheat cultivation was also assessed. Furthermore, the identified mould was classified according to the Biosafety Levels (BSL).

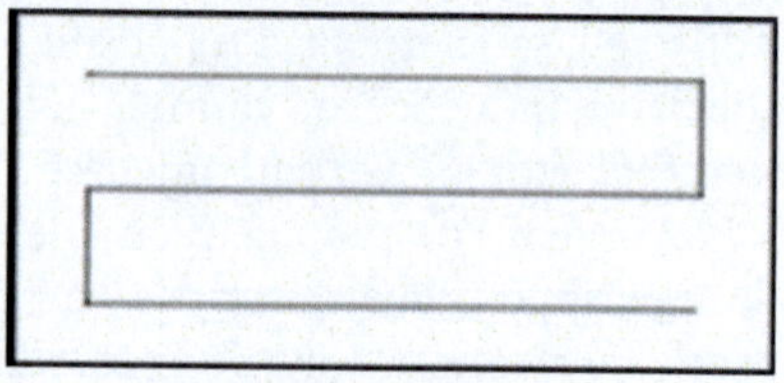

Fig. 1 The scheme of the field sampling

2 Materials and Methods

Examined material consisted of soil samples collected from wheat, rye, and barley cultivation from 30 different farms in south-eastern Poland. The research was carried out after combine harvesting of cereals cultivated under two agricultural strategies, viz. conventional and organic farming. In the course of the study mature soil was sampled. In total 30 soil samples were collected. Samples were taken from the soil surface to the depth of 20 cm, excluding field edges, ditches, recesses and hills. The mass of one test sample equalled 300 g and it was a mixture of 20 original samples taken from points evenly distributed over the test area as shown in Fig. 1.

Determination of soil pH was carried out on pH meter CP 411 combined with an EPS-1 pH electrode. The obtained soil pH results varied according to the type of cereal grown. Statistical analysis of the results obtained was performed using Statistica 13.3 software.

Microbiological tests were carried out on samples of soil used for wheat cultivation carried out under two farming strategies, i.e. organic and conventional (10 soil samples per farming strategy – 20 samples in total).

In order to determine the concentration and composition of the fungi in the collected soil samples two substrates were used, i.e. Malt Agar (MA, Becton, Dickinson and Company) and Potato Dextrose Agar (PDA, Becton, Dickinson and Company), using the plate dilution method with Tween 80 as the emulsifier. The research was conducted in two parallel repetitions. The PDA substrate was incubated at 24 °C for 144 h. The MA substrate was incubated at 30 °C for 72 h followed by another 72 h at room temperature. The fungi species composition was determined using macroscopic and microscopic methods and taxonomic keys and atlases [11–16] and expressed in CFU/g units. The isolated mould species were ranked according to the Biosafety Levels (BSL).

3 Results

The pH of the soil for wheat crops cultivated following conventional strategy ranged from 6.00 to 7.55, while in the case of the organic farming of the same cereal for 10 samples it was within the range of 6.0–7.35. It should be noted that in the case of organic strategy the soil pH showed smaller discrepancies (Fig. 2).

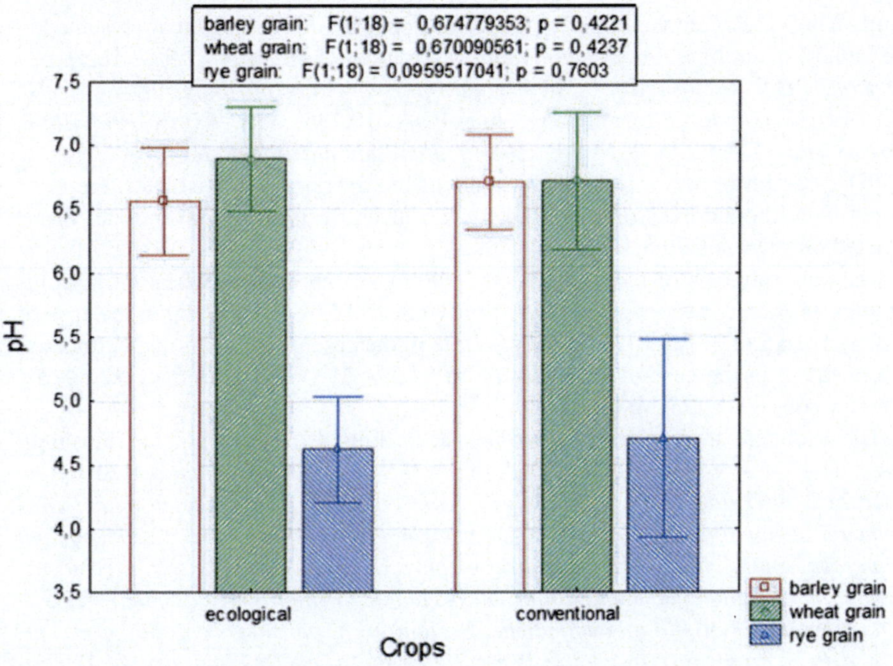

Fig. 2 Average soil pH for various species and its standard deviation

After the pH tests were carried out for the soil under conventional and organic cultivation, the obtained results were within the range of 4.15 to 6.20. In the case of the organic strategy, the results were less varied and were within the range of 4.15–5.20. The pH of the soil rye was cultivated on was significantly lower than that where wheat was grown. It was observed that for the examined cereals the pH of the soil from rye cultivation was the closest to the acidic one (Fig. 2).

In the case of barley cultivation, the pH of soil ranged between 6.0 and 7.10 for the organic strategy, and between 6.25 and 7.25 in the case of the conventional one.

Among the three examined species of cereals, the pH of the soil where barley was cultivated, was the closest to the neutral pH (Fig. 2).

Statistical analysis of variance with the significance level of 0.05 showed no significant difference regardless of the implemented strategy of agriculture (Fig. 2). However, significant differences occurred depending on the cultivated crop species. The pH of the soil was the lowest after rye cultivation (pH 4.6–4.7). There were no statistically significant differences when comparing barley and wheat, in the case of which the average pH was about 6.7

In the soil samples from the organic wheat cultivation the following types and species of mould were isolated in the highest quantities: *Penicillium tardum* PDA - 10.62%, MA - 4.9%, *Scopularipsis brevicaulis* PDA - 10.62%, MA - 20.71%, *Paecilomyces lilacinus* PDA - 4.6%, MA - 8.86%, non-spore-forming mycelium PDA - 9.73%, MA- 5.45%, *Acremonium sp.* PDA - 1.06%, MA - 0.54%. On the other hand, in the case of conventionally cultivated crops the following were represented in the highest quantities: *Acremonium murorum* PDA - 3.89%, MA - 2.9%, *Alternaria alternata* PDA - 6.29%, MA - 1.94%, *Cladosporium cladosporioides* PDA - 4.79%, MA - 1.61%, non-spore-forming mycelium PDA - 10.78%, MA - 8.06%, *Penicillium citreo-viride* PDA - 3.89%, MA - 4.19%, *Penicillium pururogenum.*

PDA - 11.08%, MA - 18.39%. In terms of biodiversity of the isolates, the most varied were the samples of soil used for wheat cultivation in conventional farms isolated on PDA substrate.

The concentration of mould in the soil samples from organic wheat crops cultivation was equal: in the case of PDA from 10 to 100.5 $\times$ 10^3 CFU/g, in the case of MA from 16.5 to 177.5 $\times$ 10^3 CFU/g. While in the case of conventional cultivation it was equal: in the case of PDA from 5 to 54.5 $\times$ 10^3 CFU/g, in the case of MA from 6 to 65 $\times$ 10^3 CFU/g (Table 1).

The isolated strains were classified according to the respective biosafety levels (BSL). The species classified to level 1 include: *Acremonium strictum, Alternaria alternata, Aspergillus Niger, Aspergillus versiolor, Chaetomium atrobruneum, Chaetomium funicola, Cladosporium cladosporioides, Cladosporium sphaerospermum, Fusarium sacchari, Geomyces pannorum, Mucor racemosus, Paeciliomyces lilacinus, Penicillium chrysogenum, Penicillium citrinum, Penicillium commune, Penicillium decumbens, Penicillium expansum, Penicillium griseofulvum, Penicillium purpurogenum, Penicillium rugulosum, Rhizopus oryzae, Trichoderma viride, Ulocladium chartarum* (23 species). Whereas the species classified to level 2 are: *Aspergillus fumigatus, Fusarium solani, Fusarium verticillioides, Scopulariopsis brumptii, Scopularipsis brevicaulis* (5 species).

Table 1 The total number of mould (CFU/g $\times$ 10^3) in the soil samples taken from wheat cultivation, depending on the crop cultivation strategy (organic, conventional) and the type of substrate (PDA, MA)

Sample no.	Organic farm		Sample no.	Conventional farm	
	PDA	MA		PDA	MA
1	33.25	31	11	5	6
2	27.75	40	12	7.5	3.5
3	26.5	18.75	13	14.25	12
4	100.5	73.75	14	29.75	34.75
5	29.5	44.5	15	54.5	17
6	10	16.5	16	51.5	65
7	44.25	32.25	17	14	9
8	83.75	177.5	18	25.5	11.5
9	22.25	25.25	19	18.75	10
10	18.5	52.75	20	16.5	13
Average	39.63	51.23	Average	23.73	18.18

4 Discussion

In both organic farming and conventional crops cultivation, the good quality of yields is the primary concern, hence the soil pH and its nutrition value plays an important role in the proper growth and yield of plants. Plants reacting strongly to acidification, such as, for instance, winter wheat and barley, have optimal conditions for pH of 6.0 to 7.5. On the other hand, plants medium sensitive to acidification, such as rye and oat, have optimal conditions for pH of 5.0 to 6.5 [17]. Below these values, i.e. in an acid environment, heavy metals can be drawn in excessive quantities and accumulated in plant products. Heavy metals absorbed by plants can be reduced, for example, by liming [18, 19]. In organic farms, only carbonate lime may be used for soil deacidification. This is due to the principle of maintaining the chemical and biological balance of the soil in organic farming. Liming treatment is one of the means of regulating soil pH intended for soil deacidification to reach optimal values for a given soil agronomic category or plant species.

Soil liming in the case of wheat and barley cultivation on sandy and acidified soils showed an increase in pH both in the topsoil and underneath the plough layer, as well as a decrease in the aluminium toxicity of the soil [20].

Liming is an agrotechnical process that determines soil condition. Not only does it make the soil less acidic but it is crucial for the absorption of micro-elements by plants. It also supports organic and mineral fertilization. It improves the quality of crops indirectly. Skilful liming allows obtaining pH optimal for the soil and cultivated crops. Determination of the desired pH level for a particular field starts with determining the soil type and then the requirements of intended crop.

Acidification of soils greatly reduces their productivity and, in many cases, hinders obtaining high quality raw materials needed for the production of healthy food from organic farms. Soil pH by decreasing below the optimal value for a given element leads to a rapid decrease in the element yield efficiency. Niemiec et al. (2020) in their surveys on the sample consisting of 50 organic farms (without animal husbandry) and conventional farms, in which cattle husbandry took place additionally, have shown that the properties of tested soil indicate a high risk of chemical and biological degradation. The investigated farms showed very low organic carbon content and were characterized by very low pH. In most cases, the tested soils were strongly acidic. In the above work, as well as other studies, it was unanimously acknowledged that without implementing control measures, the pH level and the problem of low absorption of phosphorus and mineral nitrogen may lead to the disturbance in the agroecosystems homeostasis and, consequently, to soil degradation [4, 21, 22]. Conventional farms with animal production, included in the above study, have shown beneficial soil characteristics, from the point of view of cultivation, as compared to organic farms which do not involve in animal husbandry [23]. Farming based on balancing crop production and livestock husbandry that respects the principles of organic farming and implements natural resources seems to be the response to the objective of improving soil chemical properties. A system aimed at food production with minimal damage done to ecosystems, animals, and people is perfectly in line with the principle of sustainable agriculture. According to Becher et al. (2020) organic fertilizers, including, in particular, manure fertilization, have shown a significant impact on the quantity and quality of organic matter accumulated in the topsoil layer. Organic matter, especially humus, regulates soil productivity and resistance to degradation [23]. In addition, according to Jakubus and Graczyk (2020), the use of organic fertilizers can supplement soil organic matter content, improves soil fertility and increases crop yield and quality [24].

In organic crop rotation, beneficial chemical parameters of the soil are affected in particular by both the high humus and total nitrogen (N) content, as well as the content of three micronutrients (B, Cu, Mn) in the soil [25].

Frąc et al. (2011) in the course of the research on microbiological properties of the soil used for winter wheat cultivation in organic and conventional strategies, found a higher activity of soil micro-organisms in the case of organic farming, which could indicate good soil quality and making nutrients available to plants. This is probably associated with higher organic matter content and soil pH more beneficial for the development of micro-organisms [26]. Similar relationships were obtained in this work by determining the concentration of mould in the soil samples from organic cultivation of wheat crops (PDA: from 10 to 100.5×10^3 CFU/g, MA: from 16.5 to 177.5×10^3 CFU/g), while in the case of conventional cultivation the concentration of mould for the respective substrates was: PDA: from 5 to 54.5×10^3 CFU/g, MA: from 6 to 65×10^3 CFU/g (Table 1). However, it should be noted that higher genetic and species biodiversity was obtained in the case of mould isolates from conventional crop cultivation (Table 1).

Sorption properties of soil consist of the ability to absorb and retain solids, gases and, above all, various chemical compounds and ions. Sorption properties play an essential role in soil. They enable to retain the nutrients formed as a result of mineralization processes or introduced into the soil with fertilizers. Retention of fertilizing components in the soil prevents the loss thereof, affects the efficiency of fertilization and prevents water eutrophication. Sorption capacity also allows water retention and regulation of soil pH. Many studies indicate that, as pH increases, the sorption capacity of the soils increases [27].

Based on the reports in the literature of the subject, it can be unequivocally admitted that the agricultural strategy used in agriculture affects the chemical properties of the soil [28, 29]. The organic soil cultivation strategy favours soil characteristics based on the use of natural (organic) fertilizers, as well as humus, organic carbon, and micro-elements content in the soil. A conventional strategy may cause acidification of the soil if large amounts of mineral fertilizer (especially nitrogen fertilizer) are used [30, 31]. Kwiatkowski and Harasim (2020) in their studies, which consisted in defining certain parameters of the soil chemical composition, e.g. soil pH, total sorption capacity, humus content, and micronutrients content showed that organic crop cultivation contributed to the increase of magnesium, boron, copper and other elements content in the soil. In addition, organic farming practices had a positive effect on soil pH, however, this difference was not statistically significant. Moreover, a higher humus content was noted, which is a cornerstone of healthy soil. On the other hand, conventional crop cultivation showed higher levels of phosphorus and potassium in the soil as compared with organic cultivation. Finally, it was concluded that organic crop cultivation significantly improved the sorption capacity of the soil as compared to traditional (conventional) cultivation [25].

Organic and conventional farming systems differ significantly in terms of the methods of fertilisation and plant protection, hence the impact of biological factors may cause various health effects in humans. The opinions regarding the pro-health aspect of cultivation according to these two production systems still vary, not only when it comes to the quality of raw materials produced, but also in regards to the health risks faced by the people working with these crops. One theory assumes that different method of fertilisation characterizing organic farming causes a change in the metabolism of cultivated crops. A larger amount of secondary metabolites produced by them makes these plants cope better with challenges such as fungal diseases and other pests. In this case, the pH of the soil has labile value sensitive to modifications of the soil environment, which is related to the reaction of fertilisers components with the soil. Changes in the soil pH are the key factor in the development and vegetation of many soil microorganisms, such as bacteria and fungi [32, 33]. According to the data from the literature of the subject, the results of studies concerning the content of filamentous fungi in the soil advocate for the change to ecological agriculture. Adherence to the principles of organic farming is more efficient and soil-friendly due to the biodiversity and concentration of filamentous fungi. According to the data from the literature of the subject, in the case of soil samples from conventional cultivation the concentration of filamentous fungi equalled 124.65×103, and thus was more than two times higher compared to the results obtained for the samples from organic cultivation, viz. 48.55×10^3 (CFU/g) [32]. In the research carried out, the concentration of mould from the samples of soil obtained from organic wheat farms was: for PDA: from 10 to 100.5×10^3 CFU/g, for MA: from 16.5 to 177.5×10^3 CFU/g. While in the case of conventional cultivation it equalled: for PDA: from 5 to 54.5×10^3 CFU/g, for MA: from 6 to 65×10^3 CFU/g (Table 2).

Organic farming accepts only the use of reproductive material derived solely from organic farming and chemical treatment of raw materials is forbidden. The composition of fungi present in the grain depends on the conditions present during the vegetation period, such as, for example, weather conditions, cultivation method, use or non-use of synthetic fertilisers and plant protection products. The results of the existing studies confirm that climatic conditions have a major impact on the development of particular plant pathogens, for instance, the presence of fungi of the Fusarium genus depends on temperature. According to the sources, the most favourable conditions for the development of species of the Fusarium genus are high humidity lasting for more than 24 h, and air temperature above 20 °C [32, 33].

From the data obtained from the Statistical Office (GUS) and representing the studies conducted in the years 2009 - 2010 it is known that in Poland most of the producers involved in agricultural production are organic farmers (98% of all producers) [34]. According to data provided by the Agricultural and Food Quality Inspection in recent years, organic farming in Poland has been developing dynamically. The continued support of the European Union in this area and the growing interest in organic products are the factors conditioning this development. In the years 2009 and 2010, meadows and pastures had the largest share in the organic agricultural area (42.3%), with fodder crops cultivation being on second place (20.6% of arable land). These were followed by cereals, which accounted for 19.6% of organic farmland.

The pH of the soil is associated with a number of endogenous and exogenous factors occurring in the soil. Depending on the anthropogenic soil conditions, sandy, neutral and alkaline soils are present in the environment of agricultural work.

Sandy soils are characterised by a pH which is very acidic, i.e. 3.0, acidic, i.e. 4.0, and moderately acidic, i.e. 5.0. Neutral soils are slightly acidic (pH 6.0) and slightly alkaline (pH 7.0). In the case of alkaline soils the pH is around 8.0, while for strongly alkaline – it is even 9.0. According to the literature of the subject, many biochemical processes depend on the soil pH, which is called the main variable of the soil. In agriculture, controlling the soil pH is the determinant of the availability of nutrients for plants. Therefore, soil productivity and yield levels are directly proportional to the soil pH value. In agriculture, soil pH value ranges, on average, between 5.5 and 7.5. On the basis of data from the Natural Resources Conservation Service of the United States Department of Agriculture the soil pH ranges are classified as follows: ultra-acidic (<3.5), extremely acidic (3.5–4.4), very strongly acidic (4.5–5.0), strongly acidic (5.1–5.5), moderately acidic (5.6–6.0), slightly acidic (6.1–6.5), neutral (6.6–7.3), slightly alkaline (7.4–7.8), moderately alkaline (7.9–8.4), strongly alkaline (9.0), and very strongly alkaline (9.0) [35, 36].

The soil pH is associated with the properties of solutions present in the soil and it does not apply to the solid phase of the soil, viz. to the elementary particles. According to several authors (38,39,40] the processes between the solid phase of the soil and the water contained by it account for the acidic or alkaline character of the soil. Soil acidity is the key parameter for the availability of nutrients for arable crops. According to the data from the literature of the subject, low pH levels prevent plants from obtaining nutrients. This is related to the transformation of minerals into forms that are difficult to absorb, e.g. Mo, B. Decline in organic matter, as well as the increase in heavy metal activity (Mn, Zn, Pb and others) are other factors negatively affecting the nutrient content of the soil [37]. In this own research, the soil pH for wheat cultivation in accordance to conventional strategies ranged from 6.00 to 7.55, while in the case of organic farming of the same grain it ranged from 6.10 to 7.35 (for 10 samples tested). In the case of rye cultivation harvested with a combine harvester for both organic and conventional strategy, the observed range was between 4.15 and 6.20. In the case of organic strategy smaller variation was recorded, i.e. within the range 4.15–5.20. When it comes to barley cultivation, the soil pH was within the range of 6.0–7.10 for organic farming and 6.25–7.25 for conventional farming. The examination of soil pH carried out in New Zealand gave the pH result of 6.29 for conventional strategies and pH 6.10 for organic farming [38–42].

Biosafety levels classification (BSL)

In 1996, the working party of the European Confederation of Medical Mycology developed the classification of fungi in terms of their biosafety [43]. The Biosafety

Level (BSL) classification shows the safety scale for fungi potentially pathogenic to humans and animals and distinguishes between 3 hazard levels represented by various fungal species. The BSL-1 includes saprophytes and plant pathogens that cause superficial, non-invasive or mild hazards. The BSL-2 consists of species characterized by a relatively high capacity to survive in vertebrate tissues, and in patients with severe immune disorders they may cause deep, opportunistic infections. The BSL-3 includes pathogens that are potentially able to cause severe, deep fungal infections in generally healthy persons [44].

The species classified as BSL-1 include 23 species, while the species classified as BSL-2 include 5 species. This data shows the important role of moulds in the pathomechanisms of infection. In addition, they are valuable in the in vitro studies on the methods of assessing exposure to harmful mycotic agents and confirm the utilitarian potential of these techniques in preventive healthcare.

5 Conclusions

On the basis of the conducted study, the following conclusions are formulated:

1. Cereal crops cultivation systems, i.e. organic or conventional, do not significantly affect the pH of the soil.
2. Significant differences in soil pH were observed depending on the crop cultivated.
3. The lowest soil pH (4.6) was recorded after rye cultivation, while after wheat and barley cultivation soil pH was within the range of 6.7 to 6.8.
4. Higher biodiversity of mould was observed in the isolates coming from the conventional cultivation as compared to those from organic farming.
5. The concentration of mould from the soil samples from organic wheat cultivation was higher (for PDA from 10 to 100.5 $\times 10^3$ CFU/g, for MA from 16.5 to 177.5 $\times 10^3$ CFU/g) as compared with that for the soil samples taken from conventional farming (for PDA from 5 to 54.5 $\times 10^3$ CFU/g, for MA - 6–65 $\times 10^3$ CFU/g).

References

1. Rozporządzenie MRiRW z dnia 14 maja 2002 r. w sprawie szczegółowych warunków wytwarzania produktów rolnictwa ekologicznego [Dz. U. 2002, Nr 77, poz. 699]. (in Polish)
2. Wittczak T (2012) Choroby alergiczne i zagrożenia toksykologiczne w środowisku pracy rolników. Alergia 1:12–14 (in Polish)
3. Filipek T, Skowrońska M (2013) Aktualne dominujące przyczyny oraz skutki zakwaszenia gleb użytkowanych rolniczo w Polsce. Acta Agrophisica 20(2):283–294. (in Polish)
4. Pietrzak S (2002) Priorytetowe środki zaradcze w zakresie ograniczania strat azotu i fosforu z rolnictwa w aspekcie ochrony jakości wody. Instytut Technologiczno-Przyrodniczy w Falentach. (in Polish). ISBN 978-83-62416-42-4

5. Rutkowska A (2018) Ocena przestrzennego zróżnicowania odczynu gleb w Polsce w latach 2008–2016. Studia i Raporty IUMG-PIB 56(10):9–20. https://doi.org/10.26114/sir.iung.2018.56.01
6. Hołubowicz-Kliza G (2006) Wapnowanie gleb w Polsce. Instrukcja upowszechnieniowa nr 128, Wydawnictwo IUNG-PIB, Puławy. ISBN-83-89576-31-7
7. Szatanik-Kloc A (2004) Wpływ ph i stężenie wybranych metali ciężkich w glebie na ich zawartość w roślinach. Acta Agrophysica 4(1):177–183. (in Polish)
8. Karlen DL, Obrycki JF (2018) Measuring rotation and manure effects in an Iowa farm soil health assessment. Agron J 110:63–73. https://doi.org/10.2134/agronj2018.02.0113
9. Lopatka A, Siebielec G, Smreczak B (2020) Zasobność gleb w podstawowe składniki oraz zanieczyszczenie gleb i wód azotanami. Studia i Raporty 2020 IUNG-PIB 64(18):77–90. https://doi.org/10.26114/sir.iung.2020.64.05
10. Tarczyński C (2015) Właściwości chemiczne gleby lekkiej w ekologicznym systemie produkcji. Biuletyn Instytutu Hodowli i Aklimatyzacji Roślin 2015:275 (in polish)
11. Ramirez C (1982) Manual and atlas of the Penicillia. Elsevier Biomedical Press, Amsterdam
12. Baran E (1998) Zarys mikologii lekarskiej. VOLUMED. (in Polish)
13. Larone DH (2011) Medically Important Fungi. A Guide to Identification, 5th edn. ASM Press, Washington
14. Kwaśna H, Chełkowski J, Zajkowski P (1991) Grzyby Tom XXII. Instytut Botaniki PAN. (in Polish)
15. Krzyściak P, Skóra M, Macura AB (2011) Atlas grzybów chorobotwórczych człowieka. MedPharm Polska. (in polish)
16. Samson RA, Hoekstra ES, Frisvad JC, Filtenborg O (2002) Introduction to food- and airborne fungi. Sixth Edition. Centraalbureau voor Schimmelcultures- Utrecht, The Netherlands
17. Filipek T, Skowrońska M (2009) Optymalizacja odczynu gleby i gospodarki składnikami pokarmowymi w rolnictwie polskim. Postępy Nauk Rolniczyhch 1:25–23 (in Polish)
18. Sacco D, Moretti B, Monaco S, Grignani C (2015) Six-year transition from conventional to organic farming: efects on crop production and soil quality. Eur J Agron 69:10–20. https://doi.org/10.1016/j.eja.2015.05.002
19. Cima S, Luik D, Reintam AE (2015) Organic farming and cover crops as an alternative to mineral fertilizers to improve soil physical properties. Int Agrophys 29:405–412. https://doi.org/10.1515/intag-2015-0056
20. Tang C, Rengel Z, Diatloff E, Gazey C (2003) Responses of wheat and barley to liming on a sandy soil with subsoil acidity. Field Crop Res 80(3):235–244. https://doi.org/10.1016/S0378-4290(02)00192-2
21. Niemiec M, Chowaniak M, Sikora J, Szeląg-Sikora A, Gródek-Szostak Z, Komorowska M (2020) Selected properties of soils for long-term use in organic farming. Sustainability 12(6):2509. https://doi.org/10.3390/su1206250
22. Abit MJM, Lofton J, Arnall DB (2018) Effect of diammonium phosphate as in-furrow starter fertilizer on winter canola grain yield and quality. Crop Forage Turfgrass Manag 4(1):1–6. https://doi.org/10.2134/cftm2017.11.0078
23. Becher M, Pakuła K, Czapliński K (2020) Soil organic matter quality in soils with different levels of manure fertilisation. Environ Prot Nat Resour 31 2(84):17–23. https://doi.org/10.2478/oszn-2020-0007
24. Jakubus M, Graczyk M: Assessment of chemical composition of plants cultivated on two different soils amended with sewage sludge compost using MANOVA. J Plant Nutr 14:1–10. https://doi.org/10.1080/01904167.2020.1766064
25. Kwiatkowski CA, Harasim E (2020) Chemical properties of soil in four-field crop rotations under organic and conventional farming systems. Agronomy 10(7):1045. https://doi.org/10.3390/agronomy10071045
26. Frąc M, Lipiec J, Rutkowska A, Oszust K, Półtorak M (2011) Właściwości mikrobiologiczne gleby pod uprawą pszenicy ozimej w systemach ekologicznym i konwencjonalnym. Acta Agrophys 18(2):245–254 (in Polish)

27. Šimanský V, Jonczak J (2019) Sorption capacity of sandy soil under long-term fertilisation. Agriculture 65(4):164–171. https://doi.org/10.2478/agri-2019-0017
28. Sainju UM, Alasinrin SY (2020) Changes in soil chemical properties and crop yields with long-term cropping system and nitrogen fertilization. Agrosyst Geosci Environ 3:1. https://doi.org/10.1002/agg2.20019
29. Seufert V, Ramankutty N, Foley J (2012) Comparing the yields of organic and conventional agriculture. Nature 485:229–232. https://doi.org/10.1038/nature11069
30. Araújo ASF, Santos VB (2008) Responses of soil microbial biomass and activity for practices of organic and conventional farming systems in Piauí State Brazil. Eur J Soil Biol 44(2):225–230. https://doi.org/10.1016/j.ejsobi.2007.06.001
31. Schroder JL, Zhang H, Girma K, Raun WR, Penn CHJ, Payton ME (2011) Soil acidification from long–term use of nitrogen fertilizers on winter wheat. Soil Sci Soc Am J 75(3):975–964. https://doi.org/10.2136/sssaj2010.0187
32. Żukiewicz-Sobczak W et al (2012) Pathogenic fungi in the work environment of organic and conventional farmers. Postep Derm Alergol XXIX 4:256–262. https://doi.org/10.5114/pdia.2012.30463
33. Żukiewicz-Sobczak W, Cholewa G, Krasowska E, Chmielewska-Badora J, Zwoliński J, Sobczak P (2013) Rye grains and the soil derived from under the organic and conventional rye crops as a potential source of biological agents causing respiratory diseases in farmers. Postep Derm Alergol XXX 6:373–380. https://doi.org/10.5114/pdia.2013.39436
34. Socha K, Waszewska M (2012) Producenci ekologiczni w 2012 r. Wiedza i Jakość Inspekcja Jakości Handlowej Artykułów Rolno-Spożywczych, 3(26):14–15. (in Polish)
35. Soil Survey Staff (2014) Soil survey laboratory methods manual. In: Burt R (ed) Soil Survey Investigations Report No. 42, Version 5.0. 5th ed. U.S. Department of Agriculture, Natural Resources Conservation Service, pp 276–279
36. Bloom PR, Skyllberg U (2012) Soil pH and pH buffering. In: Huang PM, Li Y, Sumner ME (eds) Handbook of Soil Sciences: Properties and Processes, 2nd ed, pp 19–14. CRC Press, Boca Raton, FL. ISBN: 9781439803059
37. Handzel A, Królczyk JB, Latawiec AE, Pluta K, Malina D, Sobczak-Kupiec A (2017) Determination of element contens and physicochemical properties of selected soils. Infrastructure and Ecology of Rural Areas, I/2, PAN, Oddział w Krakowie, pp 419–432. http://dx.medra.org/. https://doi.org/10.14597/infraeco.2017.1.2.031
38. Condron LM et al (2000) A comparison of soil and environmental quality under organic and conventional farming systems in New Zealand. N Z J Agric Res 43(4):443–466. https://doi.org/10.1080/00288233.2000.9513442
39. Doran JW, Zeiss MR (2000) Soil health and sustainability: managing the biotic component of soil quality. Appl Soil Ecol 15(1):3–11
40. Karlen DL, Ditzler CA, Andrews SS (2003) Soil quality: Why and how? Geoderma 114(3):145–156. https://doi.org/10.1016/S0016-7061(03)00039-9
41. Doran JW, Coleman DC, Bezdicek DF, Stewart BA (1994) Defining soil quality for a sustainable environment. Madison, Wis.: Soil Science Society of America. https://doi.org/10.2136/sssaspecpub35.c1
42. Oshunsanya SO (2018) Introductory Chapter: Relevance of Soil pH to Agriculture, Soil pH for Nutrient Availability and Crop Performance, Suarau Oshunsanya, IntechOpen, 18 December 2018. https://doi.org/10.5772/intechopen.82551. https://www.intechopen.com/books/soil-ph-for-nutrient-availability-and-crop-performance/introductory-chapter-relevance-of-soil-ph-to-agriculture
43. De Hoog GS (1996) Risk assessment of fungi reported from humans and animals. Mycoses 39:407–417. https://doi.org/10.1111/j.1439-0507.1996.tb00089.x
44. Nowicki R, Korting HC (1995) Różnice w aktywności hydrolitycznej dermatofitów. Mikol Lek 4:209–213. (in Polish)

Phosphorus Balance Trend in Albania Agriculture During the Period 1950–2019

Fran Gjoka, Enkeleda Shkurta, Liri Miho, and Thoma Nasto

Abstract In Albania, although there is some evidence of declining phosphorus-poor soil surface, an in-depth study on the inflows and outflows of this nutrient from the soil has not been conducted. Therefore, this study aimed to quantify soil phosphorus balance at national level during the period 1950–2019, using the OECD approach. Phosphorus use efficiency is calculated only for arable land, as only it has received fertilizers. In cases of positive phosphorus balance, freshwater quality is endangered by eutrophication, while in cases of negative balance the yield of plants may decline. The results showed that for most of the study period (1970–2000), the phosphorus balance in Albania was positive, reaching a peak of about 22 kg/ha in 1990. During the period 2000–2019, the phosphorus balance returned in negative (about –3 and –9 kg/ha in 2010 and 2019, but significantly lower compared to 1950 (about –21.0 kg/ha). In terms of phosphorus deficit, Albania is included in the group of countries with the largest deficit compared to other EU countries (on average 1.7 times higher) and the OECD. The value of phosphorus use efficiency, as an average for the entire study period, was 23%, or about 3.7 times lower compared to the average value of this indicator in the EU-28, suggesting the immediate need to improve plant production practices including manuring. Data of this study can be used in the management of phosphorus sources to agricultural to increase production and protect the environment.

Keywords Phosphorus balance · Use efficiency · OECD approach · Albania

F. Gjoka (✉) · L. Miho · T. Nasto
Agricultural University of Tirana, Rruga Paisi Vodica 1029, Tirana, Albania
e-mail: fgjoka@ubt.edu.al

E. Shkurta
National Environmental Agency, Rruga Sami Frashëri Nr. 4, Tirana, Albania

S. Pascuzzi and F. Santoro (eds.), *Farm Machinery and Processes Management in Sustainable Agriculture*, Lecture Notes in Civil Engineering 289,
https://doi.org/10.1007/978-3-031-13090-8_17

1 Introduction

After nitrogen, phosphorus is the most important soil nutrient that determines the quality and quantity of crop yields, but has adverse environmental effects when part of it is removed from agricultural soils that have received phosphatic fertilizers and accumulates in aquatic systems above the maximum permissible limits. For these reasons, this plant nutrient is in the focus of current agri-ecological research [2, 8, 11, 17, 18].

Despite the improvements in production technology, yields for most crops in Albania are much lower compared to the EU average [4]. And this is mainly due to the application of small amounts of fertilizers. On the other hand, a previous study showed that in soils with low nutrient-holding capacity (sandy soils) there is a potential risk of environmental pollution from nitrate leaching [3].

Although there is some evidence of declining phosphorus-poor soil surface [1], and data on soil nutrient (NPK) balances for agricultural land [9], an in-depth study on the inflows and outflows of phosphorous in Albania agriculture has not been conducted. Therefore, this study aimed to quantify soil phosphorus balance at national level during the period 1950–2019, using the OECD approach. Phosphorus balance serves as a useful tool to evaluate the effectiveness of using this essential plant nutrient and promoting environmental protection.

2 Materials and Methods

2.1 Study Area

The study area includes agricultural land, which is the main category of land use in Albania. In fact, only cropland is fertilized with livestock manure and mineral fertilizers, while pastures are included in the study as grazing by livestock is considered an important cause of nutrient export from the soil. The dynamics of agricultural land area for the study period is given in Fig. 1.

The territory of the country is dominated by hilly mountainous relief (77%), and the rest (23%) has plain relief. Agricultural land is spread in different categories of relief, making the Mediterranean climate (with mild and wet winters and hot and dry summers) have significant variations of its main elements such as rainfall and temperature.

The main soil types according to WRB [7] are Cambisols, Regosols, Luvisols, Leptosols, Phaeozems, Fluvisols. The main land cover categories are forest land (37%), arable land (14.5%), permanent crops (9.5%), pastures and meadows (16%) and others (23%) [10].

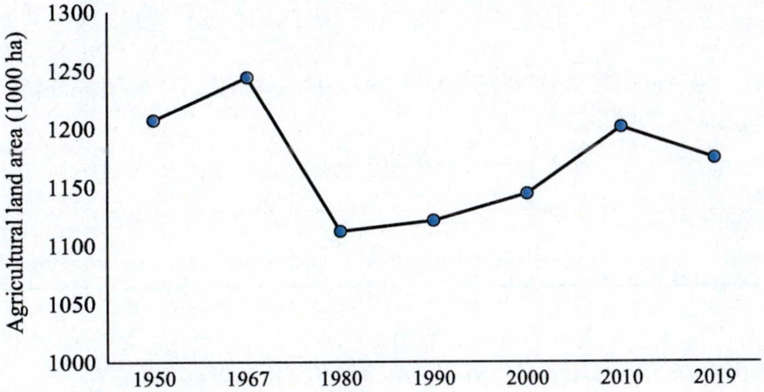

Fig. 1 Dynamics in years of agricultural land area

2.2 *Phosphorous Balance Approach*

To calculate the balance of nutrients in agriculture, two calculation methods can be used, farm-gate balance and soil surface balance. Although they use different data sources, both methods could be used when this assessment is done on a national scale [15].

To determine the phosphorus balance and to evaluate its trend in agricultural land, the OECD gross phosphorus balance approach (OECD/EUROSTAT, 2003, quoted by Panten et al. [16]) was implemented considering the Albanian conditions. Secondary data sources were used in the calculation, which are described in detail by Gjoka et al., [9]. The data was collected nationally and belong to a 69-year period (1950–2019). They include agricultural land use categories, types of cultivated plants, land area occupied by each plant and corresponding yields. For this purpose, inflows, and outflows of phosphorous from the soil per hectare of agricultural land are determined by using the following equation:

$$P - Balance = P - inflows - P - outflows \tag{1}$$

where: P-inflows (kg/ha/year) include mineral fertilizers, livestock manure and seeds and planting materials, and P-outflows (kg/ha/year) include removal of P by harvesting plants and by grazing and mowing fodder.

Phosphorous balance can be positive (>0) indicating surplus, negative (<0) indicating deficit, and zero (0) indicating equilibrium. Amounts of phosphorous export from soils via crop production or added to the soil from livestock manure are obtained by multiplying crop production (ton) and livestock (number) by the corresponding phosphorous coefficients (kg P/ton/year and (kg P/head/year).

Phosphorus use efficiency (PUE) was calculated only for crop land that received fertilizers, using the equation proposed by Syers et al., 2008, quoted by Medinski et al. [13] corresponding to the balance method:

$$PUE = [P - Outflows/P - Inflows] \cdot 100 \quad (2)$$

P-outflows - Removals of P through crop yield; P-inflows - Entrances of P with mineral and organic fertilizers.

Data were processed using the IBM SPSS 22.

3 Results and discussion

3.1 Inflow and Outflow Trends in Albania Agriculture

Figure 2 shows the average relative contribution of different sources of phosphorus, including mineral fertilizers (46.5%), livestock manure (53%), and seeds and planting material (0.5%), to phosphorus inflow to cropland. Livestock manure is obviously the main source of phosphorus in Albania agriculture.

While, annual data show a marked variation of these sources throughout the study period, where two-time segments are distinguished, the first from 1950–1990 and the second from 2000–2019. Total phosphorus inflow ranged from 35.6 kg/ha in 1950 to 63.1 kg/ha in 1990, with phosphorus sources contributing to varying degrees. Thus, the contribution of chemical fertilizers increases significantly from 6.9% in 1950 to 54.3% in 1990, and increases less from 2000 (40.2%) in 2019 (46.3%), the contribution of animal manure decreases significantly from 92, 8% in 1950 to 45.2% in 1990, and decreases less from 2000 (59.4%) in 2019 (53.5%), the contribution of seed and planting material increases significantly from 0.3% in 1950 to 0,5% in 1990, and also decreases significantly from 2000 (0.4%) to 2019 (0.2%) [9]. This variation can be explained by significant changes in the amount of chemical fertilizers used, the number of livestock heads, and the structure of the crops planted.

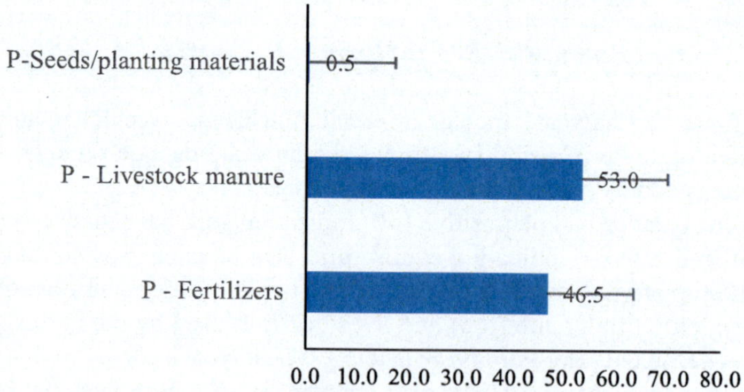

Fig. 2 The average relative contributions (1950–2019) of the P sources within the P inflows to Albania cropland. Error bars are ±1 standard error (n = 7)

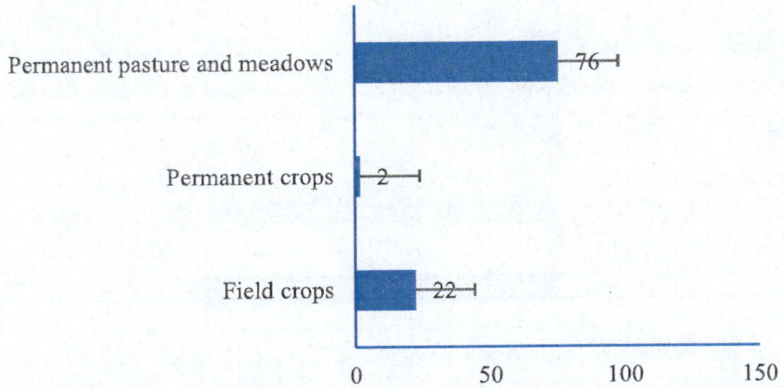

Fig. 3 The average relative contributions (1950–2019) of different mechanisms to P removal from agricultural land. Error bars are ±1 standard error (n = 7)

To determine the amount of phosphorus removed from the soil by grazing and mowing, the P coefficients proposed by FAO [7] were used. The results showed that the main mechanism of phosphorus removal from the soil was livestock grazing and fodder mowing contributing on average 76% of the total phosphorus removed, followed by field crops by 2% and permanent crops by 22% (Fig. 3).

As described by Gjoka et al. [9], phosphorus removal rates varied greatly from year to year, ranging from 53.7 kg/ha (94.9%) in 1950 to 33.5 kg/ha (66.9%) in 2019. The rest of this nutrient is removed by the harvest of field crops (14.2 kg/ha or 28.3% in 2019 and 2.8 kg/ha or 4.9% in 1950) and permanent crops (2.4 kg/ha or 4.8% in 2019 and 0.1 kg/ha or 0.2% in 1950).

This annual variation can be explained by variations in yields and phosphorous contents per ton of plant products, while the latter, among others, depends on plant cultivation practices such as fertilization, irrigation, mowing or grazing, etc.) [5]. Comparing phosphorus outflows with inflows, it results that the latter exceed the outflows for most of the study period (1970–2000), suggesting a surplus of phosphorus in agriculture.

On the contrary, in the last two decades there has been a phosphorus deficiency. According to MacDonald et al. [12], phosphorus deficiencies are more common in soils planted with forage crops, and globally, total phosphorus inflows on agricultural land exceed phosphorus removals from harvested crops.

3.2 *Phosphorous Balance Trend in Albania Agriculture*

Phosphorus balance is calculated for the entire surface of agricultural land as from the pastureland, although they have not received fertilizers, nutrients are removed through grazing and mowing. If the amount of total phosphorus outflows is greater than the amount of its total inflows to the soil, then we have a deficit, and on the

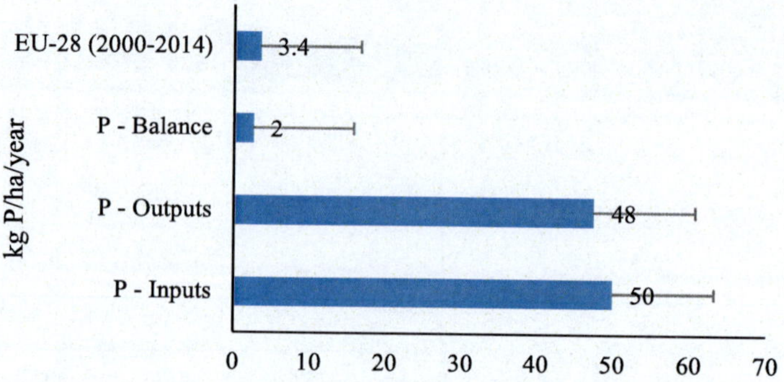

Fig. 4 Average P balance (1950–2019) for all Albania agricultural land. Error bars are ±1 standard error (n = 7)

contrary, we have a surplus. Phosphorus deficit and surplus are expressed in kg per hectare of agricultural land per year.

As shown in Fig. 4, the average value of phosphorus balance in Albanian agriculture was positive for most of the 69-year study period, by around 2.0 kg/ha/year. Compared to the value of the phosphorus balance in agriculture of the European Union (EU), it was lower, 2 kg/ha/year versus 3.4 kg/ha/year [14].

According to Gjoka et al. [9], the phosphorus balance in the last two decades resulted to be negative (about –3 and –7 kg/ha in 2010 and 2019 respectively, but significantly lower compared to 1950 (about –21 kg/ha). The data suggests that, currently, there are no environmental disturbances caused by phosphorus used in agriculture, and that these disturbances must have existed in the time segment from 1970 to 2000, marking a peak in 1990 (about +22 kgP/ha).

3.3 *Phosphorus Use Efficiency*

To determine the phosphorous uptake efficiency by crops, the phosphorus use efficiency (PUE) is calculated, which is the ratio as a percentage of phosphorus removed by the crops divided by the phosphorus inflows with fertilizers. Based on the concept, phosphorus use efficiency (PUE) was calculated only for arable land, as only this category of land use has received fertilizers.

The data of this study showed that the value of the PUE as an average for the study period (1950–2019), was 23%, or about 3.7 times lower compared to the average value of this indicator in the EU-28 (200–2014) [6]. Generally, it resulted that with the increase in the amount of fertilizers applied in agriculture, the values of PUE have also increased (Fig. 5), suggesting a deterioration in the use of nutrients applied in agriculture in Albania.

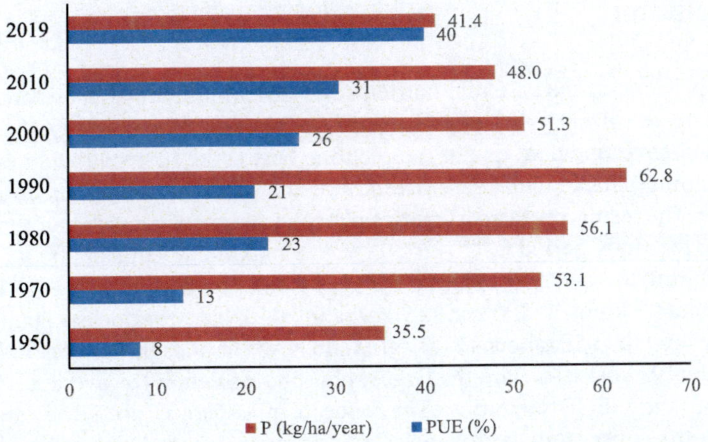

Fig. 5 Phosphorus use efficiency (%) in Albania agriculture (arable land)

The correlation between phosphorous use efficiency (PUE) and phosphorous application rate was not significant (R2=0.191) (Fig. 6). According to the determination coefficient (R^2), the phosphorous application rate explains about 19% of the variance of the phosphorus use efficiency in Albania agriculture. This means that other plant cultivation practices have a more significant impact than the use of fertilizers. Therefore, significant improvements need to be made in plant cultivation technologies. Opposite results have been reported by Medinski et al. [13].

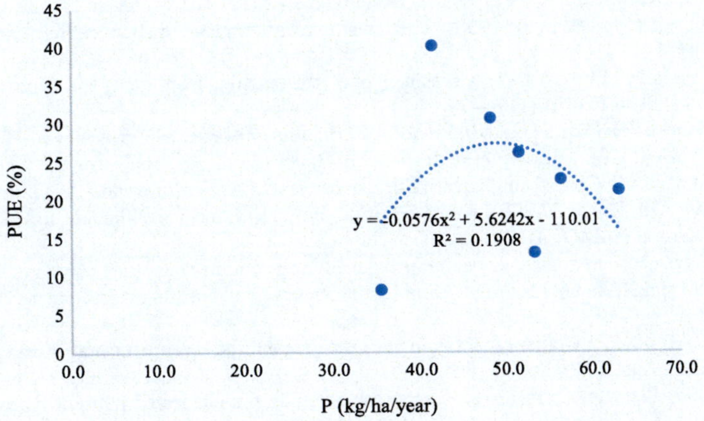

Fig. 6 Relationship between PUE and phosphorous application rate. The polynomial model was used as it was the most suitable of the others

4 Conclusion

Phosphorus is one of the essential nutrients for agricultural crops, therefore the use of phosphate fertilizers is an effective tool for obtaining high and quality yields of crops and environmental quality protection. This is the first study that evaluates the phosphorus balance and the efficiency of the use of this nutrient in Albania agriculture. The results revealed a positive phosphorus balance in Albania agriculture for most of the study period (1970–2000), with a maximum value of about 22 kg/ha in 1990. During the last two decades, the phosphorus balance turned negative with values around –3 kg/ha in 2000 and –9 kg/ha in 2019, but nevertheless significantly lower compared to 1950 (about –21.0 kg/ha). In terms of phosphorus deficit, Albania is among the EU countries with the largest phosphorus deficit (on average 1.7 times higher), and the OECD countries. The value of phosphorus use efficiency, as an average for the entire study period, was 23%, or about 3.7 times lower compared to the average value of this indicator in the EU-28, suggesting the immediate need to improve plant production practices including manuring.

Acknowledgements This research was supported by the Agricultural University of Tirana.

References

1. Binaj A, Veizi P, Beqiraj E, Gjoka F, Kasa E (2014) Economic losses from soil degradation in agricultural area in Albania. Agric Econ Czech 60(6):287–293
2. Carmo M, García-Ruiz R, Ferreira MI, Domingos T (2017) The N-P-K soil nutrient balance of Portuguese cropland in the 1950s: The transition from organic to chemical fertilization. Sci Rep 7(1):1–14
3. Ceci O, Gjoka F, Miho L (2017) Assessment of nitrate leaching from sandy soil in potato-carrot rotation. J Balkan Ecol 20(1):69–75
4. Cela R, Marku SH, Imami D (2010) Review of agriculture and agricultural policy in Albania. In: Volk, T (Ed.) (2010): Agriculture in the Western Balkan countries, Studies on the Agricultural and Food Sector in Central and Eastern Europe, Review of Agriculture and Agricultural Policy in Albania, vol 57, pp 37–62, Leibniz Institute of Agricultural Development in Central and Eastern Europe (IAMO), Halle (Saale)
5. EUROSTAT (2011) Archive: Phosphorus balance in agriculture. https://ec.europa.eu/eurostat/statisticsexplained/index.php?title=Archive:Phosphorus_balance_in_agriculture&oldid=66744
6. EUROSTAT (2022) Statistics Explained. https://ec.europa.eu/eurostat/statisticsexplained/. Accessed 11 Apr 2022
7. FAO (2006) Plant nutrition for food security: A guide for integrated nutrient management. FAO fertilizer and plant nutrition bulletin, p 16. http://www.fao.org/fileadmin/templates/soilbiodiversity/Downloadable_files/fpnb16.pdf
8. Gaj R, Bellaloui N (2012) Evaluation of phosphorus and nitrogen balances as an indicator for the impact of agriculture on environment: a comparison of a case study from Poland and Mississippi US. Agric Sci 3(2):317–329
9. Gjoka F, Miho L, Vata N, Spaholli A (2021) Temporal variability of environmental indicators for agriculture in Albania. Environ Monit Assess 193:829

10. INSTAT (2019) Use of chemical ferftilizres. http://databaza.instat.gov.al/pxweb/sq/DST/START%20BU__AM/NewBU0012/?rxid=b5069c81-9c75-4560-905a-2cb719af3ada
11. Kopiński J, Jurga B (2016) Managing phosphorus in Polish agriculture–production and environmental aspects. Pol J Environ Stud 25(6):2451–2458
12. MacDonald GK, Bennett EM, Potter PA, Ramankutty N (2011) Agronomic phosphorus imbalances across the world's croplands. Proc Natl Acad Sci 108(7):3086–3091
13. Medinski T, Freese D, Reitz T (2018) Changes in soil phosphorus balance and phosphorus-use efficiency under long-term fertilization conducted on agriculturally used Chernozem in Germany. Can J Soil Sci 98:650–662
14. OECD (2020) Agri-Environmental indicators: nutrients, Phosphorus balance. https://stats.oecd.org/index.aspx?queryid=79765
15. OECD/Eurostat (2007) Gross Phosphorus Balances Handbook. https://www.oecd.org/greengrowth/sustainable-agriculture/40820243.pdf. Accessed 18 Jan 2021
16. Panten K, Rogasik J, Godlinski F, Funder U, Greef JM, Schnug E (2009) Gross soil surface nutrient balances: the OECD approach implemented under German conditions. Agric For Res 1(59):19–28
17. Uusitalo R et al (2007) Finnish trends in phosphorus balance and soil test phosphorus. Agric Food Sci Finland 16:301–316
18. Van Eerdt MM, Olsthoorn CSM, Fong NPK, Van Veller M (2003) The nutrient balance as an agri-environmental indicator on three scales, National balances, regional balances and farm balances. In: Proceedings of the International Conference Agricultural Statistics in the New Millennium: The Challenge of Agri-environmental Indicators as a Tool for the Planning of Sustainable Development for Agriculture. Chania-Crete, Greece. 13–15 November 2002. National Statistical Service of Greece, Athens

Applicability and Efficiency of Remote Monitoring of Agricultural Crops

Asparuh Atanasov, Radko Mihaylov, and Gallina Mihova

Abstract The object of research in this article is the applicability and effectiveness of a photo sensor for remote monitoring of crops. The analysis and processing of the information contained in a large amount of photographic material showing the vegetation of wheat in the spring of 2021 was made. The images show how over time, under different weather conditions, plants reflect solar radiation in the frequency ranges of visible light. The photos were taken with a digital camera Hasseblad L1D-20, equipped with a 20 MP 1″ CMOS sensor, which is equipped with a drone model DJI Mavic 2 Pro. Through the methods of mathematical statistical processing, it is established how controllable and uncontrollable factors affect the process of image acquisition with the digital sensor, which is built into the camera. The regression equations are obtained, which give the relationship between the factors of the shooting process during the flight of the drone and the parameters that show the change of numerical values for the colours of the visible solar spectrum: red, green, and blue at the time of digital image acquisition for the condition of the crop. Conclusions have been made on how to set the f-stop and exposition values for shooting at a given value of solar radiation when working with a digital camera.

Keywords Photo sensor · Statistical processing · Regression equations · f-stop and exposition values · Remote monitoring of crops

1 Introduction

The object of the study is the reflection of solar radiation from wheat crops. Drone model DJI Mavic 2 Pro is equipped with a digital camera Hasseblad L1D-20, equipped with a 1 "20.1 MP CMOS sensor, which is registered red, green and blue, ie

A. Atanasov (✉) · R. Mihaylov
Technical University of Varna, ul. "Studentska" No. 1, Varna 9000, Bulgaria
e-mail: asparuh.atanasov@tu-varna.bg

G. Mihova
Dobrudhza Agriculture Institute, Agricultural Academy, General Toshevo 9521, Bulgaria

S. Pascuzzi and F. Santoro (eds.), *Farm Machinery and Processes Management in Sustainable Agriculture*, Lecture Notes in Civil Engineering 289,
https://doi.org/10.1007/978-3-031-13090-8_18

RGB (red, green, blue) the rays reflected by the plants. The sensor matrix: 1 "20.1 MP CMOS, consists of photodetectors, the number of which determines the resolution of the image. Each pixel converts the light on it into an electric charge. The pixel in the raster map is an object characterized by a specific color, brightness and transparency recorded in digital form - 10-bit Dlog-M color profile. Each pixel describes only one color. The three colors correspond to reflections with different wavelengths. Naturally, these three colors have different energies, different propagation frequencies, which is the main reason why the photosensor records different numerical values in each pixel of the image.

The aim of the study is to determine how the factors that determine the function of the photosensor to convert the reflected solar radiation in the visible area, affect its work.

2 Related Works

A common approach in research practice is the use of mathematical apparatus for statistical processing of data obtained from various experimental studies. In crop production or more generally, in agriculture, the quantities, the factors that determine the development of processes are observed over a relatively long period of time, during which they acquire variable, random, in nature values. To identify trends in change, which is very important for predicting and managing processes, researchers use the capabilities and methods of mathematical statistics. In many publications the object of application is the regression analysis of an array of data, which in a certain way characterize the course of a natural process. Such an approach was applied in [3] to study a platform for remote monitoring of unmanned aerial vehicles which maps the values of the normalized differential vegetation index (NDVI). High resolution has been achieved, which allows to assess the impact of the parameters of the drone on the quality of the data obtained. A study by, [4] developed a simple method based on the light efficiency model to estimate the growth and yield of irrigated winter wheat under semi-dry conditions. The originality of the proposed method consists in modifying the expression of the conversion factor by integrating at an appropriate voltage threshold to trigger irrigation, replacing the product of two maximum interception coefficients and conversion by a single parameter. In the work of [9] regression analysis is used to establish a relationship between a set of variables: annual precipitation, cultivation area and food price index and their effects on rice yield. In this study, regression analysis evaluates the analysis of environmental factors and their impact on crop yields. A multivariate analysis technique is applied, which determines which factors are important variables for decision making. They work with a sample of environmental factors, considering a period of 10 years from 1990–2000. Linear regression is applied to establish a relationship between independent variables and crop yield as a dependent variable. In her study [2] she used a methodology for determining the condition of winter oilseed rape before and after overwintering and estimating the duration of flowering. It has been successfully tested to determine the state of the culture and is coded by the author of the Python programming language. The three conditions: good, average, and bad for each parameter (duration of flowering) are determined expertly by the author and the agronomist. Ten variables corresponding to the Sentinel-2 channels used with a spatial resolution of 10 and 20 m were used; all input data that are determined to be abnormal with a 5% significance level are excluded in subsequent treatments. The best regression model for the considered process, both parametric and nonparametric, is determined.

2.1 Overview

A central compositional plan has been drawn up, [1], which consists of three parts: core plan of the matrix for complete factorial experiment (CFE), star points, on the coordinate axes and zero points, experiments, symmetrically located in the center of the plan. The central compositional plan is applied to obtain a regression model and to describe local areas of the factor space with the number of factors $n = 3$. They reveal the essence of the state and functioning of the object of study. This is expressed by several conditions under which it takes place (factors) and the results obtained (parameters). The optimization of the mathematical model was performed with the optimization program Solver included in Microsoft Office 2006. Microsoft Graph chart software package was used for the graphical interpretation of the models. Graphical interpretation is performed by the method of two-dimensional sections. In a specific case [6], the ranges of change in volume, water level pressure and turbine speed are determined after conducting preliminary one-factor experiments, summarized graphically. The planning of the experiment and the processing of the obtained results were performed according to the known methods. When describing the object, a water turbine, which has static characteristics with extremum, and when planning the experiment, a second-degree polynomial was chosen for the regression equation, which is easy to systematize and study for extremum. A numerical matrix of the second-order partial derivatives is constructed. In this case, the numerical matrix is negatively defined and has a maximum that accurately characterizes the mode of operation of the turbine. In June 2019 [5], an aerial image was recorded using a drone equipped with an RGB camera to collect information to determine the condition of an experimental field sown with wheat, at intervals of seven days three times a day. The information on meteorological conditions was collected within one day (June 12, 2019) and eight measurements were made every hour. The values of the following vegetation indices were studied: VARIgreen and ExG, observed for a period of one month. For the purposes of the study, a regression model was created for a complete factorial experiment of type 2^3. The interaction of the factors that characterize the conditions of video recording was evaluated. Relevant conclusions have been formulated regarding the influence of the factors under which the observations were made. A recommendation was made for the best time interval during the day for filming and video surveillance of the wheat condition.

3 Real Content

The current study was conducted in the spring of 2021 in a field with coordinates: (43.656963N, 28.023110E). This is a plateau with an average altitude of about 230 m (Fig. 1). The climate is temperate continental. Critical for the winter months are low temperatures in the absence of snow cover. Due to the frequent influx of ground cooling air from the sea, spring comes later. Summer is cool and autumn is long with gradual cooling. Winds are common, with the northern component predominating.

The input values or factors that determine the course of the process of recording the reflection of solar radiation from wheat crops are two groups. In one are the height and speed of the drone, and in the other are the values that characterize the very acquisition of the image, namely: f-stop, exposition, and shutter speed. They are measurable and manageable. Meteorological factors (wind speed and direction, air temperature, relative humidity, precipitation, solar radiation) determine various disturbances during image acquisition. They are important for the reflection of solar radiation from plants but are practically unmanageable. At the time of filming, the meteorological factors have specific values, but vary greatly during the growing season. Information about their level and dynamics is collected from the respective meteorological stations located near the surveyed areas.

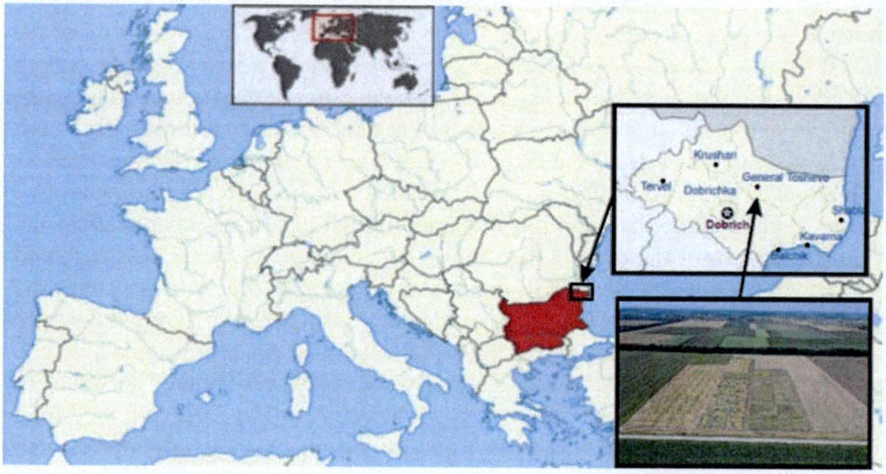

Fig. 1 Location of the experiments

These factors and disturbances form a large amount of information, which is difficult to statistically process. In accordance with the aim of the present study, a selection of three factors was made, the importance of which is essential for obtaining numerical values for the three colors from which vegetation indices for the condition of wheat are calculated. From the second group of controllable factors these are: f-stop and exposition (Exposition time) of the shooting, and from the distracting, i.e., uncontrollable factors, that is: meteorological information about the solar radiation at the time of shooting. Table 1 gives the natural data for the two controllable factors and for the solar radiation according to the dates on which the wheat crops images were taken, as well as the numerical data for the reflection of red, green, and blue color. The numerical values for the colors are obtained after processing the RGB images with the software product ImageJ [9] is a Java-based image processing program developed at the National Institutes of Health and the Laboratory for Optical and Computational Instrumentation (LOCI, University of Wisconsin, USA) [11].

To describe the areas of optimum of the parameters Y_j, $j = 1, 2, 3$ or to describe the response surfaces, a regression model more complex than the linear one is applicable in a larger area of the factor space. Such a model cannot be built with a first-order plan, i.e., complete factorial experiment (CFE) of type 2^m and fractional replicas of it. To describe the area of the optimum, a polynomial of the second degree was used, according to the methodology of [6].

In this case a polynomial of the type:

Table 1 Natural data on the factors and parameters of the regression model

Datas 2021	№ of experiment	Factors: (F) X_1, (E) X_2, (W) X_3			Parameters: (R) Y_1, (G) Y_2, (B) Y_3		
		F-stop [mm]	Exposition [s]	Solar radiation [w/m^2]	Red colour [-]	Green colour [-]	Blue colour [-]
05.03	1	4,5	1/120	93,7	194,54	193,34	141,31
19.03	2	3,5	1/80	202,2	190,49	196,8	155,5
02.04	3	4,0	1/120	134,52	188,08	209,0	159,25
09.04	4	4,0	1/80	105,5	207,8	234,3	184,92
16.04	5	4,0	1/100	127,8	197,9	218,2	173,22
23.04	6	3,2	1/50	165,2	188,81	235,08	179,93
30.04	7	3,2	1/30	190,3	209,09	247,74	203,96
14.05	8	3,5	1/80	203,3	206,69	234,38	193,05
22.05	9	3,2	1/50	286,3	209,92	241,98	192,03
24.05	10	3,5	1/50	224,0	218,18	235,35	207,63
30.05	11	4,0	1/100	262,1	238,15	245,8	184,33
05.06	12	4,0	1/120	170,9	233,83	234,71	162,23
14.06	13	3,5	1/60	163,4	234,75	230,1	179,04
25.06	14	5,0	1/250	140,9	143,74	141,07	120,99

$$\hat{Y}_j = b_0 + \sum\nolimits_{i=1}^{m} b_i \cdot \dot{x}_i + \sum\nolimits_{i,k=1}^{m} b_{\substack{i,k \\ 1<k}} \cdot \dot{x}_i \cdot \dot{x}_k + \sum\nolimits_{i=1}^{m} b_{ii} \cdot \dot{b}_{ii}^2 + \qquad (1)$$

where $\hat{\boldsymbol{Y}}_j$ is the estimate of the parameter $\boldsymbol{Y}_j$; $\boldsymbol{b}_i$ coefficients of the regression equation; $\boldsymbol{i}$ is an index of factors from 1 to 3; $\dot{\boldsymbol{x}}_i$ coded values of the factors from Table 3; $\boldsymbol{m}$ - number of factors. The plan that allows you to find the values of the quantities involved in this polynomial is called a second-order plan. Here the factors must vary at least on three levels. Table 2 shows the variation and coding of factors at five levels.

The description of the area of the optimum is first planned with a first-order plan, then the center of the new plan is chosen as the center of the CFE or a fractional replica of it. Appropriately selected points are added to the plan. A rotatable central

Table 2 Levels of variation and coding of the factors of the regression model

Levels	F-stop (F) - X_1 [mm]		Exposition (E) - X_2 [s]		Solar radiation (W) - X_3 [w/m^2]	
	Natural values	Code values	Natural values	Code values	Natural values	Code values
Down $\boldsymbol{X_{0i}} + \boldsymbol{\alpha_i}$	3,2	+1,682	0,004	+1,682	93,7	+1,682
$\boldsymbol{X_{0i}} + \boldsymbol{J}$	3,65	+1	0,0113	+1	141.77	+1
Average $\boldsymbol{X_{0j}}$	3,8	0	0,0186	0	189,85	0
$\boldsymbol{X_{0i}}$-$\boldsymbol{J}$	4,25	−1	0,0259	−1	237,93	−1
Upstairs $\boldsymbol{X_{0i}}$-$\boldsymbol{\alpha_i}$	5,0	−1,682	0,0333	−1,682	286,0	−1,682
StepJ	0,45	–	0,0082	–	48,075	–

Table 3 Rotatable central composition plan for three factors

X1	X2	X3	X12	X13	X23	X11	X22	X33	Y1	Y2	Y3
1,000	1,000	1,000	1,000	1,000	1,000	1,000	1,000	1,000	194,540	193,340	141,310
-1,000	1,000	1,000	-1,000	-1,000	1,000	1,000	1,000	1,000	190,490	196,800	155,500
1,000	-1,000	1,000	-1,000	1,000	-1,000	1,000	1,000	1,000	188,080	209,000	159,250
-1,000	-1,000	1,000	1,000	-1,000	-1,000	1,000	1,000	1,000	207,800	234,300	184,920
1,000	1,000	-1,000	1,000	-1,000	-1,000	1,000	1,000	1,000	197,900	218,200	173,220
-1,000	1,000	-1,000	-1,000	1,000	-1,000	1,000	1,000	1,000	188,810	235,080	179,930
1,000	-1,000	-1,000	-1,000	-1,000	1,000	1,000	1,000	1,000	209,090	247,740	203,960
-1,000	-1,000	-1,000	1,000	1,000	1,000	1,000	1,000	1,000	206,960	234,380	193,050
1,683	0,000	0,000	0,000	0,000	0,000	2,829	0,000	0,000	209,920	241,980	192,030
-1,683	0,000	0,000	0,000	0,000	0,000	2,829	0,000	0,000	218,180	235,350	207,630
0,000	1,683	0,000	0,000	0,000	0,000	0,000	2,829	0,000	238,150	245,800	184,330
0,000	-1,683	0,000	0,000	0,000	0,000	0,000	2,829	0,000	233,830	234,710	162,230
0,000	0,000	1,683	0,000	0,000	0,000	0,000	0,000	2,829	234,750	230,100	179,040
0,000	0,000	-1,683	0,000	0,000	0,000	0,000	0,000	2,829	143,740	141,070	120,990
0,000	0,000	0,000	0,000	0,000	0,000	0,000	0,000	0,000	0,000	0,000	0,000
0,000	0,000	0,000	0,000	0,000	0,000	0,000	0,000	0,000	0,000	0,000	0,000
0,000	0,000	0,000	0,000	0,000	0,000	0,000	0,000	0,000	0,000	0,000	0,000
0,000	0,000	0,000	0,000	0,000	0,000	0,000	0,000	0,000	0,000	0,000	0,000
0,000	0,000	0,000	0,000	0,000	0,000	0,000	0,000	0,000	0,000	0,000	0,000
0,000	0,000	0,000	0,000	0,000	0,000	0,000	0,000	0,000	0,000	0,000	0,000

compositional plan (RCCP) is obtained [1, 6]. The newly added points lie on the coordinate axes at a distance α from the center of the plan. These are the star points, and α is the star arm. The value of $\alpha = \mathbf{1{,}683}$ was selected, (from the source, [6]) for the number of factors $m = 3$ according to the formula $\alpha = 2^{m/4}$. A symmetrical plan is obtained, which is orthogonal to the central composition, as shown in Table 3. The points of the rotatable plan are located so that when rotating the coordinate axes, the distribution of the variance does not change. With an appropriate choice of the number of re-observations in the center of the plan, an almost uniform distribution of the variance over the whole area can be achieved. This area does not depend on the distance to the center of the plan. Such planning is called uniform planning. The number of observations in the center of the plan is chosen to obtain uniform planning.

With the statistical analysis program IBM SPSS Statistics v26 [10] was performed statistical analysis and was determined the coefficients of the second-degree model, Eq. (1). The results of statistical processing are given in Table 4, where F (k_1, k_2) is the value of the Fisher test; k_1 and k_2 are degrees of freedom of the respective sums, and R^2 is the coefficient of certainty.

The obtained values for the Fisher criterion [6] $F_{(9,\,10)}$, compared with that for the degrees of freedom $k_1 = 9$ and $k_2 = 10$ according to Table 1 of Annex 2 of [6], which is 3,155 and its corresponding probability at a significance level $\alpha = 0.05$ is much smaller than calculated, which means that the null hypothesis is rejected, and model Eq. (1) is adequate. The coefficients of determination R^2 are very close to one, and it follows that 98% of the changes in the parameters $Y_j, j = 1, 2, 3$ are described by the model Eq. (1) of the second degree. The assessment of the significance of the obtained coefficients according to Student's criterion, [6] at a significance level $\alpha = 0{,}05$, respectively $\gamma = 1\text{-}\alpha = 0{,}95$, number of degrees of freedom $k = n\text{-}1 = 13$, (where n is the volume of the sample (i.e., number of experiments) shows that $t_\gamma, k = 2{,}16$, (Table 2, [6]). The comparison with the calculated coefficients b_{ij}, Table 4 and the values for t_i from Table 5 shows that which are the insignificant coefficients, namely those that are less than ***2.16***, i.e., the uncolored values in Table 4.

Table 4 The results of the regression analysis

	Y_1			Y_2			Y_3		
	b_{ji}	$F_{(9,\,10)}$	R/R^2	b_{ji}	$F_{(9,\,10)}$	R/R^2	b_{ji}	$F_{(9,\,10)}$	R/R^2
Const	0,902	**35,08**	**0,985/0,969**	0,444	**34,39**	**0,984/0,969**	0,355	**30,57**	**0,982/0,965**
x1	−1,343			−1,546			−4,531		
x2	−2,409			−4,635			−3,954		
x3	9,61			3,504			0,84		
x12	3,841			−1,05			−0,768		
x13	−3,361			−3,155			−5,507		
x23	2,311			−3,04			−0,437		
x11	69,752			81,456			68,311		
x22	77,508			82,018			58,926		
x33	60,984			62,693			50,702		

Table 5 Values of the quantity t_i, against which the significant coefficients before the respective factors are determined

Factors	t_i (Y_1)	t_i (Y_2)	t_i (Y_3)
Const	0,093	0,042	0,04
x1	−0,209	−0,22	−0,77
x2	−0,375	−0,659	−0,672
x3	1,498	0,498	−0,143
x12	0,458	−0,114	−0,1
x13	11,166	11,887	11,925
x23	0,276	−0,33	−0,57
x11	11,166	11,887	11,925
x22	12,407	11,969	10,287
x33	9,762	9,149	8,851

4 Results

The graphs (Fig. 2) show the maximum values of the factors compared to each other. In all combinations, they have two maxima and are most clearly expressed where the factor X1 - f-stop.

The regression equations of the parameters take the following form, considering the significant coefficients b_{ij}, Table 4 and after moving to the natural values of the factors, Table 3:

$$\hat{Y}_1 = -3.361FW + 69.752F^2 + 77.508E^2 + 69.984W^2 \tag{2}$$

$$\hat{Y}_2 = -3.155FW + 81.456F^2 + 82.018E^2 + 62.0693W^2 \tag{3}$$

$$\hat{Y}_3 = -5.507FW + 68.311F^2 + 58.926E^2 + 50.702W^2 \tag{4}$$

Figures 3, 4 and 5 show the influence of individual factors, the effect of second-order interaction on responses, parameters: Y_1, Y_2 and Y_3. Here, too, more than one maximum is obtained.

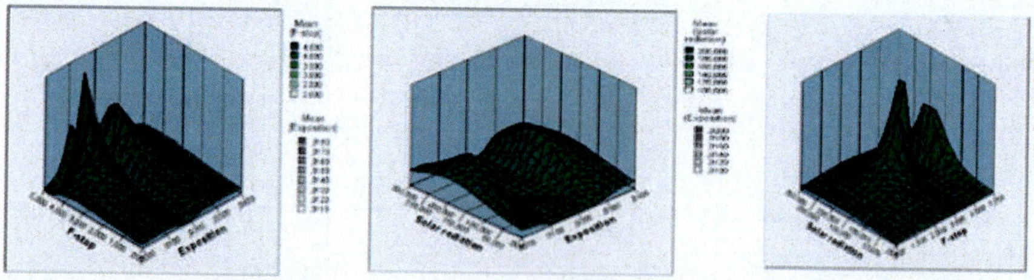

Fig. 2 Dependence between factors (effect of interaction)

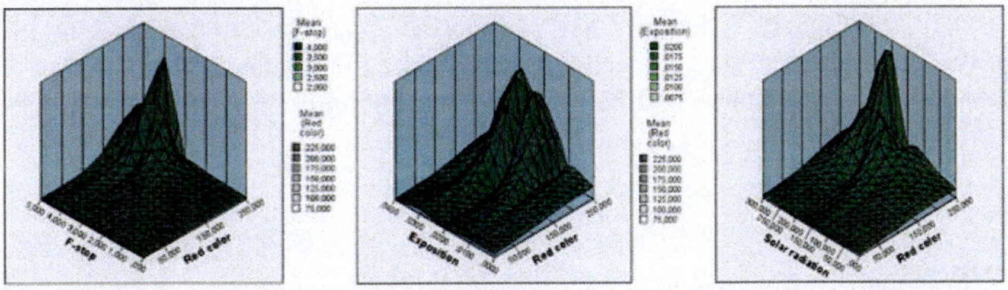

Fig. 3 Maximum response Y_1 "red" to factors

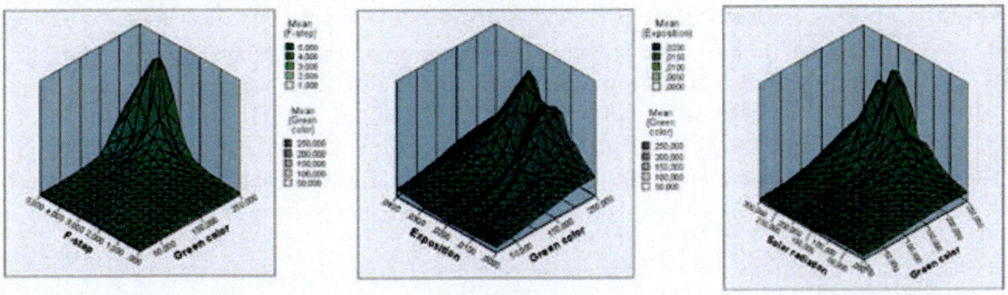

Fig. 4 Maximum response Y_2 "green" to factors

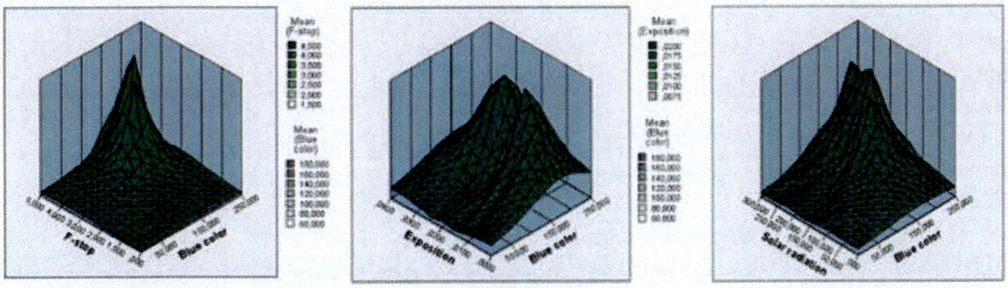

Fig. 5 Maximum response Y_3 "blue" to factors

5 Conclusion

An active–passive experiment was performed to study the properties of a photosensor. A regression model was obtained using a polynomial of the second degree, which provides information on the influence of three factors: two controlled F-stop and E-exposition and one uncontrolled W-solar radiation on the model parameters, which represent the numerical values of the three colors: $\boldsymbol{Y_1}$ - red, $\boldsymbol{Y_2}$ - green and $\boldsymbol{Y_3}$ - blue. Useful information is given in Fig. 2, which shows the interaction of the factors with respect to each other. The obtained maximum values are important to consider when determining the mode of image capture with the photo sensor, which is the subject of this study. The total result of the obtained graphs (Figs. 3, 4 and 5) shows that the maxima of the colours, as the number of pixels with which they are registered by the photo sensor are obtained when the factors are in the upper close to their maximum values. It should be noted that this is not a prerequisite for a better image!

The findings of this study will be used in planning and conducting surveys of cereal crops in the 2022 season, with NIR digital camera.

References

1. Andonova-Vakarelska T (2016) Approbation of the methodology for experimental research and processing of results in automated production. In: Collection of Reports, XXVI ISTC ADP-2016, pp 416–422. (in Bulgarian)
2. Ganeva-Kiryakova DG (2020) Assessment of the condition of winter rapeseed crops in north-eastern Bulgaria by satellite and terrestrial data, Dissertation for obtaining the educational and scientific degree Doctor Field of higher education, no 4. Natural science. (in Bulgarian)
3. Jiang R et al (1850) Assessing the operation parameters of a low-altitude UAV for the collection of NDVI values over a paddy rice field. Remote Sens 12(11):2020. https://doi.org/10.3390/rs12111850
4. Khabba S et al (2020) A simple light-use-efficiency model to estimate wheat yield in the semi-arid areas. Agronomy 10(10):1524. https://doi.org/10.3390/agronomy10101524
5. Mihajlow R, Ivanova A (2019) Drone video capture–a new method in precision agriculture. Ann J Tech Univ Varna Bulgaria (AJTUV) 3(2). ISSN 2603-316X. https://doi.org/10.29114/ajtuv.vol3.iss2.141
6. Mitkov A (2011) Theory of Experiment. Danube Press, Ruse. (in Bulgarian). ISBN 968-954-612-464-5
7. Stefanov N (2012) Application of the factorial experiment for determining the efficiency of a water turbine. In: Scientific Papers of the University of Ruse, vol 51, Series, pp 138–142. (in Bulgarian)
8. Sellam V, Poovammal E (2016) Prediction of crop yield using regression analysis Indian. J Sci Technol 9(38). ISSN (Print): 0964-6846, ISSN (Online):0964-5645. https://doi.org/10.16485/ijst/2016/v9i38/91614
9. ImageJ, Homepage. https://imagej.nih.gov/ij/
10. IBM SPSS Statistics v28.0.1.1, Homepage. https://www.ibm.com/products/spss-statistics
11. National Institutes of Health and the Laboratory for Optical and Computational Instrumentation, Homepage. https://eliceirilab.org/software/imagej/

Image Analysis in the Assessment of Homogeneity of Fine-Grained Mixtures

Dominika B. Matuszek, Łukasz A. Biłos, and Jolanta B. Królczyk

Abstract Digital image analysis can successfully substitute traditional methods of quality assessment—e.g. sieve analysis or other manual methods that consist dividing samples. In case of granular mixtures in which the components have grain dimensions measured in microns the assessment of mixture quality received after mixing process is crucial. This work presents the results of mixing micro-grained materials used in food industry. The process was carried out in turbulent mixer in which 90 ml of mixing vessels was used. The mixtures were prepared using the 3D Turbula® mixer by 30 and 60 s of mixing time and the quality was verified using a digital image analysis. The main aim of the study was to test the computer image analysis for the analysis of the mixing process in different conditions by determining the fraction of the key component (algae).

Keywords Turbulent mixer · Homogeneity · Computer image analysis · RGB scale · Fine-grained mixtures

1 Introduction

The significance of mixing granular materials becomes more relevant over the past decades. The development of computer technologies has introduced new analytical possibilities to the subject, especially in the field of evaluation the quality of mixtures [1]. The granular material mixing is visible in many branches of the industry. The popularity of the process can be observed in the ceramics, fertilizers and even in plastics or glass production. It is an important unit process in the food and pharmaceutical industries and the powder metallurgy industry as well [2, 3]. The methods

D. B. Matuszek (✉) · Ł. A. Biłos
Department of Biosystem Engineering and Chemical Processes, Faculty of Production Engineering and Logistics, Opole University of Technology, 45-758 Opole, Poland
e-mail: d.matuszek@po.edu.pl

J. B. Królczyk
Department of Machine Technology and Materials Science, Faculty of Mechanical Engineering, Opole University of Technology, 45-758 Opole, Poland

S. Pascuzzi and F. Santoro (eds.), *Farm Machinery and Processes Management in Sustainable Agriculture*, Lecture Notes in Civil Engineering 289,
https://doi.org/10.1007/978-3-031-13090-8_19

and techniques have a great influence on the final form of the mixtures, like the solid mixers type that are available in the market [4]. Efficient powder mixing technique can influence the quality by increasing homogeneity of the final product. In this case a great role plays the time of mixing as well, e.g. not sufficient time can cause visible inhomogeneity and in effect make the mixture low quality [5]. The major obstacle with obtaining perfectly homogenous mixture is the diversity of the granular materials—the shape, size or the surface type, irregularity of the grains. The same we can observe if it's about the elasticity, hardness texture and along with other factors [6]. The other factors are also the speed, the kind of vessel used in the mixing tool an e.g., the flowability of the powders [7]. Hence, the mixing of fine-grained materials is not unambiguous. The results can frequently be incorrect, unreliable. The assessment of the mixture quality is significant for the final product form. Provided that we can put traditional way of assessment in practice if it's about the coarse grained mixtures e.g. sieve analysis, in case of fine-grained powders the digital image analysis is often the only possible method. Digital image analysis as an advanced method of assessing the quality of mixtures has also found application in simulating the movements of granular mixtures [8].

The main aim of the study was to test the computer image analysis for the analysis of the mixing process in different conditions by determining the fraction of the key component (algae).

2 Materials and Methods

Prior to testing, the RGB (decimal RGB) scale standard was made for a cornmeal—algae mixture with a set algae content from 0 to 10% (in 1% increments). Example images of samples with a given share of the key component are shown in Fig. 1.

From vessel 3 samples were taken, which were placed in the chamber used for taking pictures in repetitive conditions. The detailed parameters of the chamber have been presented in other work of authors [9, 10]. The obtained photos were analyzed in the dedicated Patan® application [11, 12], determining for each sample three values respectively for the RGB scale. Then, the color number (as R*65536+G*256+B) was calculated based on this data. Due to the objective—comparing the images, it was necessary to transform the data's received from the RGB scale, into other scale which enables the numerical determination of the color hue. In the RGB scale, attributable

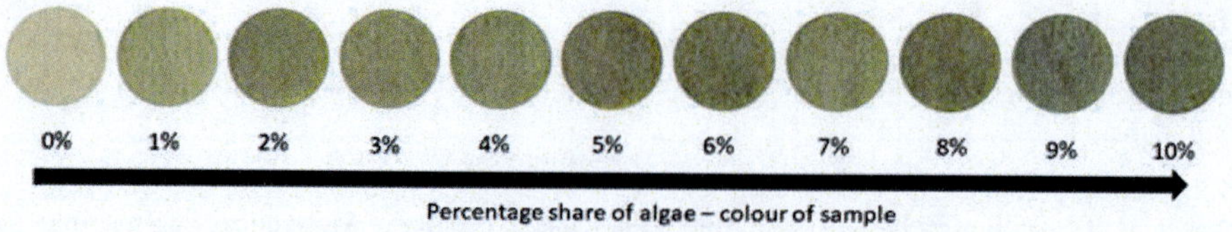

Fig. 1 Photos of samples of the mixture cornmeal—algae with algae 0–10%

colors in the bands 0–255 create a model that allows to replicate 16 777 216 different colors. Despite such a significant value, in order to compare the colors, the RGB model was converted to the HSL model thanks to which the components of hue, saturation and lightness were received in compared images. In Fig. 2, the RGB and HSL models are shown. The conversion was made with conversion rate (1–4) based on.

$$\begin{cases} R' = \frac{R}{255}; G' = \frac{G}{255}; B' = \frac{B}{255} \\ C_{max} = Max(R', G', B'); C_{min} = Min(R', G', B') \\ \Delta = C_{max} - C_{min} \end{cases} \quad (1)$$

$$H = \begin{cases} 60°x\left(\frac{G'-B'}{\Delta} mod6\right), C_{max} = R' \\ 60°x\left(\frac{B'-R'}{\Delta} + 2\right), C_{max} = G' \\ 60°x\left(\frac{R'-G'}{\Delta} + 4\right), C_{max} = B' \end{cases} \quad (2)$$

$$S = \begin{cases} 0 & , \Delta = 0 \\ \frac{\Delta}{1-|2L-1|} & , \Delta <> 0 \end{cases} \quad (3)$$

$$L = \frac{C_{max} + C_{min}}{2} \quad (4)$$

Except transformation there were also needed to represent RGB color in percentage RGB %. In this way the decimal color, % RGB, Hue, Saturation and Lightness parameters representing the mixtures color was defined.

The values were averaged and an exponentional function formula was determined based on the variables: color number (decimal), algae proportion (Fig. 3). This formula was used to estimate the share of algae in the further part of this study.

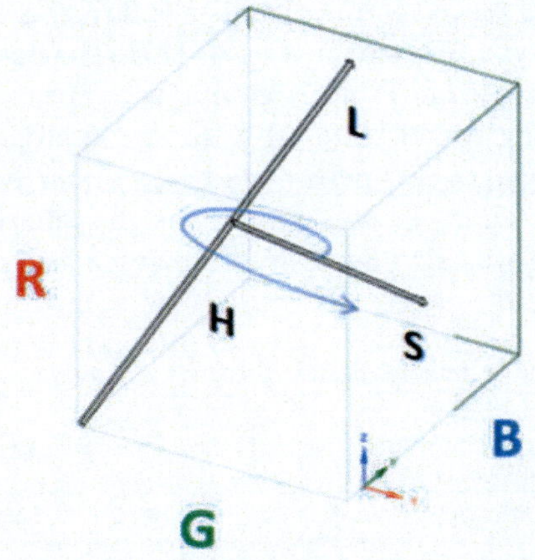

Fig. 2 Color RGB and HSL graphic interpretation with conversion relationships [14]

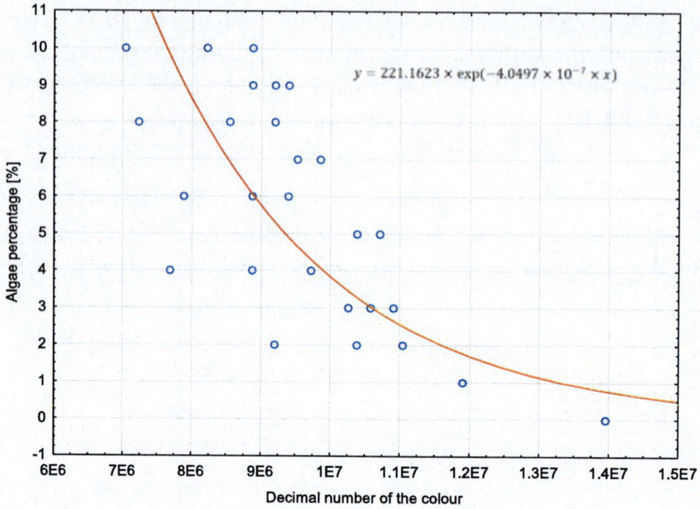

Fig. 3 Standard specimen—graph of algae proportional dependence on decimal color number

Entire process of acquiring the photos and digital analysis of the image was carried out in the same exact way in every test of the mixtures (composed of cornmeal and algae) during which the tracer fraction was assessed.

For the mixing process the vessel with capacity 90 ml was used. Identical procedure was applied—the components were placed in mixing vessels in the same manner (lower part of the container was filled with cornmeal, upper part with algae). The role of the key component in the mixture was played by algae (*chlorella*). Based on the contribution of the key component, it is possible to determine the homogeneity of the granular mixture [10, 15, 16]. The set amount of algae was 4% of the mix. The percentages and masses of the ingredients are presented in Table 1. The vessel prepared in this way was placed in the chamber and mixed with oloid movement characteristic—Turbula type 2TC mixer (Fig. 4) [17]. The parameters of the mixer used for the tests are presented in Table 2. Mixing was repeated in three series.

The duration of mixing was 30 and 60 s. After the process was complete all procedure of obtaining images was performed based on methodology presented in [10]. In this work after obtaining images of mixtures, the pattern interpreted in graph (Fig. 3) was used. There was also necessity to upgrade analysis to different scale

Table 1 Characteristics of the ingredients used in the study and mixture compositions

Parameter	Algae	Cornmeal
Geometric weighted mean particle size, mm	0.16	0.18
Humidity, %	4.61	10.35
Mixing ratio, %	4	96
Mass of constituents, mg	1.56	37.44

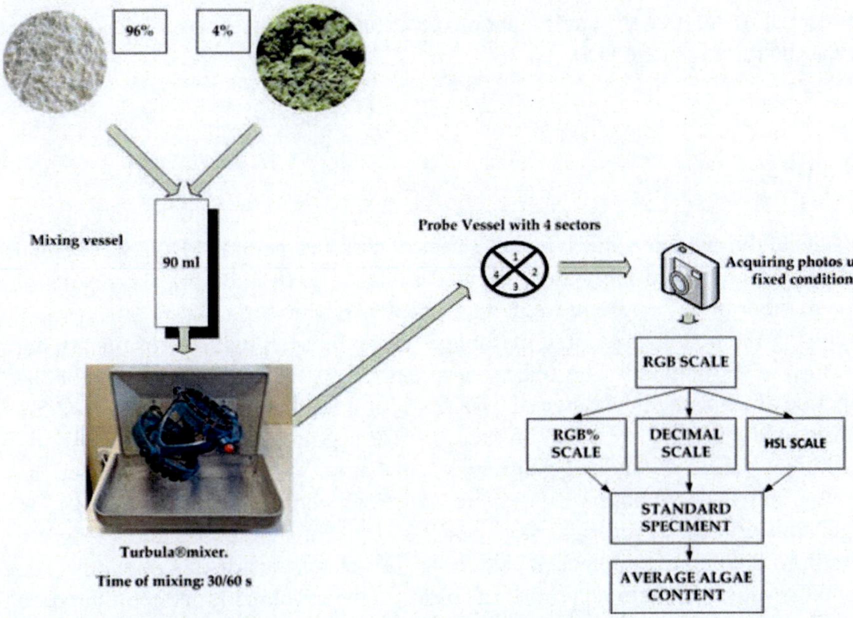

Fig. 4 Set up from research

Table 2 Mixer parameters Turbula® Type T2C

Parameter	Turbula mixer
Volume of the vessel, ml	90
Diameter of the vessel, mm	37
Length of the vessel, mm	57
Rotational speed, rpm	67
Mixing time, s	30/60

of colors representation than RGB model. The %RGB model was used in research because of the possibility of displaying average value of color in decimal scale and after that the HLS model was applied considering Hue, Lighting and Saturation of pixels. On this basis, the content of the key ingredient (algae) color representation was obtained for each series of tests depending on the time of mixing. The whole procedure is summarized in Fig. 4. Each test was performed in three repetitions. Additionally, the homogeneity of the mixtures was calculated using the coefficient of variation formula CV (as 100*standard deviation/mean). This parameter is the official standard used to assess the homogeneity of marketed drugs and its acceptable level in this case is CV < 6% [15]. In the feed industry this indicator is also often used and its acceptable level, depending on the intended use of the feed, is within the range $CV \leq 10\%$ or $CV \leq 15\%$ [18]. For food mixtures there are no similar standards, but in practice the coefficient of variation is often used to ensure adequate quality at the mixing stage [15].

3 Results

The results of the average values (from 3 series of measurements) of RGB%, decimal number of the color and HSL are presented in Table 3. In addition, the results of decimal number of colors are presented graphically in Fig. 5.

Analysing the results presented in Table 3, it can be seen that the results are not significantly differentiated. The obtained values of the standard deviation for the mixing time of 30 s are in the range of 0.42–2.62, and for the time of 60 s, 0.42–2.53. Moreover, analyzing Fig. 5, it can be seen that for segments 1 and 3 of the mixing vessel, higher values of the decimal number of colors were obtained than for the segments 2 and 4. These observations were statistically confirmed by distinguishing homogeneous groups (a) for segments 1 and 3 and (b) for segments 2 and 4 (Fig. 5). It is worth to take up the subject of searching for the reasons of this phenomenon. Based on the values of decimal number of color and the developed specimen standard (Fig. 3), the obtained values were converted into the share of algae. These results are presented in Table 4 and graphically in Fig. 6.

Analyzing the results from this part of the research (Table 4, Fig. 6), it can be seen that mixing cornmeal with algae allowed to obtain results close to the set values of algae (4%). The obtained values are in the range of 2.50–4.34% for the standard deviation in the range of 0.06–0.27. And the obtained average share of algae in the mixture was 3.42% (SD = 0.88) for the mixing time of 30 s and 3.36% (SD = 0.86) for the mixing time of 60 s. With the set amount of algae 4%, it can be

Table 3 Results of computer image analysis

Segment	%R[a]	%G[a]	%B[a]	Decimal number of color	H°	S%	L%
30 s							
1	66 ± 0.47	65 ± 0.42	46 ± 1.70	11,052,661	57	23	56
2	58 ± 0.68	57 ± 0.87	38 ± 1.91	9,736,545	57	21	48
3	66 ± 0.72	65 ± 0.57	46 ± 2.06	11,052,661	57	23	56
4	58 ± 1.03	57 ± 1.25	38 ± 2.62	9,736,545	56	21	48
60 s							
1	66 ± 0.57	65 ± 0.42	46 ± 2.11	11,052,661	57	23	56
2	58 ± 0.57	57 ± 0.31	44 ± 1.66	9,736,560	56	14	51
3	66 ± 0.0.72	65 ± 1.10	46 ± 2.53	11,052,661	57	23	56
4	58 ± 0.79	56 ± 1.10	38 ± 2.22	9,736,033	54	21	48

[a]mean of three series ± standard deviation

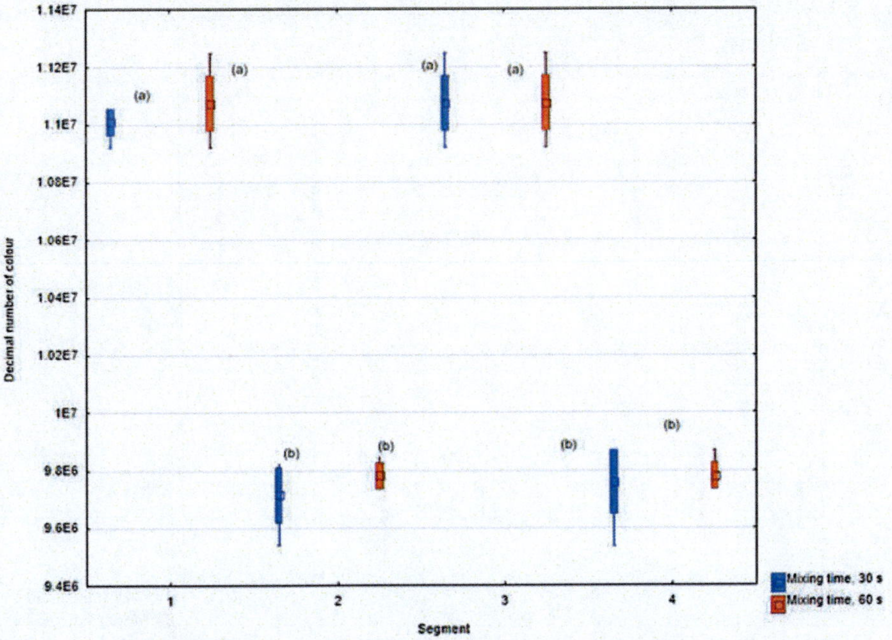

Fig. 5 Box plot graph of decimal number of colors obtained in individual segments of the mixing vessel for the times 30 and 60 s. Point—mean, frame—standard error, whiskers—min, max. (…)—statistically homogeneous groups

Table 4 The results of the algae share and the coefficient of variation for mixing times of 30 and 60 s

Segment	Average share of algae, %[a]	Share of algae in the mixture, %[b]	CV, %
Mixing time, 30 s			
1	2.56 ± 0.06	3.42 ± 0.88	25.73
2	4.34 ± 0.24		
3	2.50 ± 0.14		
4	4.26 ± 0.27		
Mixing time, 60 s			
1	2.50 ± 0.14	3.36 ± 0.86	23.89
2	4.22 ± 0.10		
3	2.50 ± 0.14		
4	4.22 ± 0.11		

[a]mean of three series ± standard deviation
[b]mean share of algae in the whole mixture ± standard deviation

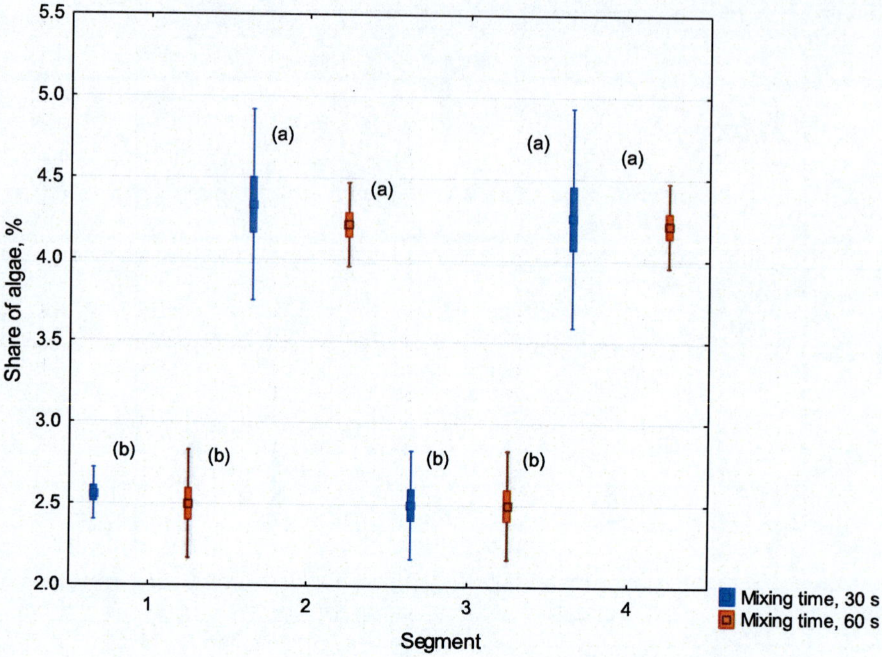

Fig. 6 Box plot graph of the share of algae for individual segments of the mixing vessel for the times 30 and 60 s. Point—mean, frame—standard error, whisker—standard deviation. (…)—statistically homogeneous groups

concluded that the obtained results are satisfactory. Slight deviations from the mean value (results of standard deviation) allow to conclude on high repeatability and accuracy of the proposed method. Moreover, the obtained results indicate that the mixing time (extension of the time from 30 to 60 s) did no influence on the homogeneity of the mixture. It is true that the coefficient of variation decreased from 25.79% for the time of 30 s to 23.89% for the time of 60 s. Obtaining significant values of the coefficient of variation and thus unsatisfactory homogeneity of the mixtures (CV > 15%) results certainly from the observed higher share of algae in segments 2 and 4 and a lower in segments 1 and 3. This differentiation was also confirmed statistically (as in the case of the decimal number of color results). The presented methodology and the obtained results indicate the possibility of applying the proposed solution in the assessment of the share of the key component in the two-component fine-grained system. The material described in the manuscript is a preliminary element of the planned series of tests.

4 Discussion

In this study the results of the attempt of correlation the actual share of key ingredient in micro-grained mixture with the digital image analysis indications was made. Thanks to the application of extended color scale with one hue, saturation and lightness it is possible to place the analysed samples in the color space in which they are actually located. In digital image analysis as the analytical reference method, the calibration is particularly noteworthy factor. The calibration has to be congruent with the type of the mixed substance we evaluate. It can take place in various actions. The idea of calibration is to correlate the key ingredient concentration with the representation of the colors in the image received. In particular, we can mention e.g. the camera settings, lighting conditions and so on that can be significant for the further work [19]. Despite the stable conditions and setting unfortunately there is still a risk of calibration problem, claim. Grasa and Abanades in their work [20] they analyse the correlation between the grey scale view and the solids concentration. Hence the most recommended way to receive the highest possible reliable image is to pay attention to substantial environmental influences that could cause a distortion of the view. The process allows to appraise the level of homogeneity of the ready mixture by comparing the view of several samples received on the digital images using the grey scale. In most of the works focused on this subject the grey scale value is the crucial view factor, the color scale is not this popular estimation method. Although most of the samples of the mixture are mostly colored [19]. There are the methods which consist in color analysis of the substance added to the mixture, that has the luminescence properties. In such methods, the fluorescent substance is a tracer in describing the quality of the mixture [9]. The phase of acquiring the digital images is equally important as the subsequent photo editing (image filtration, representation of the colors in the selected model and so on) [21]. Applying different color scales can be used in the assessment of the powder's quality. The objective of the grayscale (0–255), even assisted by pixelation, is to compare the value of the hue. The RGB scale, despite its wide range of color creating possibilities, quite often requires transformation to another color model [22]. By this we mean the model that indicates numerically characteristic elements of the additional dimensions, e.g., saturation, hue, lighting. An example of a transition with color models is the RGB -> HSL path in mathematical relation [13]. Similar kind of studies were performed previously by Realpe and Velázquez [23]. The scientists also compared the powder samples by using grayscale method in digital image analysis. The components in their work were different combinations of powder substances—lactose, blue lactose, chocolate and cellulose. The grayscale method allowed to assess the concentration of the mixtures. Mayer-Laigle et al. [7] carried out the research in which they were analysing the images of mixtures composed of black and white semolina (size 100–250 μm). Parallel to the authors of this paper they could assess a homogeneity level of the mixed particles during the work process. The methods and techniques have a great influence on the final form of the research, like the device type applied in solids mixing. The research presented in this work consists in determining the quality of mixtures composed of cornmeal and algae (with 4% algae share in the substance tested) in two configurations of the mixing time and in vessel capacity 90 ml used in the Turbula® mixer. Presented methodology requires further improvement and more tests of correlation of digital image analysis with actual key ingredient share in the mixture. The discrepancy observed in the work (share of algae in 1 and 3 segment connected to 2 and 4 segment) can have connection with the conditions during acquiring the photos needed for assessment. Another reason may be the tendency of algae to occupy a specific place in the mixed bed. Therefore, further tests are planned to optimize the methodology.

The quality of the final product depends significantly on the steps of production—among other things mixing the ingredients. In their work Mishra and Bhatt [24] determine highest quality pasta as the product with features such as strong consistency, elasticity and slight loses during the heat-treating. In turn, Spinelli et al. [25] claim that that fortification applied in pasta production is the method that allows to enhance the quality as regards color, texture and sensory properties. The color of food products not only presents their quality but can also be an indicator of their safety [26]. Currently the industry pays increasingly more attention to the nutritional values of the food [24]. Thus, fortification becomes more popular as the key to create the adequate products enriched with healthy ingredients that affects the consumers well-being. More broadly, considering further industries connected to the subject, low quality of final product would be serious obstacle in placing it on the market. Which is why the authors of the work decided to use algae as the component of the analysed mixtures along with the cornmeal.

5 Conclusions

Conducted experiments and presented literature allow to present conclusions:

- The described methodology allowed to obtain information on the percentage share of tracer (algae) in the mixture with cornmeal. The methodology is based on image analysis and relating the results with the created pattern.
- Mixing cornmeal with algae allowed to obtain results close to the set values of algae (4%). The average share of algae in the mixture was 3.42% (SD = 0.88) for the mixing time of 30 s and 3.36% (SD = 0.86) for the mixing time of 60 s. Thus, the mixing time was not observed to affect on the quality of the mixture.
- It was observed that the samples taken from the segments 2 and 4 of the mixing vessel had a higher proportion of the key component than the samples from the segments 1 and 3. This significance was confirmed statistically. Therefore, it is necessary to conduct additional observations to determine the reasons of this phenomenon.
- Computer image analysis has been successfully applied in the assessment of homogeneity of granular mixtures with the use of various mixers. The obtained

results indicate its applicability to food mixtures. However, the methodology requires further clarification and elimination of possible measurement errors.

References

1. Zuki SAM, Rahman NA, Yassin IM (2014) Particle mixing analysis using digital image processing technique. J Appl Sci 14:1391–1396
2. Królczyk JB (2016) Metrological changes in the surface morphology of cereal grains in the mixing process. Int Agrophys 30(2):193–202
3. Krolczyk JB (2014) An attempt to predict quality changes in a ten-component granular system. Tehnički vjesnik - Technical Gazette 21(2):255–261
4. Marigo M, Cairns DL, Davies M, Ingram A, Stitt EH (2012) A numerical comparison of mixing efficiencies of solids in a cylindrical vessel subject to a range of motions. Powder Technol 217:540–547
5. Obadele BA, Masuku ZH, Olubambi PA (2012) Turbula mixing characteristics of carbide powders and its influence on laser processing of stainless steel composite coatings. Powder Technol 230:169–182
6. Bauman I, Ćurić D, Boban M (2008) Mixing of solids in different mixing devices. Sadhana 33:721–731
7. Mayer-Laigle C, Gatumel C, Berthiaux H (2015) Mixing dynamics for easy flowing powders in a lab scale Turbula® mixer. Chem Eng Res Des 95:248–261
8. Tahvildarian P, Ein-Mozaffari F, Upreti SR (2013) Circulation intensity and axial dispersion of non-cohesive solid particles in a V-blender via DEM simulation. Particuology 11:619–626
9. Matuszek DB (2020) Ultraviolet fluorescence in the assessment of quality in the mixing of granular material. Sustainability 12:1546
10. Matuszek DB, Biłos ŁA (2021) Computer image analysis as a method of evaluating the quality of selected fine-grained food mixtures. Sustainability 13:3018
11. Tukiendorf M, Krótkiewicz M, Boss J (2003) The application of image analysis as a method of evaluation of granular blend's quality. Polish J Food Nutr Sci 53(2):27–30
12. Matuszek DB, Królczyk JB (2021) Fluorescence optical analysis method for assessing homogeneity of granular mixtures. Metrol Meas Syst 28(1):41–54
13. Saravanan G, Yamuna G, Nandhini S (2016) Real time implementation of RGB to HSV/HSI/HSL and its reverse color space models. In: International conference on communication and signal processing, ICCSP. IEEE, India, pp 0462–0466
14. Loesdau M, Chabrier S, Gabillon A (2014) Hue and saturation in the RGB color cpace, conference paper in lecture notes in computer science. In: Elmoataz A, Lezoray O, Nouboud F, Mammass D (eds) Image and signal processing. International conference on image and signal processing, ICISP. Springer, Cham, pp 203–2012
15. Cuq B, Berthiaux H, Gatumel C (2013) Powder mixing in the production of food powders. In: Handbook of food powders, processes and properties. Woodhead Publishing Series in Food Science, Technology and Nutrition, pp 200–229
16. Matuszek DB, Bierczyński K, Jędrysiak A, Kraszewska A (2021) Homogeneity of the selected food mixes. Czech J Food Sci 39(3):197–207
17. Turbula® Shaker Mixer Type T2C Operating Instruction. Willy A. Bachofen AG, Maschinenfabrik, Utengasse 15–17, CH-4005 Basel. https://www.novatechuk.com/content/products/f/turbula-t2c-manual.pdf. Archived 28 Apr 2021
18. Groesbeck CN, Goodband RD, Tokach MD, Nelssen JL, Dritz SS (2004) Using mixer efficiency testing to evaluate feed segregation in feed lines. Kansas agricultural experiment station research reports 0(10). https://doi.org/10.4148/2378-5977.6870

19. Emady HN, Wittman M, Koynov S, Borghard WG, Muzzio FJ, Glasser BJ, Cuitino AM (2015) A simple color concentration measurement technique for powders. Powder Technol 286:392–400
20. Grasa G, Abanades JC (2001) A calibration procedure to obtain solid concentrations from digital images of bulk powders. Powder Technol 114:125–128
21. Tadeusiewicz R, Korohoda T (1997) Komputerowa analiza i przetwarzanie obrazów. Wydawnictwo FPT, Kraków (in Polish)
22. Ibrahim AWS, Sartep HJ (2017) Transfer color from color image to grayscale image. Int J Eng Res Adv Technol 03(3):18–25
23. Realpe A, Velázquez C (2003) Image processing and analysis for determination of concentrations of powder mixtures. Powder Technol 134:193–200
24. Mishra P, Bhatt DC (2016) A Study on development of fortified pasta with ginger powder. J Environ Sci Toxicol Food Technol 10(8):14–18. Ver. I
25. Spinelli S, Padalino L, Costa C, Del Nobile MA, Conte A (2019) Food by-products to fortified pasta: a new approach for optimization. J Clean Prod 215:985–991
26. Sobol Z, Jakubowski T, Nawara PR (2020) The effect of UV-C stimulation of potato tubers and soaking of potato strips in water on color and analyzed color by CIE L*a*b*. Sustainability 12(18):7473

The Use of Farm Digital Tools in Crop Production in Wallonia: Situation, Key Points and Actors

Thi Minh Hop Ho and Philippe Burny

Abstract This paper shows that the use of farm digital tools in crop production is already well spread in Wallonia. This evolution is supported by the public authorities, as well as by many private and public actors involved in this field, leading to a large supply of tools available to the farmers. However, 80% of available tools are not used, while farmers' needs are far to be met. This is due to several constraints: the profitability of specific digital tools, the legal protection of personal data, the time consumption to learn how to use the new tools efficiently, the complexity of some tools, and the lack of interconnection among the new tools and also with the already existing ones. It finally appears that there is a need for a better adequation between supply and demand of farm digital tools, and also for better training facilities, though several structures are already active in this field.

Keywords Digital tools · Agriculture · Crop production · Training · Wallonia

1 Introduction

New technologies, and more specially digital technologies, are now spreading in all fields of human activity, including agriculture [3]. Animal raising and breeding, agricultural crops, horticultural production… are all concerned today with digital technologies and "big data" management [5]. The potential use of these technologies is rather broad, as they could really contribute to face the main problems of agriculture: environmental negative impacts on air, water and soil quality, biodiversity reduction, and also farm profitability, as well as the disappearance of a lot of farms and the decline of the agricultural population [2, 6].

T. M. H. Ho
Gembloux Agro-Bio Tech, University of Liege, 5030 Gembloux, Belgium

P. Burny (✉)
Walloon Agricultural Research Centre, 5030 Gembloux, Belgium
e-mail: p.burny@cra.wallonie.be

S. Pascuzzi and F. Santoro (eds.), *Farm Machinery and Processes Management in Sustainable Agriculture*, Lecture Notes in Civil Engineering 289,
https://doi.org/10.1007/978-3-031-13090-8_20

Digital technologies, robotization, the artificial intelligence can help to find appropriate answers to the challenges of the future. However, their adoption by farmers is not obvious and there are several obstacles to their spreading among farms, and family farms more specifically.

This paper is one of the results of the ITFARM (IT for interconnection of social, economic and environmental aspects in agribusiness) Erasmus + international project led by the Czech University of Life Sciences in Prague and gathering seven partners from seven EU countries.

This paper is only a small part of the ITFARM project and focuses on Wallonia (South of Belgium) on one hand, and crop production, on the other hand.

Its objectives are to give answers to the following questions:

- What is the general framework for agricultural digitalization?
- What are the key actors involved in agricultural digitalization and technology transfer?
- What is the present situation of the use of digital tools in crop production?
- What are the key questions for a larger adoption of digital tools by crop farmers?

2 The General Context

2.1 The European Framework

The Internet of Things (IoT) is at the centre of the digitalization of the world economy and is one of the priorities of the European Union [1]. Since 2015, the European Commission developed several strategies and actions in order to strengthen the use of IoT in Europe, like: the digital single market (DSM-a framework for actions), advancing the internet of things in Europe and IoT cluster study (sector program), AIOTI (actors gathering), Horizon 2020 (support for research and innovation)... These priorities create a dynamics of innovation about the IoT implementation in given key sectors, like smart farming.

Several pilot projects have been selected within Horizon 2020:

- Internet of Food and Farm 2020 (IoF 2020)
  This project concerns 19 "good agricultural practices" of the IoT in different sectors of agricultural activities: dairy production, fruits and vegetables, arable land. There are 73 partners throughout Europe involved in this project. Among these good practices are concerned the optimization of wine production, data processing about weeding actions, large farm management and table grapefruit production channels.

  Belgian actors are involved in two good practices:

  - "Grazing Cows Monitor": cow identification and traceability by GPS

  - "Big farm management": optimization of pig production management thanks to captors

- Agricultural Interoperability and Analysis System (ATLAS)
  The use of good practices of the IoT is illustrated with 13 demonstrating professional and experimental farms. They concern animal behaviour, soil data management and the use of captors for irrigation.
- The DEMETER project
  This project focuses on advanced interoperability technologies, involving several sectors, actors, production channels and farms of different sizes. The goal is to offer the farmers a choice among many IoT tools more adapted and easier to combine with their machines.

  Another goal is to prove the role of IoT solutions in the development of sustainable agricultural systems and food production channels.

  The concerned sectors are: arable crops, precision agriculture, fruits and vegetables, cattle raising and supply chain. In Belgium, the fruits and vegetables sector is involved. More than 5,700 farmers and 318,000 ha are involved in this project.
- AgROBOfood
  This project supports pilot initiatives developed in the field of robotization. Belgium is concerned with the creation of an autonomous greenhouse using interconnected robots able to evaluate the maturity of the food products, to harvest and to transport them without any human intervention.
- SmartAgrihubs
  This project includes 140 digital innovation hubs spread within 9 different European regions.

2.2 *The Regional Policy About Digitalization*

On December 6th, 2018, the regional government of Wallonia defined the new strategy called "Digital Wallonia" for the five-year period 2019–2024. This strategy describes the framework within which the ways to explore the socio-economic opportunities brought by the digitalization process are identified (Digital Wallonia).

The digital strategy of the Walloon government is ambitious, innovative and inclusive. Digitalization is a global project which concerns the regional territory, its economy and its society in order to enhance its attractivity, its competitiveness and the well-being of all its citizens, based on values like transversality, transparency, coherence, openness and ability.

The program "Digital Wallonia 2019–2024" is based on eight transversal challenges:

- Governance: the specific governance of the phenomenon of digitalization is established in order to master the rapid evolution of this sector and to make it profitable for all sectors of activity
- Digital society: the process of digitalization must be accompanied in order to avoid the possible societal fractures among citizens
- Ecosystem: specific fields of activity will be chosen and will concentrate the public and private efforts
- Giga region: the whole territory must be connected with the most efficient system
- Competences: all citizens must get a sufficient knowledge of the digital tools used both at work and at home
- Data: the collection and processing of data must be at the heart of all actions and their follow up
- Platform: the platform "digitalwallonia.be" must identify, activate and promote the actors of the digital evolution
- Label: the label "Digital Wallonia" must be strong and attractive in order to represent the digital region of the South of Belgium

The program "Digital Wallonia 2019–2024" defines five specific fields of action:

- Digital competences
- Digital territory
- The digital sector
- The digital economy
- Digital administration

In all these, more specific projects are implemented. One of them directly concerns the Walloon agriculture and is called "Smart Farming".

2.3 *Smart Farming Projects*

Walloon farmers, who are mainly cattle raisers (bovine milk and meat) and crop producers (cereals, potato, sugar beet, rapeseed…) use today a large set of digital tools: captors, robots, connected objects, decision making tools, monitoring tools… in order to make farming more profitable and more environment-friendly.

The goals of Smart Farming are:

- To optimize the agricultural production process: for example the use of Nitrogen can be precisely determined on each small part of a plot, according to the development level of the plants, which can guarantee the maximal yield while using the minimal quantities of fertilizers, so reducing the environmental risk (water nitrification) while improving the farmers' income
- To improve the agricultural practices: the large data collection and processing can lead to the implementation of better practices
- To strengthen the contacts between farmers and consumers: the consumers can have a better knowledge of the conditions in which agricultural goods are

produced and farmers can know more precisely what are the main concerns for the consumers, so that they can adapt their supply to the demand. Farmers and consumers can know each other and direct sales are easier.
- To develop the commercial activities of the farms: the digital tools can help farmers to reduce their production costs while e-commerce can boost their sales

The Smart Farming project has identified two main challenges:

- To increase the number of farmers who use digital tools
- To harmonize and secure the collected data

Two data management systems have been set up for the access to and use of data concerning crops (WalDigiFarm) and animal raising (WalleSmart). These systems collect and protect the data.

3 Methodology

The methodology is based on a qualitative approach. Secondary data are collected from the website of public institutions and private organizations supposed to be involved in the development of digital tools, the agricultural statistics publications, surveys made by private companies and the relevant literature review.

In addition, primary data are collected through the interview of important actors of the supply side of digital tools used in the field of agriculture, of the demand side of these tools, and also from the in-between organizations dealing with education, training and extension services.

4 Key Actors

WalDigiFarm is a non profit association which is now financially supported by the Walloon government. In 2021, 90 members were registered, including farmers and agricultural entrepreneurs, companies which produce inputs, food processing entreprises, food distributors, agricultural management and consultancy organizations, research institutions, teaching and training institutions... [7].

The work of this association is done according to the following guidelines:

- To federate the actors of agricultural digitalization
- To promote/strengthen the use of digital tools in agriculture
- To be the Walloon think tank for the digital transition of agriculture
- To conceive the tools and structures of the future

The Walloon Agricultural Research Centre is the main institution involved in agricultural research in Wallonia and many regional, national and international research projects deal with new digital technologies both in animal and vegetal production and processing. The researchers are in permanent contact with the users of these new technologies.

The Universities are also involved in research and in teaching (the University of Liege and the Catholic University of Louvain deal with agronomy and veterinary medicine; the University of Mons has a specialization in computer science, software and artificial intelligence).

The Centres for Advanced Technologies play an important role in the field of professional training and extension services. One is specialized in the agricultural sector (Huy) and another one in the "green sectors"—horticulture (Gembloux). They are based in high schools and are in daily contact with digital tools suppliers and with students and professionals, for whom they organize regular trainings and among whom they spread the relevant information.

Farmers' unions are also well organized and financially supported by the regional authorities, notably in order to implement trainings and conferences.

The "Walloon public service" (Ministry) is of course involved in all fields dealing with agriculture, including extension services.

Some more specific organizations, like "**Eleveo**", the Walloon association of animal raisers and breeders, are cofinanced by the regional government and are deeply concerned with the spreading of new technologies and extension services.

Of course, as Belgium is very open to import/export activities, all **the international companies** which produce and sell agricultural machines, are present on the national market and propose new tools to the farmers.

Private companies which sell inputs like fertilizers to the farmers are also dealing with field trials and information spreading.

The regional Agency for Digitalization has been established to promote the use of digital tools in all fields of activity, including agriculture.

Several **Walloon entreprises** have also developed some specific digital tools for the use of farmers.

A complete list of actors is impossible to establish, as they are too numerous.

5 The Current Situation

According to the Agency for Digitalization (see Fig. 1), Walloon farmers are well equipped as far as information technologies are concerned: 60% have a desktop; 61% have a laptop, 28% have an iPad and 63% have a smartphone.

Farmers are also buyers and sellers on line:

- 20% of Walloon farmers sell their products on line (against 47% for all sectors together)

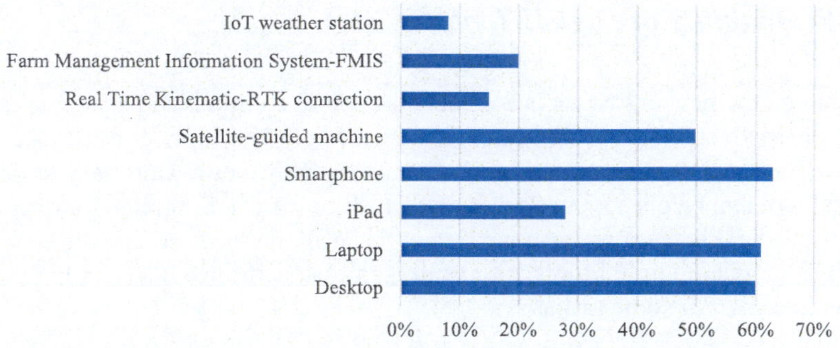

Fig. 1 Percentage of farmers equipped with different kind of digital tools

- 65% of Walloon farmers buy their inputs on line (against 48% for all economic sectors together)

According to surveys and extrapolations made by private companies, 50% of Walloon farmers have satellite-guided machines, 15% with Real Time Kinematic-RTK connection, and 20% have a Farm Management Information System-FMIS. Around 8% of Walloon farmers use IoT weather stations [7].

There are around 250 applications which are available, with a mean of 4.2 apps per farmer.

The fertilization with satellite/drone is implemented on around 1% of the area of arable land.

In general, the situation seems rather good, though there is still room for a significant progress. However, an important problem must be stressed: it can be estimated that around 80% of the available digital tools are not used by farmers, while most of farmers' needs are not met. It is clear that significant efforts have to be made in order to adapt the supply of digital tools to the demand.

6 Key Questions for a Larger Adoption of Digital Tools by Crop Farmers

Digital tools seem to be very useful in the field of agriculture and their future seems to be bright. However, this adoption by relatively small family farms (the average area in Wallonia is around 60 ha) does raise problems which deserve to be taken into account and overcome. The main ones have been identified as follows:

6.1 Profitability of Digital Tools

Even though the new digital tools are becoming more common and available on the market at reasonable prices, or sometimes even freely, the farmers must of course calculate the return on investment, which is, by the way, not always easy to define. A digital tool can be cheap, but does it really reduce the use of inputs, for example? Another problem is the small average size of the Walloon farm, so that the economy of scale is reduced. Digital tools are also linked to specific machines, which could be too expensive for some farmers.

6.2 Personal Data Protection

The use of new technologies in general can lead to an unwanted spreading of personal data. The farmers want to keep their data for their personal use only, or at least that the data stay anonymous, notably for fiscal reasons. With the evolution of the technologies, new regulations must be adopted and implemented in order to guarantee the protection of personal data.

6.3 Time Needed for Training

The farmers must of course know how to use the technologies they would like to buy. However, this is not always so easy to access and it is always time consuming, even if there are several possibilities for the farmers to know how to master the new technologies, offered by public and private organizations.

Time management is already a problem for farmers, especially during some specific periods.

6.4 The Complexity of Digital Tools

Some tools can appear to be complex for unexperienced and also elder farmers who, for example, do not have a computer at home. Farmers can feel that the use of digital tools is too difficult compared to their potential benefit.

6.5 *The Lack of Interconnection Between the Existing Tools/Machines*

Many software and digital tools are available on the market, but their use is sometimes limited to a specific machine from a given company. Several softwares can be useful, but are not interconnected and so there is a waste of time and information.

Today digital tools suppliers try to make them as interconnected as possible.

6.6 *Adequation Between the Tools Available on the Market and Farmers' Needs*

It appears that 80% of digital tools available on the market are not used by the farmers, while their needs are far to be satisfied. It is clear that a permanent dialogue between digital tools suppliers and the farmers is necessary to define tools which are really useful for diverse reasons (better profitability, gain of time, easy use...).

6.7 *Existence of Good Training Facilities*

In the present situation, it clearly appears that good training facilities are very important in order to promote the use of agricultural digital tools.

7 Conclusions

Thanks to a favourable global framework, including the fast global evolution of new digital technologies in all fields of human activities, among which agriculture and agricultural products processing, and the European policy and its implementation in the Walloon region, it appears that the use of digital tools among crop producers is already widely spread. Many private and public actors are involved in the supply and development of new technologies among farmers: multinational companies producing agricultural machines, local small enterprises producing software, research and teaching institutions, farmers' unions and associations, the Ministry of Agriculture, the Agency for Digitalization... However, there are still some problems preventing an even larger use of digital tools in crop agriculture: their cost and profitability, their complexity, the lack of interconnection between tools and machines, the time needed to learn how to use the new tools, and also the legal aspects linked to personal data protection, the relative lack of training facilities. All these reasons lead to the inadequation between supply and demand: around 80% of the available tools are not used, while farmers' needs are not satisfied. In such a situation, intermediary structures between the suppliers and the farmers, which can organize regular up-to-date training and spread the relevant information, can play a significant role to improve the use of new digital tools. The non-profit association "WalDigiFarm", of which all kinds of stakeholders are members, can significantly contribute to clearly define the problems and to find the appropriate solutions. The Centres for Advanced Technologies can also help to reach a better equilibrium between supply and demand.

References

1. Alliance for Internet of Things Innovation-AIOTI. European IoT challenges and opportunities 2019–2024. https://aioti.eu/wp-content/uploads/2019/09/AIOTI-Priorities-2019-2024-Digital.pdf. Accessed 10 Apr 2022
2. Batter MT, Ehsani MR (2006) The economics of precision guidance with auto-boom control for farmers-owned agricultural sprayers. Comput Electron Agric 53(1):28–44
3. Birner R, Daum T, Pray C (2021) Who drives the digital revolution in agriculture? A review of supply-side trends, players and challenges. Appl Econ Perspect Policy 43:1260–1285
4. Digital Wallonia (2022) Bilan 2015–2018 vision and actions 2049–2024. https://content.digitalwallonia.be/post/20190626100319/Bilan-strategique.pdf. Accessed 10 Apr 2022
5. Trendov NM, Varas S, Zeng M (2019) Digital technologies in agriculture and rural areas-Status report, Rome
6. Van Es H, Woodard J (2017) Innovation in agriculture and food systems in digital age. In: Global Innovation index 2017. World Intellectual Property Organisation, Geneva
7. WalDigiFarm (2019) 1st association between the agricultural and digital sectors in Wallonia. https://www.waldigifarm.be/fr/accueil/. Accessed 7 Apr 2022
8. WalDigiFarm (2022) Valoriser et favoriser l'usage du numérique en productions végétales en Wallonie

Effect of Digestate Fertilization on the Quality of Beans

Milan Koszel, Stanisław Parafiniuk, Sławomir Kocira, Atanas Z. Atanasov, Artur Przywara, and Artur Kraszkiewicz

Abstract The possibility of agricultural use of the digestate as a fertilizer was determined. The digestate collected from an agricultural biogas plant was tested for the content of macroelements and heavy metals. The research covered the "Kontra" variety of snap bean. Four variants of the research were applied—25,000 l * ha^{-1}, 37,500 l * ha^{-1}, 50,000 l/ha^{-1} and a control object. The analysis showed an increase in the yield depending on the applied dose of digestate. The best results of mean values were obtained during the application of the dose of 50,000 l * ha^{-1}. Average height—84.00 cm. The seed yield was 2.60 t * ha^{-1}. Number of pods per plant—11 pcs. Number of seeds per plant—44 pcs. Number of seeds in a pod—4.4 pcs. Thousand seed weight—184 g. The average length of the pods was 11 cm and the width was 0.7 cm. Applying the digestate dose of 50,000 l * ha^{-1}, the following nutritional values were found in 100 g of beans: protein—21.4 g, fat—0.6 g, carbohydrates—54.0 g, calcium—124 mg, magnesium—137 mg, phosphorus—387 mg, potassium—1394 mg.

Keywords Digestate · Snap beans · Sustainability agriculture · Fertilization

1 Introduction

Anaerobic digestion is one of the most promising technological solutions related to the production of fuels based on renewable resources. Biogas plants operate on the basis of various types of organic substances [1].

Biogas is produced by anaerobic microbial decomposition of organic compounds and contains two main components: methane and carbon dioxide. A positive aspect

M. Koszel (✉) · S. Parafiniuk · S. Kocira · A. Przywara · A. Kraszkiewicz
Department of Machinery Exploitation and Management of Production Processes, University of Life Sciences in Lublin, Lublin, Poland
e-mail: milan.koszel@up.lublin.pl

A. Z. Atanasov
Department of Agricultural Machinery, University of Ruse Angle Kanchev Bulgaria, Ruse, Bulgaria

S. Pascuzzi and F. Santoro (eds.), *Farm Machinery and Processes Management in Sustainable Agriculture*, Lecture Notes in Civil Engineering 289,
https://doi.org/10.1007/978-3-031-13090-8_21

of anaerobic digestion is that it reduces pathogens, kills viruses, fungi, Listeria, Salmonella, and Escherichia coli bacteria, and inactivates plant seeds. The post-fermentation mass is a secondary product of this wet fermentation [2, 3].

Anaerobic decomposition is a technology which enables the conversion of food industry or municipal waste into renewable energy sources. The process of anaerobic decomposition consists of a number of metabolic reactions performed by a wide range of microorganisms in anaerobic conditions. As a result, biogas and post-digestion matter are formed [4].

The major factors determining the way of digestate utilization include its quality, local conditions and legal regulations. In Poland the factor which determines the utilization of biogas plant's by-products is legal norms, which do not facilitate digestate management [5].

The digestate is a heterogeneous liquid formed as a by-product in the production of biogas from organic matter with a significant share of undecomposed solid organic fraction. The dry matter content in the digestate is 7–12% and is comparable to the liquid manure. Nitrogen content in fresh matter is similar to manure (0.2–1%), but has a higher pH value (7–8). The use of digestate as organic fertilizer on agricultural land is already considered the standard method of its application [6–8].

In terms of the speed of action (nutrient uptake by plants), digestate is similar to mineral fertilizers, because the N, P and K components are easily available to plants. It also contains some organic matter that has a positive effect on the physical and chemical properties of the fertilized soils [9, 10].

Two species are cultivated in the Polish climate, the common bean *Phaseolus vulgaris* L. and the multiflora bean *Phaseolus multiflorus* Wild. Within the common bean we distinguish dwarf varieties with a stiff stem 25–40 cm long and pole varieties with a slack stem up to 3 m long. There is also an intermediate snap bean form, with a stem 60–120 cm, without the ability to wrap around a support.

The usable part of the beans are unripe pods and dry, ripe seeds. Beans are considered a rich source of protein with high biological value, potassium, phosphorus and iron. The seeds contain phytic compounds that lower the level of cholesterol in blood serum and tissues, and inhibit the development of cancerous diseases, especially of the large intestine.

The aim of the research is to select the digestate dose and analyze the response of the "Kontra" variety to its various doses.

2 Method

The digestate was collected from the Piaski biogas plant (Lubelskie Province) and applied to the experimental plots for the cultivation of "Kontra" variety of snap beans. The area of the experimental plots was 75 m^2. The polyphoska 6 mineral fertilizer was sown before sowing (with a content of 6% nitrogen in ammonium form, 20% phosphorus, 30% potassium). The dose was 350 kg * ha^{-1}. Plant protection treatments were performed with a mounted field sprayer Pilmet 312 LM. Liquid dose during the treatments was 270 l * ha^{-1}, the working speed was 5.4 km * h^{-1}, and the working pressure was 3 bars. The height of the field boom above the sprayed surface was 50 cm.

The scope of the field research covered 4 variants of the experiment, on which the digestate was poured with liquid manure with hoses on the soil surface:

- Variant I—a dose of digestate 25,000 l * ha^{-1},
- Variant II—post-fermentation dose 37,500 l * ha^{-1},
- Variant III—post-fermentation dose 50,000 l * ha^{-1},
- Variant IV—the control object, which consisted of sown seeds not fertilized with digestate.

The above doses were determined on the basis of soil samples and studies by other authors [11, 12].

The digestate samples were tested for the content of heavy metals. The tests were carried out at the Regional Chemical and Agricultural Station in Lublin according to the PN-EN ISO/IEC 17,025: 2005 standard. The research on the quality of seeds after harvesting the snap bean was based on the determination of seed moisture, thousand seed weight (TSW), protein, fat, carbohydrate, calcium, magnesium, phosphorus and potassium content. The above research was carried out at the Central Agroecological Laboratory of the University of Life Sciences in Lublin.

3 Results

The biogas plant (a biogas combined heat and power plant) is located in Piaski Commune, Lubelskie Province. The electric power is 0.99 MW, and the thermal power—1.1 MW. The annual electricity production—approximately 8 400 MWh. The generated biogas is desulfurized, dewatered, cooled and pumped by means of an underground gas pipeline into a cogeneration engine which generates electricity and heat in a combined process. The following are used as an input into the digestion process: green waste matter, maize silage, beet pulp, stillage, whey.

Before application, the digestate was tested for the content of macroelements and heavy metals (Table 1). The pH of the digestate used for the cultivation of snap beans—8.50 and is similar to the pH of cattle liquid manure (7.90).

When analyzing the obtained test results, the content of heavy metals was below the detectability of the measuring equipment. Moreover, the digestate contains significant amounts of macronutrients, therefore it has been found possible to use the digestate as a fertilizer.

The average height (from 10 measurements) of plants after the application of digestate in the BBCH 78-BBCH 79 phase was 41 cm for variant I, 81 cm for variant II, 84 cm for variant III, and 73 cm for variant IV.

Table 2 shows the changes in the length and width of the pods, the number of pods per plant and the number of seeds per plant in each variant of the experiment.

Table 1 Comparison of selected macroelements and heavy metals in the post-product

Examined feature	Content in digestate
Phosphorus [g/l]	0,13
Potassium [g/l]	5,08
Calcium [g/l]	0,33
Magnesium [g/l]	0,08
Cadmium [mg/l]	<0,43
Lead [mg/l]	<0,43
Nickel [mg/l]	<0,43
Chromium [mg/l]	<0,43
Copper [mg/l]	0,47
Zink [mg/l]	1,99
Manganese [mg/l]	2,02
Iron [mg/l]	75,01

Table 2 Changes in the length and width of pods, the number of pods per plant and the number of seeds per plant

Variant	Average length of pods [cm]	Average width of pods [cm]	Average number of pods per plant [pcs]	Average number of seeds per plant [pcs]
I	9,70	0,66	8	31
II	10,40	0,67	10	39
III	11,00	0,70	11	44
IV	8,80	0,61	7	25

Analyzing the obtained research results, slight discrepancies were found in the average length and width of the pods, the average number of pods per plant and the average number of seeds per plant. The lowest values were recorded in variant IV (control object) and the highest for plants harvested from a plot with a dose of 50,000 $1 * ha^{-1}$. The average length of the pods ranged from 9.70 (variant I) to 11.00 cm (variant III), while the average width ranged from 0.66 cm (variant I) to 0.70 cm (variant III). For the control object, these values were 8.80 cm and 0.61 cm, respectively. The average number of pods per plant ranged from 8 (variant I) to 11 (variant III), and for variant IV—7 pieces. Average number of seeds per plant—31 (variant I) to 44 (variant III). In turn, for the control object—25 pcs. Based on the data presented in Table 2, the average number of seeds in a pod was determined, which was as follows: variant I—3.8 pcs; variant II—3.9 items; variant III—4 pieces, and for variant IV—3.6 pieces.

The humidity of the bean seeds of the "Kontra" variety for the control object was 12.94%, for the variant I—12%, for the variant II—11.98%, and for the variant III—11.87%.

Table 3 presents changes in the obtained yield values and the thousand seed weight in individual variants of the experiment.

Table 3 Changes in the yield size and the thousand seed weight in different variants of the experiment

Variant	Yield [t * ha^{-1}]	Thousand seed weight [g]
I	2,05	172
II	2,34	179
III	2,60	184
IV	1,80	167

Analyzing the obtained results, the obtained yield was found to be divergent depending on the digestate dose from 2.05 t * ha^{-1} to 2.60 t * ha^{-1}, for the control object—1.80 t * ha^{-1}. The obtained thousand seed weight ranged from 172 g (variant I) to 184 g (variant III), while for variant IV this value was 167 g.

Table 4 shows the results of the research on the content of protein and fat in 100 g of snap bean seeds in individual variants of the experiment.

By analyzing the results obtained, an increase in the protein content in 100 g of seeds of the snap bean seeds, depending on the dose of digestate used, was found—19.41 g (variant I) to 21.40 g (variant III). In the control object, the protein content in 100 g of snap bean seeds was 18.87 g.

The fat content in 100 g of the seeds of snap bean was from 0.48 g (variant I) to 0.60 g (variant III), and for the control object—0.41 g.

Table 5 presents the results of the research on the content of carbohydrates, calcium, magnesium, phosphorus and potassium in 100 g of snap bean seeds.

Analyzing the obtained research results, an increase in nutritional value was found in the seeds of the snap bean of the "Kontra" variety, depending on the dose of digestate used. The percentage increase in the content of carbohydrates in variants

Table 4 Changes in protein and fat content in 100 g of bean seeds in different variants of the experiment

Variant	Protein content in 100 g of beans [g]	Fat content in 100 g of beans [g]
I	19,41	0,48
II	20,81	0,52
III	21,40	0,60
IV	18,87	0,41

Table 5 Changes in the content of carbohydrates, calcium, magnesium, phosphorus and potassium in 100 g of bean seeds in different variants of the experiment

Variant	Carbohydrate content in 100 g beans [g]	Calcium content in 100 g beans [mg]	Magnesium content in 100 g beans [mg]	Phosphorus content in 100 g beans [mg]	Potassium content in 100 g beans [mg]
I	49,76	109,23	129,88	367,00	1341,18
II	51,37	119,56	132,00	372,88	1357,00
III	54,00	124,00	137,00	387,00	1394,00
IV	47,00	103,12	119,96	361,23	1325,16

I to III was 8.5%, calcium—13.5%, magnesium—5.5%, phosphorus—5.4%, and potassium—4.2%.

4 Discussions

In the presented article, an attempt was made to evaluate the effect of the dose size on the yield and quality of bean seeds.

In the conducted research, the following factors were adopted as determinants of seed quality: the thousand seed weight and the nutritional value in 100 g of seeds (e.g. the content of protein, fat, carbohydrate, calcium, magnesium, phosphorus and potassium). Dostatny et al. [13] and Wondołowska-Grabowska [14] classify also the thousand seed weight and nutritional value per 100 g of seeds as the basic quality characteristics of bean seeds.

In the research, the average number of pods per plant was obtained in the range of 7 to 11 pcs depending on the variant of the experiment. On the other hand, the average number of seeds per plant, depending on the experiment variant, was 25–44 pcs. Similar values of the average number of pods per plant and the average number of seeds per plant were obtained by Dostatny et al. [13]. In the studies of Dostatny et al. [13] the number of pods per plant was 11 pcs, and the number of seeds per plant was 41 pcs. Higher values of the number of pods per plant were obtained by Peeters et al. [15].

The thousand seed weight in the studies of other authors was similar. In the studies of Dostatny et al. [13] it was 175 g and it is a value lower than the obtained thousand seed weight in variant III.

In the conducted studies, the effect of the dose of digestate fertilization on the nutritional value of 100 g of bean seeds was found. Montemurro et al. [16] and Pan et al. [17] state that by using digestate, macronutrients, in particular nitrogen, phosphorus and potassium, are delivered to the soil. The results of studies by Panuccio et al. [18] indicate an alternative use of digestate instead of mineral fertilizers. In turn, Kapusta [19] found that protein is the basic ingredient in the bean seeds, which contain about 23% of it. Dostatny et al. [13] state that bean seeds are a good source of protein, starch, fiber, minerals and vitamins. Carbohydrates in beans, unlike other high-carbohydrate products, are characterized by a low glycemic index, and the high fiber content, including the soluble fraction, reduces the LDL cholesterol fraction in the blood. Regular consumption of legumes reduces the risk of cancer, type 2 diabetes and coronary heart disease. The authors provide the following nutrient content in 100 g of beans: protein 23.6 g, fat—0.8 g, carbohydrate—60.0 g, calcium—143.0 mg, magnesium—140.0 mg, phosphorus—407 0.0 mg and potassium—1406.0 mg. In the conducted research, the protein content was from 19 to 21 g, and the fat content from 0.41 g to 0.60 g. The carbohydrate content in 100 g of bean seeds was from 47 g. up to 54 g, calcium—103.12–124 mg, magnesium—119.96 to 137 mg, phosphorus—361.23–387 mg and potassium 1325.16 to 1394 mg. The obtained values are lower than in the studies other authors, however, they testify to the good suitability of the digestate for fertilization.

5 Conclusions

Changing economic conditions and social preferences cause a reorientation in farming systems. Production systems that ensure obtaining raw materials of socially acceptable quality while maintaining ecological safety are becoming more important.

Based on the research, it was found that the digestate contains large amounts of macronutrients. The content of heavy metals was below detectable by laboratory equipment.

The size of the digestate dose affects the size of the obtained yield and the thousand seed weight. In addition, the size of the digestate dose determines the nutritional content of 100 g of snap bean seeds. Of the three doses of digestate, it is recommended to use the highest one.

The digestate from biogas plants can be used as fertilizer. The use of digestate as a fertilizer brings measurable benefits in agricultural production, but it is also a product the use of which may reduce the negative effects of mineral fertilization and contribute to the development of sustainable agriculture.

References

1. Przygocka-Cyna K, Grzebisz W (2018) Biogas digestate – benefits and risks for soil fertility and crop quality – an evaluation of grain maize response. Open Chem 16(1):258–271
2. Ditl P, Nápravník J, Šulc R (2017) Chemical pre-treatment of fugate from biogas stations. Biomass Bioenergy 96:180–182
3. Rasi S, Veijanen A, Rintala J (2007) Trace compounds of biogas from different biogas production plants. Energy 32:1375–1380
4. Comparetti A, Febo P, Greco C, Orlando S (2015) Italian potential biogas and biomethane production from OFMSW. In: IV international conference on safety, health and welfare in agriculture agro-food and forestry systems, HyblaCampus, Ragusta, Sicily – Italy, 2–5 September 2015, pp 1–9
5. Czekała W, Pilarski K, Dach J, Janczak D, Szymańska M (2012) Analiza możliwości zagospodarowania pofermentu z biogazowni. Technika Rolnicza Ogrodnicza Leśna 4:13–15
6. Koszel M, Lorencowicz E (2015) Agricultural use of biogas digestate as a replacement fertilizers. Agric Agric Sci Procedia 7:119–124
7. Lijó L, González-García S, Bacenetti J, Negri M, Fiala M, Feijoo G, Moreira MT (2015) Environmental assessment of farm-scaled anaerobic co-digestion for bioenergy production. Waste Manage 41:50–59
8. Makádi M, Tomócsik A, Orosz V (2012) Digestate: a new nutrient source–review. Biogas 14:295–312
9. Comparetti A, Febo P, Greco C, Orlando S (2013) Current state and future of biogas and digestate production. Bulgarian J Agr Sci 19(1):1–14
10. Kowalczyk-Juśko A, Szymańska M (2015) Poferment nawozem dla rolnictwa. FnRRPR, Warszawa

11. Muśnicki C (1989) Charakterystyka Botaniczno-Rolnicza Rzepaku Ozimego i jego Plonowanie w Zmiennych Warunkach Siedliskowo-Agrotechnicznych. Roczniki Akademii Rolniczej w Poznaniu, Rozprawy Naukowe. Zeszyt 191, Wydawnictwo Akademii Rolniczej w Poznaniu: Poznań, Poland, pp 5–130
12. Wysocka-Czubaszek A (2019) Dynamic of nitrogen transformations in the soil fertilized with digestate from agricultural biogas plant. J Ecol Eng 20:108–117
13. Dostatny DF, Tyburski J, Żurek M (2020) Vademecum dawnych roślin uprawnych. Monografie i rozprawy naukowe IHAR – PIB 52:146–154
14. Wondołowska-Grabowska A (2001) Wpływ terminu i sposobu zbioru fasoli zwyczajnej (*Phaseolus Vulgaris* L.) na przyrost masy nasion i ich wartość użytkową. Część II. Masa 1000 nasion, energia i zdolność kiełkowania oraz wigor. Zeszyty Naukowe Akademii Rolniczej we Wrocławiu 415, Rolnictwo LXXX:117–128
15. Peeters NPA, Willick IR, Nabben RHM, Waterer DR, Verhoeven TMO, Tanino KK (2020) Effect of location on dwarf French bean (*Phaseolus vulgaris* L.) seed production and seedling vigour. Acta Agric Scandinavica Sect B Soil Plant Sci 70(3):224–232
16. Montemurro F, Ferri D, Tittarelli F, Canali S, Vitti C (2010) Anaerobic digestate and on-farm compost application: effects on lettuce (*Lactusa sativa, L.*) crop production and soil properties. Comput Sci Utilization 18:184–193
17. Pan Z, Qi G, Adriamanohiarisoamanana FJ, Yamashiro T, Iwasaki M, Nishida T, Tangtaweewipat S, Umetsu K (2018) Potential of anaerobic digestate of dairy manure in suppressing soil-borne plant disease. Anim Sci J 89:15212–21518
18. Panuccio MR, Papalia T, Attinà E, Giuffrè A, Muscolo A (2019) Use of digestate as an alternative to mineral fertilizer: effect of growth and crop quality. Arch Agron Soil Sci 65:700–711
19. Kapusta F (2012) Rośliny strączkowe źródłem białka dla ludzi i zwierząt. Nauki Inżynierskie i Technologie Eng Sci Technol 1(4):16–32

Online Pest Warning System—Support for Agriculture and Transfer from Science to Practice

Tratwal Anna, Jakubowska Magdalena, and Baran Marcin

Abstract The primary aim of integrated control and *The Polish National Action Plan for Mitigating the Risks Associated with the Use of Plant Protection Products* is to promote the general principles of integrated pest management and eliminate hazards linked to the use of plant protection products. This aim has been pursued in a number of projects with the active involvement of consultancies, research institutes, public administration bodies, and sectoral organizations. Essential support in pursuit of the objectives and measures of the National Action Plan comes, among others, from multiannual programmes carried out by research institutes and supervised by the Ministry of Agriculture, which additionally ensure knowledge transfers into practical applications. A key new plant-protection-related task performed in the multiannual programmes is to develop and launch an Online Pest Warning System (www.agrofagi.com.pl) and monitor agricultural plant pests of economic significance. The system is operated by the Institute of Plant Protection – the NRI in close cooperation with the Institute of Horticulture—NRI, Institute of Soil Science and Plant Cultivation—NRI, Plant Breeding and Acclimatization Institute—NRI and all units of the Agriculture Advisory System.

Keywords Integrated control · Pest warning · Monitoring

1 Introduction

As of 1 January 2014, by virtue of Directive 2009/128/EC of the European Parliament and of the Council establishing a framework for Community action for the sustainable use of pesticides, all professional users of pesticides in the Republic of Poland have to comply with the general principles of integrated pest management. Integrated pest management is defined as the careful consideration of all available pest control techniques and subsequent integration of appropriate measures that discourage the development of pest populations and keep pesticides and other interventions to levels

T. Anna (✉) · J. Magdalena · B. Marcin
Institute of Plant Protection – National Research Institute, Poznan, Poland
e-mail: A.Tratwal@iorpib.poznan.pl

S. Pascuzzi and F. Santoro (eds.), *Farm Machinery and Processes Management in Sustainable Agriculture*, Lecture Notes in Civil Engineering 289,
https://doi.org/10.1007/978-3-031-13090-8_22

that are economically justified and reduce or minimize risks to human health and the environment. It should be emphasized that measures and principles of integrated pest management are nothing new in Polish agriculture. For many years farmers and agricultural consultants have been using various methods to prevent the development of large populations of pests and the spread of pathogens. In Poland, the principles and methods of integrated pest management rely on interdisciplinary activities requiring the cooperation of specialists in many areas, such as entomology, phytopathology, agriculture, soil science and others [2]. Integrated pest control strongly emphasizes the achievement of yields anticipated by the farmer (free from pesticide residues, damage, symptoms of infestation by pests, etc.) with the least possible disruption to agro-ecosystems, and encourages natural pest control mechanisms. Relevant activities are required to popularize integrated pest control programmes for different crops across the country, for example training consultants to promote and supervise integrated pest control, modification of teaching curricula at all levels of education, and above all change in the approach to the protection of plants and the agricultural environment [3, 4]. Directive 2009/128/EC specifies measures to be implemented by professional users and research institutions dealing with plant protection. According to the Directive on integrated pest management, harmful organisms and diseases must be regularly monitored in the field using appropriate methods and tools. The range of tools includes systems for alerting, forecasting and early detection, based on solid scientific evidence. Moreover, for harmful organisms threshold levels defined for the region, specific areas, crops and particular climatic conditions must be taken into account before plant protection treatments.

2 Material and Methods

The Online Pest Warning System (www.agrofagi.com.pl) (Figs. 1 and 2) launched in 2016, operated by the Institute of Plant Protection—the NRI in Poznań in close cooperation with the Institute of Horticulture in Skierniewice, the Institute of Soil Science and Plant Cultivation in Puławy, is an important tool supporting the implementation of objectives and measures related to compliance with guidelines on integrated production and protection of crops. All units of the Agriculture Advisory System, the Agricultural Advisory Centre in Brwinów, Research Centre for Cultivar Testing in Słupia Wielka, the Plant Breeding and Acclimatization Institute in Radzików, and several private businesses from the agricultural sector have also joined this initiative. The most important role of the Online Pest Warning System is public sharing of results from the monitoring of harmful organisms in crops, which is carried out in about 500 locations across Poland. Crops covered by monitoring include winter wheat, winter rape, maize, potato, sugar beet and large seeded legumes. Short guides on all diseases and pests have been prepared, and they contain descriptions of methods, timing and tools useful for the process of monitoring.

The home page of the Online Pest Warning System is arranged in six modules. The first one, Alerts on Plant Pests, provides information from field monitoring,

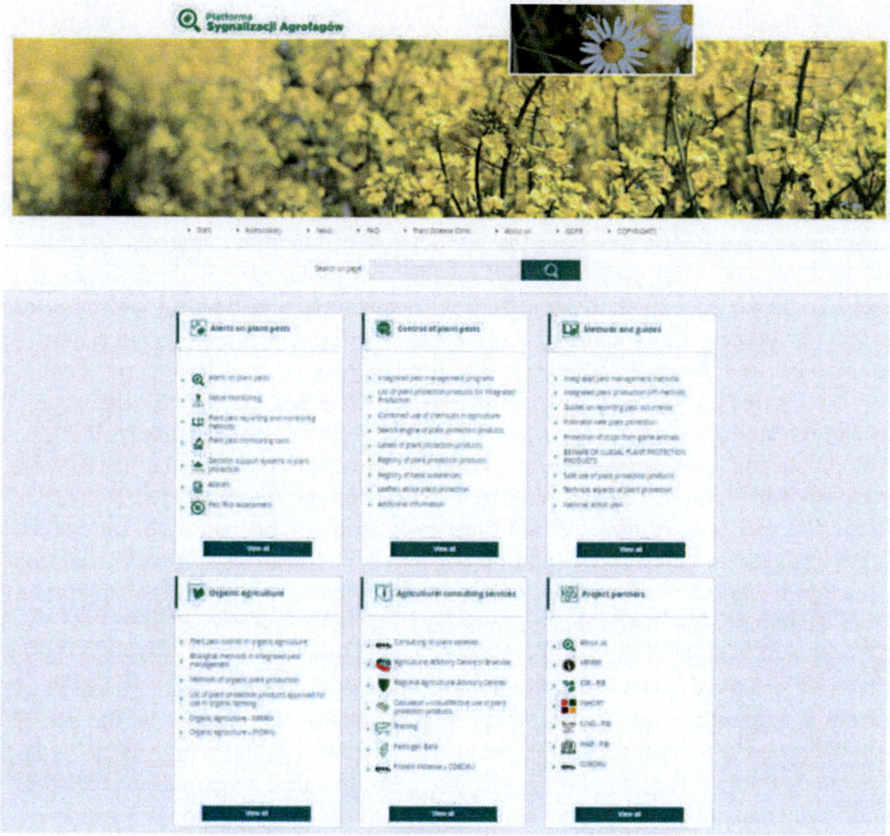

Fig. 1 On line Pest Warning System—web page

Fig. 2 On line Pest Warning System—logotype

descriptions of tools useful for pest monitoring, methodology of pest alerts and monitoring, and systems supporting the decision-making process about chemical control treatments if needed [5] Regular plant pest monitoring is a vital part of integrated pest management. This basic activity is aimed at identifying any risks to crops posed by pests with a view to determining the phytosanitary status of crops.

Plant pest monitoring helps determine the current phytosanitary status of crops for the purpose of selecting the best date to apply a protective treatment and issuing treatment alerts. By making adequate use of observational data on the emergence and rise in pest populations, it is possible to develop pest control strategies, minimize potential damage, and prevent the excessive and often unnecessary use of chemicals, as required under the Integrated Pest Management Directive [1].

An essential part of integrated pest management other than the choice of preventive measures are decisions about the application of chemical control. Integrated pest management, crop protection and production guidelines, methodologies and programs are no substitute for regular field inspections and due consideration of weather conditions and economic harm levels by farmers facing the decision to use chemical treatments. In recent years, advanced research has been carried out to provide scientific background knowledge for short-term pest emergence forecasting. Much of this research centered on determining correlations between disease and pest development and weather conditions. Its findings have been used to establish support systems to, for example, forecast pathogens breakouts and the developmental stages of pests as a way to determine the best time for treatment. The instructions produced on that basis are to help farmers and plant protection consultants identify the need for plant protection treatment based on ecological principles, with due consideration of cost–benefit analysis and climate conditions.

The second module, Control of Plant Pests, includes publications, product leaflets, posters, etc. related to Plant Protection Programmes and other important issues. Plant protection programmes are developed to help farmers make choices on the use of chemical protection products for the control of harmful organisms in different types of crops (including fruit plants, vegetables, and industrial crops). Chemical pest control is increasingly challenging and complicated because of the ongoing changes in the register of plant protection products. This module also contains a list of plant protection products, their labels, and a search engine of plant protection products. There is also a wide range of different leaflets related to the control of pests in specific groups of plants. Additional materials provide information on plant protection that satisfies third-country requirements, and plant succession following rape and cereals after tillage.

The third module, Methods and Guides, presents methods of integrated production and integrated pest management in a wide range of crops, including fruit plants, vegetables and industrial plants. Valuable information on the correct monitoring of pests is provided on this website in "Guides on reporting pest occurrence", which have been prepared for a number of crops (cereals, maize, small and large seeded legumes, vegetables, fruit plants, and others, such as hops, tobacco and ornamentals). This section also contains essential information regarding pollinator-safe plant protection, and advice and publications on the safe use and storage of plant protection products and handling their unused remains. Honey bees and other pollinating insects are an integral part of ecosystems. They play a crucial role in pollinating crops and wild plants. An estimated 78% of all plant species on Earth (including more than 200 crop species) are pollinated by insects. Therefore, insect pollinators ensure the survival of most species living on our planet. Most plants in Poland are pollinated by the honey bee (that applies to over 90% of flowers of entomophilous plants). Other flowers are pollinated by bumblebees, solitary bees, flies, butterflies, beetles and other insects. In Poland honey bees pollinate some 50 crop species, about 60 vegetable species as well as ca. 140 horticultural plant species, including 15 species of fruit trees and shrubs, and over 60 species of medicinal plants. Insects increase crop yields and improve their quality, particularly for such plant species as rapeseed (up to 30%), apple (up to 90%), gooseberry (up to 70%) and strawberry (up to 20%). However, used incorrectly, the chemical plant protection products that are vital in farming may pose risks to human and animal health and harm the environment. It is therefore essential to apply plant protection products properly to either reduce or fully eliminate such risks.

Another problem described in this section is protection of crops from game animals. The damage caused to crops by some game species in Poland has increased dramatically over the last decade becoming a serious plant protection concern. the key factors contributing to this problem are: rapid increases in the country's populations of wild boar, red deer, fallow deer, elk and beaver, climate change causing faster maturation and increased reproduction of wild boar (year-round), deer and other species, as well as increased survival rates owing to mild winters, changes in the biology and behaviour of wild boar, deer and other crop-damaging game species (resistance to odor-based and other repellents). changes in species makeup and crop acreage (maize, rape, wheat, and other cereals), reduced biodiversity in field and forest ecosystems, significant rise in crop yields (new, high-yield varieties, effective plant protection, intensive fertilization).

This section also addresses the important issue of technical aspects of plant protection. The proper use of equipment for the application of plant protection products, including its regular maintenance, is of crucial importance for both the efficacy and safety of treatments. To reduce the risk of using defective (hazardous) sprayers, a system of mandatory periodic inspections of the equipment used to apply plant protection products has been established in Poland. Such equipment is inspected by registered institutions supervised by the National Inspectorate of Plant Health and Seed Inspection.

The fourth module of the Online Pest Warning System concerns problems related to organic agriculture. Organic agriculture is the most environmentally-friendly method of farming, it is a very important sector, but requires continuous changes in order to increase its share in the total agricultural production volume. The environmental awareness of society continues to grow, which can be clearly seen in the increasing demand for organic products. Organic crops are protected mainly by means of appropriate treatments and cultivation techniques. The key measures used for that purpose are crop rotation and keeping soils biologically active and highly fertile.

This section also describes biological methods in integrated pest management. In the practice of using the biological methods, plant protection relies on three main strategies: classical control, introduction and conservation. The strategies vary in application and the biological agents they employ. Classical biological control involves the introduction of useful, predominantly non-native, biocontrol agents into the environment to control or significantly suppress the population of either native or non-native pests. Such organisms are usually obtained from the areas of their natural distribution and are released in new areas, where they are expected to establish and effectively keep the pest population down to a safe level. The biological control targeted at augmentation involves the periodic introduction of beneficial micro- and/or macro-organisms obtained from large-scale cultures into crop fields. These organisms are appropriately formulated and delivered in the form of biopesticides. Under this strategy, the biological agent should act immediately whenever possible, and is not expected to be established in the new area. This strategy is most popular in commercial crops (cultivation under cover, mushroom farms, orchards, and some field and forest crops). Conservation control relies on such landscape features as are naturally occurring or have been specially created in agricultural and forest areas to foster the populations of beneficial insects that naturally live in these habitats. The main objective of these efforts is to improve habitat quality for these organisms by diversifying the landscape, thus creating shaded places and hides, suitable wintering sites, and securing a necessary diverse food base for naturally occurring entomophages. A key to this strategy is the rational use of selective chemicals to reduce their deleterious effect on beneficial organisms.

In the fifth module, Agricultural Consulting Services, users can find links to the websites of all regional agricultural advisory centres, and the Agricultural Advisory Centre in Brwinów, where a lot of practical information for farmers is provided. This section also presents results from post-registration testing of cultivars (PDO COBORU experiments) and relevant recommendations.

The section on Agricultural Consulting Services also presents a training offer for all interested candidates. Courses are one of the best methods of acquiring knowledge, supplementing education, improving skills and raising professional qualifications. Many research and consulting institutions that deal with agricultural issues organize such courses for agricultural advisors, farmers, and for their own employees. Courses enable a direct transfer of knowledge from science to agricultural practice, sharing updated information with audiences in various regions of Poland, integration of people involved in agriculture, and above all sharing views and experience, discussions between scientists and specialists, as well as agricultural advisors and farmers.

The last module, Project Partners, presents information on the principles of website operation and funding sources. There also active logos of the project partners, and by clicking on them users are redirected to the website of a partner involved in the creation of the Online Pest Warning System.

3 Summary

Establishing the optimum dates for the chemical control of pests is a challenging task for every farmer. If a decision on pest control is made based on easily noticeable, high abundance of pests or very clear symptoms of pest feeding, the treatment is often too late, only partly effective, and not always justified from the economic point of view. For this reason farmers should carry out regular monitoring of plants in the field throughout the whole growing season, but also have knowledge of pest biology, regions of its occurrence and climatic preferences, and be able to estimate the size of pest populations, hence the current level of economic injury.

The Online Pest Warning System contains an extensive number of publications, reports and materials related to methods of integrated production and plant protection, guides on reporting pest occurrence, technical aspects of plant protection, pollinator-safe plant protection, risks associated with illegal and counterfeit plant protection products, and many others. The database of the Online Pest Warning System contains about 600 pdf files and about 40 links to other websites. Of note is the fact that materials available for all website users are based on solid and evidence-based knowledge from experimental studies.

References

1. Directive 2009/128/EC of the European Parliament and of the Council of 21 October 2009 establishing a framework for Community action to achieve the sustainable use of pesticides. OJ L 309, 24 November 2009. https://eur-lex.europa.eu/eli/dir/2009/128/oj. Accessed 1 Oct 2021
2. Gacek E, Behnke M (2006) Practical application of the biological progress in agriculture in conditions of the market economy. Bull Plant Breed Aclim Inst 240–241:83–89
3. Matyjaszczyk E (2013) Plant protection in Poland on the eye of obligatory integrated pest management implementation. Pest Manag Sci 69(9):991–995
4. Pruszyński S (2011) Integrowana ochrona roślin – wyzwanie dla rolników, służb doradczych i nauki. Zagadnienia Doradztwa Rolniczego 2:49–65
5. Tratwal A, Baran M (2018) The role of guidelines on pest monitoring and warning systems in integrated pest management. J Plant Prot Res 58(3):211–214

Selected Problems on Data Used in Precision Agriculture

Edmund Lorencowicz and Jacek Uziak

Abstract The paper describes data requirements in precision agriculture. Different types of data, specific in agriculture is discussed. The collection and use of data is looked in view of risks and threats it generates. Also, the rights of data and its administration is presented in view of the EU Code of Conduct for Agricultural Data Sharing by Contractual Agreement. The code, launched in Brussels in 2018, was introduced to foster trust in data sharing and can be considered as a milestone in regulating problems related to agricultural data. Code of Conduct explains contractual relations, making the actual contract between parties involved, the primary instrument in ensuring concord. It also provides guidance on the rights of access and use of the agricultural data.

Keywords Precision agriculture · Big data · Internet of Things · Agriculture 4.0 · Data ownership · Agricultural production chain

1 Introduction

Global population growth increases food production needs, which demands introduction of new solutions in agriculture. Some forecasts predict the increase in World population up to ca 11 billion by the year 2100—Fig. 1 [31]. However, it is not only the volume demand, the requirements regarding the food quality and environment safety are also changing.

In the last century, the efficiency of agriculture increased remarkably. In 1900, a farmer, performing chores by hand, using walking plough with horses, forking hay and milking by hand, produced enough agricultural product to feed a total of 10 people, whereas in 2020, that number was already 120 [4].

E. Lorencowicz (✉)
University of Life Sciences in Lublin, Lublin, Poland
e-mail: edmund.lorencowicz@up.lublin.pl

J. Uziak
College of Management and Enterprise, Wałbrzych, Poland

S. Pascuzzi and F. Santoro (eds.), *Farm Machinery and Processes Management in Sustainable Agriculture*, Lecture Notes in Civil Engineering 289,
https://doi.org/10.1007/978-3-031-13090-8_23

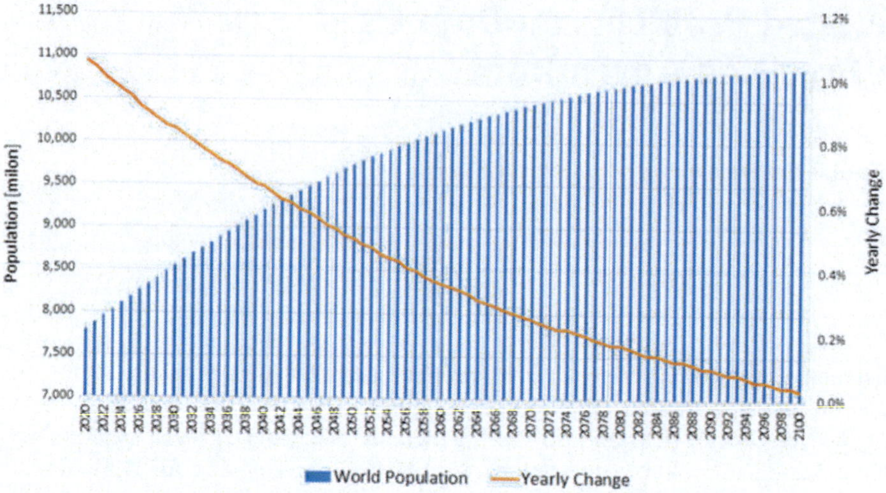

Fig. 1 World population and yearly dynamics projections to year 2100. *Source* Own study based on [31]

Rapid improved in efficiency was evident in the second half of the XX century, when during, so called 'green revolution' characterized by improved technologies in agricultural engineering, fertilization, and chemical plant protection. During that period, the global agriculture production increased and limit the world hunger.

Next step in development was related to individual treatment of a particular crop or animal. It is thanks to satellite navigation used in agricultural applications for crop and soil monitoring as well as surveying the fertility of agricultural fields to control the distribution of chemicals and fertilizers [7]. Digital techniques applied to production and management processes increased its significance and brought next steps called 'smart farming, 'digital farming' or 'agriculture 4.0' [1]. The modern digitized solutions have made its way also to developing countries [17, 29]. Despite the initial stage of digitization in agriculture, the whole agri-food sector increasingly becomes a digital industry [3, 11]. Although, it is estimated that agriculture machines and devices have reached their top parameters, it is still possible to improve their efficiency and work quality through work and management processes optimizations.

2 Objectives and Material of Study

The objective of the study is to present threats and limitations related to digitalization of agriculture. The analysis was based on the specialized literature, mainly not available to the general public. That included, beside scientific papers, also press releases, presentations and lectures performed at international exhibition for agricultural machinery and equipment Agritechnica, Hanover. Other sources included unpublished materials from the meetings of Standing Committee on Mobile Machinery of European Organisation of Agricultural, Rural and Forestry Contractors (CEETTAR) and own research [16]. A details internet query and selected research positions were also done. Since quick progress and changing conditions, the sources were limited to those from year 2017.

3 Data Requirements in Modern Agriculture

The development of integrated management systems (Fig. 2) leads to establishment of ever complicated relations between the participants of such systems and the need for the data requirements for analysis and application.

It is estimated that, the amount of data generated by farms increases every year exponentially. According to the available information it was ca 500,00 bits daily in 2020, and it is projected to reach more than 2,000,000 bits in 2030 [5].

Access to internal and external data allow, after its processing, achieved added value [14, 15].

Internal data originates from the farm, and includes information regarding type and volume of production, coordinates of the operational land and information about the equipment and agricultural processes. According to the German Mechanical Engineering Industry Association (VDMA), more than 80% of new agriculture machinery offered on the market in 2017 had Internet access, irrespective of the sector and forms of use [30]. Apart from the already common use of smartphones and tablets, automatic machine guidance with GPS, there are many new, even more sophisticated technologies, such as drone [22] or robotic systems [12]. The use of agricultural robots is bound to increase significantly spectacularly in the coming years. There are reports (Tractica and The Robot Report), which envisage imposing increase in

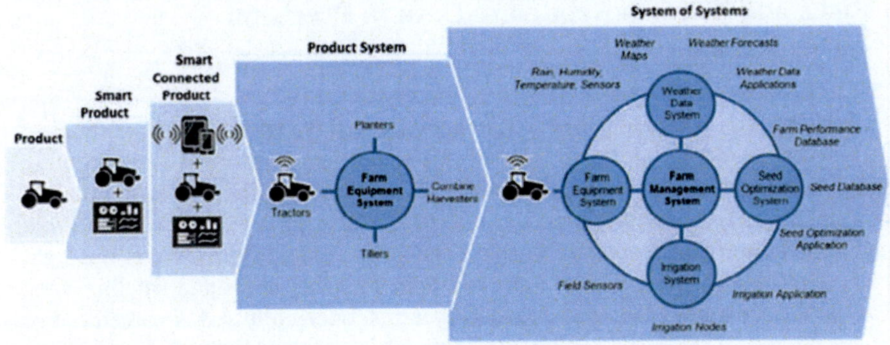

Fig. 2 Concept of expansion from 'product to product' systems to 'systems of systems' *Source* [13]

agricultural robots' shipments, from annual units of 32,000 in 2016 to 594,000 in 2024. It is also predicted that the annual revenue of that market will reach $74.1 billion in that year [10].

External data are those which are publicly available or collected information from outside of the farm, such as surveying and cadastral data, as well as satellite information regarding, for instance, weather, soil and crop conditions. Information transfer between a farmer and a provider of the production means and services improves production methods and technologies and leads to cost reduction. Transfer of the data is actually bidirectional, as the flow of information from the producers facilitates selection of the optimal technology parameters and reduction of the input costs.

There are several factors which makes automation and digitization in agriculture more difficult. Those factors are related to general issues such as disturbances in transmission and simple lack of right sensors. But also, to particular nature of agricultural material and machines, such as uncertainty and variation of inputs, lack of process knowledge, and huge variety of the machines and process execution. In comparison to typical industry the biobased industry faces challenges related to huge variation of production conditions, diversity of technologies, inevitability of working with biological materials and products, extremely reliant on external factors, such as weather [12]. However, the dynamic development of Internet of Things (IoT), in particular, improves compatibility of machines used in the technological process and their optimization [6, 26].

At the operational level, the most popular technology is automated steering and field work. The most common auto-steering used by German agriculture contractors is applied to tractors (almost 55%), combine harvesters (21%) and self-propelled forage harvesters (20%). Digital technologies are mainly used in fertilization (46%), sowing (45%), soil cultivation (39%), mowing (27%), tedding and raking (17%) and plant protection (8%) [18, 19].

4 Risks and Threats Related to Data in Precision Agriculture

Precision agriculture, due to use of IoT, industrial sensors and adapted technologies, generates different types of threats. However, also farmers themselves, are becoming aware of risks potentially generated by digitization [16, 24]. Those risks are mainly related to how and where the data is stored and used. Sharing of the sensitive data and its agglomeration from devices such as sensors on machinery (flow meters, weight sensors, and optical recognition), barn and field sensors, aerial and satellite data, as well as farm management and financial systems, once collected is passed out of farmers' control. That data security now depends on the security of the company holding it and may be used competition or even to alter commodity markets by speculating. That would effectively mean using the advantageous data from the farmers against themselves. Another possibility is the use of the data by agricultural technology providers to develop and sell new products to the actual producers of the original data. Uncertainties related to the ownership and use of the data influences farmers attitude and may affect decision making processes and, consequently, may lead to incorrect decisions and negative production results.

For years now, it has been indicated that there is need to establish general valid and accepted principles on the data exploration [2]. The analysis on data privacy and security, and its use, has been already under discussion in EU for some years. The fundamental problem is to establish range and access rights to private data received from farmers.

Whether data collected from farmer's tractors and his machines (basically private information) can be used, with no restrictions, by the producer, dealer, and machine service provider, whereas the farmer (owner of the machines) pays for the processed data, is almost a rhetoric question with no straightforward answer. Such processed data, based on information collected during machine operation, is used, among other things, in planning and control of technological processes as well as assessment of the work parameters and its quality. Data is also useful in remote steering and control of the machines, in addition to monitoring of work and energy efficiency, and profitability. Although both sides of the data equation have an interest in developing transparent property rights over agricultural production data and its use, there is a clear difference between expectations from farmers and needs of machine producers and service providers. In response to the data question a survey of German service providers indicate that data access should be given to service providers only (69% of positive answers). Only 5% of the respondents indicated that data could be freely used by the machine 'producer' and 'producer and service'. Interestingly, 21% believed nobody, apart from the farmer, could sue the data [20].

Another issue related to data rights its available to authorized dealers, repairers and independent service operators, regardless of brand and expertise, which would increase competitiveness in this sector [21].

5 Data Sharing Principles and Codes of Practice

To foster trust in data sharing, in 2018, the EU Code of Conduct for agricultural data sharing by contractual agreement was launched as a collaborative effort of farmer's cooperatives in the EU allied to Copa-Cogeca (Committee of Professional Agricultural Organisations-General Confederation of Agricultural Cooperatives) and CEJA (European Council of Young Farmers), as well as representatives of animal breeding companies and large organizations representing industries producing animal feed, fertilizers, seeds and farm machinery (like CEMA, Fertilizers Europe, CEETTAR, ECPA, EFFAB, FEFAC and ESA). As a basis, the code of conduct provides definitions for roles and processes in data sharing in agriculture, as well as support on contractual relations and guidance on re-using and sharing data. Such initiative allows for more effective use of farms' data and improves productivity as well as production profitability [8]. Future use may include plant and soil information actualization as the basis for remote mapping and forecasting of volume and quality of products.

Rapid progress of digital technologies has triggered the process of moving both data and its manipulation to the cloud (cloud computing), hence to the digital network operating in agricultural environments but covering the whole world. Modern, and in particular mobile applications, use and take advantage of cloud resources [15, 25]. The use of those resources by farmers, together with the use of software for decision making, is different in different countries. For instance, in case of Polish farmers, the most popular software was that assisting in decision making and record keeping [16].

Data used in precision agriculture are of different nature and range of use and one of the fundamental issues is to properly define it. The EU Code of Conduct defines data, in a very straightforward way, as information in all forms that is being transferred during business operation [8]. Data, according to the code, is divided into some groups [8].

1. Personal Data—leading to identification of a person.
2. Anonymized Data—information, which no longer personal as it has been stripped of any identifiable information.
3. Publicly available Data—freely used and distributed.
4. Raw Data—generated and collected without any processing.
5. Metadata—providing information on other data (e.g., date, units).
6. Primary Data—raw data processed into form identifiable by people (e.g., geological data, soil data, water data).
7. Aggregated Data—combined dataset using several sources (e.g., sensors, systems, farmers, or data platform).
8. Agricultural Data—related directly to agricultural production, including farm data and that generated in the farming processes.

Crucial agricultural data may consist of a few different types.

1. Farm data—related to farm, its operations, and management. It may consist of:
   - plant production data (agronomic data, such as yield planning, soil data, etc.);
   - control and implementation data (compliance data, which may be required by authorities);
   - animal production data (livestock data—related to the herd such as age, sex, and performance indicators, such as milk yield, animal welfare etc.).
2. Machine data—related to machine operations, which may be protected to avoid external bridge.
3. Service data—related to machine maintenance and repair.
4. Agri-supply (input) and agri-service provider data—the first one related to inputs such as fertilizers, feedstuffs, plant protection products, etc. and the second originating from agricultural services with the intention to benefit a client (e.g., farmers).

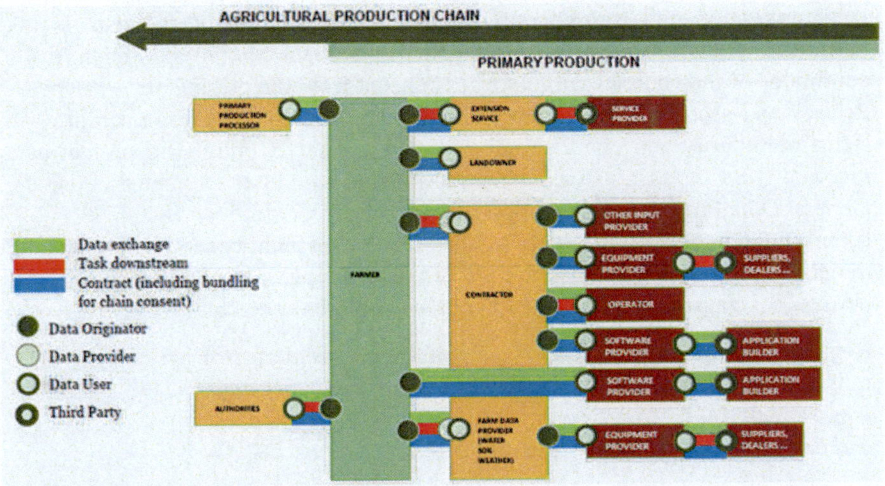

Fig. 3 Data interchange in agricultural production chain. *Source* Adopted from: [8]

Closely related to data is the problem of data users and their links and flow of information between them. The data users consist of the following main groups.

1. Contractual partners—clients (farmers), agricultural contractor and service providers.
2. Producers of machines and equipment.
3. Other contractors and service providers, which includes operators and providers of information services, software, other data users, research and development institutions, educational institutions.

Relationships between users can be considered interactive and flow of the information creates a chain for specific purposes as agreed in the contract between data originator and data users (Fig. 3).

There are also issues related to the rights of the data and its administration [8].

Data Ownership—data produced by any operator in the agricultural production chain, the operator is considered as the originator. The originator (farmer or anyone else) is recognized to benefit from and/or be compensated for the use of data created. Such benefits should be clearly stated in any contract, which can be established between the originator and other contracting parties.

- Data access, control, and portability—all manipulation of the collected agricultural data can only occur once the originator has granted an informed permission by contractual agreement. If not specified in the contract, the data originator has the right to transmit this data to another data user. Moreover, originators should be not restricted to use their data in other systems, platforms, or data storage facilities.

- Data protection and transparency—unauthorized sharing of the data is not permitted. Data users under contractual agreement should not change the conditions of the contract without prior consent from the originator.
- Privacy and security—the contract should clearly define the responsibilities in terms of security and confidentiality. The data user is obliged to protect data received from the originator. Any personal data should be protected under the General Data Protection Regulation (GDPR).
- Liability and intellectual property rights—the term of liability should be dealt with in detail by contract agreements. Furthermore, the intellectual property rights of parties in the agricultural production chain should be protected.

Regarding the principles of the legal contract agreements for agricultural data, there is a list of questions to consider when using a product or service that captures or uses agricultural data [8].

Contract check list for agricultural data.

- Is there an agreement/contract in place?
- What obligations are there? What warranties and indemnities are on each party?
- What data is collected?
- Who owns/controls access to the data?
- What services are delivered?
- Will my data be used for other goals than providing me, data originator (e.g., for instance farmer) a service? Is it clear what these are? Can I agree/disagree? What are/is the benefits/value for me (data originator)?
- Is data shared with other parties? What rules do the external parties adhere to? Can I agree/disagree with sharing data with other parties?
- Can the service provider change the agreements unilaterally?
- What happens when the service provider changes ownership?
- Can I retrieve my dataset from the system in a usable format?
- Will I be updated on security breaches?
- Can I opt out of the service and have my data deleted from the system?
- Is there a contact point to assist me with my questions?
- Do you have a need for insurance?
- What are the terms of confidentiality?

Answering the above questions should provide clear guidelines for accurate and legal use of agricultural data.

6 Conclusions

Precision agriculture is a scientific, but practical, farming concept of improving agricultural production and assisting management decisions using high technology sensor and analysis tools. Using digital techniques to monitor and optimize agricultural processes, it leads to increase production, reduce labor time, and ensure the effective management of fertilizers and irrigation processes. Applying data from multiple sources allows to improve crop yields, its quality and to increase the cost-effectiveness of management strategies including fertilizer inputs, irrigation management, and pesticide application, supporting the concept of sustainable agriculture.

However, the use of multiple sources generates a huge amount of data, with certainty that this quantity is to rapidly increase in the future. Need for information (data) sharing in agri-food sector has always occurred and has been crucial for all involved. It started by informal chatting in the local shops or conversations in the markets/agricultural fairs, where experiences were shared and information regarding seeds, new plants, new machines, etc. were exchanged. Due to the growing complexity of the agricultural system and increasing the length of supply chains, sharing of information and data between stakeholders has become both more difficult and more important. It is even more crucial in precision agriculture, and, in that respect, it becomes a valued commodity.

Sharing data by all stakeholders in the agricultural production chain creates challenges. The first, quite obvious, is a 'technical' challenge of small and large actors in the chain using different ICT systems for undertaking various tasks including accounting, invoicing etc. These systems are not compatible between each other, and the requisite infrastructure has been until now missing.

The second challenge is related to issues around data sharing for either privacy reasons, business confidentiality or suspicions concerning of their use. The ownership of data, its administration, transfer, authorization of use, all of those are at the contention, with the farmer remaining in the heart of the problem. An attempt to regulate those issue was made when the EU Code of Conduct for Agricultural Data Sharing by Contractual Agreement launched in 2018. The Code, although not binding, lays out explicitly contractual data sharing. It makes the contract between parties involved, the primary instrument in ensuring concord. Apart from signifying the contract in data handling, the document provides guidance on the rights of access and use of the agricultural data.

References

1. Albiero D, de Paulo RL, Felix Junior JC, da Silva Gomes Santos J, Melo RP (2020) Agriculture 4.0: a terminological introduction. Rev Cienc Agron 51:1–8
2. Beecher JI (2017) Deal with farm data. Future Farming 4:38–39
3. Damianos D, Cambou P (2021) Data drives Agriculture 4.0. https://www.eetimes.eu/data-drives-agriculture-4-0/. Accessed 18 Jan 2022
4. De Clercq M, Vats A, Biel A (2018) Agriculture 4.0: the future of farming technology. World Government Summit, 25. https://www.worldgovernmentsummit.org. Accessed 20 Jan 2022
5. Dealing with the Big Data Decade (2018) https://space-of-innovation.com/dealing-with-the-big-data-decade/. Accessed 20 Apr 2022
6. Wandel D (2017) Big data, internet of things, DLG Pressemitteilung 8, Agritechnica 2017, Hanower, p 4 (typescript)
7. Ekielski A, Wesołowski K (2019) Systemy agrotroniczne. PIGMiUR, Toruń

8. EU Code of conduct on agricultural data sharing by contractual agreement (2018). https://www.scc-gmbh.de/images/scc/Downloads/EU_Code_of_conduct_on_agricultural_data.pdf. Accessed 01 May 2022
9. Daheim C, Poppe K, Schrijver R (2019) Precision agriculture and the future of farming in Europe: scientific foresight study. European Parliament, Directorate General for Parliamentary Research Services. https://data.europa.eu/doi/10.2861/020809. Accessed 12 Apr 2022
10. Giles F (2017) Agricultural robots no longer science fiction. https://www.growingproduce.com/vegetables/agricultural-robots-no-longer-science-fiction/. Accessed 12 May 2022
11. Griepentrog HW (2017) Green future – smart technology: Chancen und Herausforderungen der Digitalisierung in der Landwirtschaft. DLG Pressemitteilung 3, Agritechnica 2017, Hanower, p 3 (typescript)
12. Herlitzius Th, Krzywinski J, Rückelshausen A, Weidig K (2017) Mobile Cyber Physical System concept for controlled agricultural environments. Agritechnica 2017, Hanower (typescript)
13. Bonneau V, Copigneaux B, Probst L, Pedersen B (2017) Industry 4.0 in agriculture: focus on IoT aspects. Digital transformation monitor EU, p 6
14. Jones JW, Antle JM, Basso B, Boote KJ, Conant RT, Foster I, Godfray HChJ, Herrero M, Howitt RE, Janssen S, Keating BA, Munoz-Carpena R, Porter ChH, Rosenzweig C, Wheeler TR (2017) Towards a new generation of agricultural system data, models, and knowledge products: state of agricultural system science. Agric Syst 155:269–288
15. Kamilaris A, Kartakoullis A, Prenafeta-Boldú FX (2017) A review on the practice of big data analysis in agriculture. Comput Electron Agric 143:23–37
16. Lorencowicz E, Uziak J (2019) Precision agriculture effect on farmers' information needs. In: Lorencowicz E, Uziak J, Huyghebaert B (eds) Farm machinery and processes management in sustainable agriculture. X international scientific symposium FMPMSA 2019. IN Spatium, Lublin, pp 377–382. https://doi.org/10.24326/fmpmsa.2019.1. ISBN 978-83-66017-74-0
17. Miles Ch. (2019) The combine will tell the truth: on precision agriculture and algorithmic rationality. Big Data Society 1–12
18. Noordhof J (2017) Ja, aber….. Lohnunternehmen 11:22–25
19. Noordhof J (2017) LU trend-report: datenmanagement 2017. Lohnunternehmen 3
20. Noordhof J (2017) Unser Ziel ist barrierfreier Datenaustausch. Lohnunternehmen 11:30–32
21. Open access to Repair and Maintenance Information (RMI) (2017) CLIMMAR Mag 7
22. Riley J (2017) How to fly drones within the law. Future Farming 4:32–34
23. Ruckelshausen A (2017) Green future – smart technology: field robotics in crop production – vision possible? Agritechnica 2017, Hanower (typescript)
24. Schaffner A (2017) Digitization: top value for farmers. Agrifuture 4:24–25
25. Schick M, Mersch F (2021) Trends in tractors 2021/22. Agritechnica 2021, trend reports for the media. T1, p 3. https://www.agritechnica.com/en/press/downloads. Accessed 30 Nov 2012
26. Symeonaki E, Arvanitis K, Piromalis D (2020) A context_aware middleware cloud approach for integrated precision farming facilities into the IoT toward Agriculture 4.0. Appl Sci 10:813. https://doi.org/10.3390/app10030813
27. The Future of Food and Farming. Communication from the Commission to the European Parliament, the Council, the European Economic and Social Committee and the Committee of the Regions, Brussels, 29 November 2017. COM (2017) 713, p 27
28. Trendov NM, Varas S, Zeng M (2019) Digital technologies in agriculture and rural area. Status report. FAO UN, Rome
29. Valle SS, Kienzle J (2020) Agriculture 4.0-Agricultural robotics and automated equipment for sustainable crop production. In: Integrated crop management, vol. 24. FAO, Rome, p 25
30. VDMA Landtechnik (2017) Branchenwachstum auf globalen Schlüsselmärkten verstetig sich. Pressinformation, Agritechnica, Hanower, p 3 (typescript). https://www.vdma.org/v2viewer/-/v2article/render/21845730. Accessed 01 May 2022
31. World Population Projections. Worldometer. https://www.worldometers.info/world-population/world-population-projections/. Accessed 27 Apr 2022

Estimation of Workload Level of Dynamic and Static Warehouse Assemblers in Logistic and Ergonomic Aspect

Mateusz Gawliński and Katarzyna Łyp-Wrońska

Abstract In modern logistic engineering, it is important not only to deliver the goods to the right place and time, but also to deliver, for example, details to the assembly line in the correct assembly position. This is of particular importance for companies that implement World Class Manufacturing—it is very important for process management in sustainable agriculture. The object of the research is a company, which produced farm machinery. This subject in particular concerns the pillar of security and logistics. To determine the right component configuration, many ergonomic tools can be used. Acquired skills are very often used to correct already existing workstations. In addition, the newly emerging legal regulations related to safety force the employer to apply better technical and organizational solutions. The consequence of working in such bad ergonomic conditions is a high risk of musculoskeletal disorders. This study contains a review of the effects of using two methods (OWAS and Lehmanns's tabular –chronometric method) from the field of ergonomics in modern logistic engineering in a company producing agricultural machines, cars, etc. The concepts discussed are used to determine the degree of static and dynamic severity of an employee. The results of the analysis may be useful in improving the logistics of the manufacturing process in the ergonomic aspect.

Keywords Ergonomics · OWAS · Lehmann's tabular · Chronometric method · World Class Manufacturing · Working postures

M. Gawliński—Freelance researcher.

M. Gawliński
Płock, Poland

K. Łyp-Wrońska (✉)
AGH University of Science and Technology, Krakow, Poland
e-mail: klyp@agh.edu.pl

S. Pascuzzi and F. Santoro (eds.), *Farm Machinery and Processes Management in Sustainable Agriculture*, Lecture Notes in Civil Engineering 289,
https://doi.org/10.1007/978-3-031-13090-8_24

1 Introduction

Sustainable logistics, and supply chain are one of the most important global research areas [11]. In modern factory, a specially which has approach World Class Manufacturing, is important not only to deliver the goods to the right place and time, but also to carry on the workers. The object of research is a factory, which produced a tractors, combine, cars etc. The question is how to deliver, for example, details to the assembly line in the correct assembly position. To determine the right component configuration, many ergonomic tools can be used. Acquired skills are very often used to correct already existing workstations. In addition, the newly emerging legal regulations related to safety force the employer to apply better technical and organizational solutions. The consequence of working in such bad ergonomic conditions is a high risk of musculoskeletal disorders. Many health problems among employees are caused by poor body position while working, monotypicity of movements and involvement of only one part of muscles [7, 8]. The consequences is dismissal from work and illness [1–3]

To diagnose the problem, the following in the factory is required: knowledge and ergonomic skills, motivation to improve it and knowledge of design methods.

Ergonomic research consists of observing the position and people employed there. Based on the acquired knowledge, the appropriate method of analysis is selected [9, 10]. The evaluation criteria depend on the intensity of work and the positions of the body being taken. It takes several days to diagnose the chosen workspace, the results should be averaged and applied to specially prepared tables of one of the methods. After estimating the severity of work, the changes have to be implemented in the workplace in order to improve ergonomic conditions.

The specialists in the field of ergonomic prophylaxis deal with the analysis and counteracting adverse consequences of burdens occurring in the workplace. Specialists in the field of logistic engineering also perform this task. The examination and diagnosis of the problem should be carried out at the previously selected workstation, including the relevant aspects affecting the functioning of the entire enterprise.

In the article, two methods of analysis have been used: OWAS, as well as tabular—chronometric of Lehmann. The effect of calculations is the degree of static and dynamic load on the workstation. OWAS analysis is the method used very often [4–6].

The objectives of this research is to determine the work posture risk level and suggest recommendation of improvement. The study consists in listing all the activities performer.

2 Methods

2.1 *OWAS*

OWAS method—developed by the Finnish Institute for Labor Protection in the 1970s. Its advantage is transparency, simplicity and the ability to recognize the complexity of the problem but also corrective actions. Thanks to OWAS, you can choose which body positions are the most burdened by the employee and reduce the risk of developing musculo-skeletal disorders resulting from working in poor ergonomic conditions. The entire study can be carried out in the form of observations at the workplace. The position of the human body and the external weight is taken into account, is the whole method emphasis is on identifying and estimating the risk of discomfort caused by static load.

Four factors are taken into account during the research:

- back positions,
- shoulder positions,
- leg work,
- the severity of the external leg load.

The last parameter has three compartments: mass below 10 kg, from 10 to 20 kg, over 20 kg. The different positions of different parts of the body together with the external load were classified in four variants of the assessment [2, 7].

Method of testing:

- First of all, a four-digit employee at-work position code should be determined, which uniquely defines the position of the arms, back, leg work, as well as the amount of external load.
- In the second stage, you must put on the table a combination of a four-digit code to which only one category of static load per station is assigned.
- In the last part one should carry out the analysis of the position of the body weighted during duty performance, taking into account the posture type (forced or unforced), the previously assigned category and the working time in a given position.

2.2 *Lehmann's Tabular-Chronometric Method*

Lehmann's tabular-chronometric method—developed to assess the employee's energy expenditure. It is quite easy to apply and therefore often used in factories. In comparison with other methods, the error in the assessment is small and acceptable. The test is carried out in the form of observations at the work station by analyzing all activities performed during one shift (8 h of work). The exact duration of the operation is defined in tables prepared specifically for this. The ambient temperature should also be taken into account, e.g. work performed in extremely low or high temperatures has a higher energy expenditure by about 12% compared to performing duties in thermal comfort. The amount of energy that each person has does not determine the resulting fatigue while performing tasks.

The factors illustrating fatigue are physical capacity and exercise tolerance. In addition, every living creature should provide energy to its body, depending on how much it needs at the moment to function efficiently.

It is assumed that the amount of energy consumed depends on:

- work intensity,
- operation duration,
- size of muscle groups involved in the activity,
- type of work (dynamic or static),
- workstation structure,
- environmental conditions.

24-h energy expenditure for a young man, average height (175 cm) and average body weight (75 kg) is 10,000 kJ. And the total expenditure on labor is:

- up to around 4,200 kJ/8 h—light labor,
- from 4200 to 6300 kJ/8 h—medium-heavy labor,
- from 6300 kJ to 8,400 kJ/8 h—heavy labor,
- over 8,400 kJ/8 h—heaviest labor.

After a properly performed test, the severity of the given work can be estimated in the factory conditions. When performing the timing, you can use two methods:

- gasometric—generalization of the entire work cycle,
- estimated—separation of all performed activities during 8 h of work [1].

During the observation it is worth conducting an interview with employees in order to obtain information on which position improvements considering the ergonomics and safety they would expect.

3 Results and Analysis

3.1 Characteristics of the Position

In the ergonomic aspect, the most common within enterprises are the assembly positions. The job of a fitter is a physical job, mainly with the use of manual skills. This specialization is characterized by a large monotypicity of movements, forced body positions and the involvement of only one part of muscles. These aspects negatively affect the musculoskeletal system and cause mental and physical fatigue. At the beginning of the observation, it is necessary to determine the number of employees in the examined position, working time, rest breaks and the frequency of line travel.

The employees' duties most often include:

- assembly of individual elements while maintaining technological principles,
- compliance with quality standards,
- compliance with health and safety standards,
- sorting the tools used and preparing the station for the assembly of the next machine,
- carrying out the assessment of the work carried out,
- confirmation of a correctly completed assembly by signing on a specially prepared form

One of the points of ergonomic diagnosis is the position of the body. Most often during the work, the assemblers assume a standing position with the hands raised above the shoulder and the standing position inclined. This positioning of the figure influences many health ailments.

Workers of the assembly line are exposed to loads:

- static associated with the assumed body position during work,
- dynamic associated with energy expenditure.

3.2 Ergonomic Diagnosis in the Aspect of Logistic Engineering

By means of ergonomic diagnosis, it is easier to indicate directions of changes or technical solutions that can be introduced to improve working conditions at the workstation. The whole research concerns the connection between a human being and a technical object.

Diagnosing the problem should be based on:

- developing a table with the activities performed by the assemblers on the assembly line (photograph of a working day),
- measuring the time of individual operations,
- OWAS method calculations,
- Lehmann's tabular-chronometric method calc
- analysis and consultaion of results.

3.3 Photograph of a Working day Assembler

The photograph of the working day consists in observing the examined position. Using a stopwatch, the duration of one operation is measured. Each activity should

Table 1 A photo of the working day, *Source* own research based on [2]

lp	Activity description	Time length of singular activity [min]	Number of repetitions in 8 h	Total time for 8 h [min]
1	Coming to work	5	1	5
2	Changing into work clothes	5	1	5
3	Preparing workstation	8	1	8
4	Organizing accessories	3	7	21
5	Bringing and putting back instruments	1	21	21
6	Drilling	4	7	28
7–18	…	…	…	…
19	Bathing and changing into personal clothing	10	1	10
Total				480

be noted in a specifically prepared table. The test should last about fifteen business days. Table 1 show working day.

The results of the analysis should be averaged and a workstation chronometric report should be made.

A photograph of the working day of the position is necessary for:

- estimation of level of work organization,
- calculation of energy expenditure,
- rating of static load.

3.4 *Analysis of the Lehmann's Tabular-Chronomatric Method*

The position of the body and the unit energy expenditure of the activity are added to the table with the description and duration of the individual operation. The analysis should take into account the number of muscle groups participating in the activity, with greater number making longer activity acceptable. In addition, at the time of one operation for more than 3 h without changing the position of the figure, the Lehmann method cannot be used to estimate the nuisance of work. If a certain time is exceeded in a particular position of the body, the work takes a static character and ceases to have a dynamic one. The method of lifting and moving heavy elements used for assembling the agricultural press also has an influence on the final result. In

Table 2 Energy expenditure for individual body positions [2]

Body position	Energy expenditure	
	kJ/min	kcal/min
Sitting	1,25	0,3
Kneeling	2,10	0,5
Crouching	2,10	0,5
Standing	2,51	0,6
Standing inclined	3,35	0,8
Walking	10,89	2,6

Table 3 Energy expenditure for a particular muscle load [2]

Muscle load range	Energy expenditure (depending on the intensity of the muscle load)	
	kJ/min	kcal/min
Work of the fingers and forearm	2,10–4,20	0,3
Work of one arm	4,20–8,40	0,5
Work of both arms	8,40–12,6	0,5
Whole body work (muscles, limbs, torso)	12,6–41,9	0,6

the case of exceeding the endurance level of an employee, the evaluation criterion should be increased to a higher grade. The test position is in the assembly line, which results in a large monotypicity of movements. Subsequent systematically repeated activities by fitters cause greater fatigue and reduced concentration.

Based on the photograph of the working day, the body position with the range of muscle load and the unit energy expenditure of the operation a simplified tabular chronometric of the workstation should be developed. Table 2. shown energy expenditure for individual body positions, and Table 3 for particular muscle load.

The most frequently accepted by employees is the position:

walking, standing, standing inclined.

The biggest range of muscle strain is the work of both arms.

3.5 Calculations

Equation (1) defines Energy expenditure for a separate operation over time one work shift and Eq. (2) defines total energy expenditure.

Duration of operation [min] [A].

Organization of work at the workplace:

	Time [min]
coming to work	5
changing into work clothing	5
bathing and changing into personal clothing	10
SUM	20

Remaining of the calculation done by the scheme: [A].

Energy expenditure for a separate operation over time one work shift (8 h) [kJ] [B]

$$q = t \cdot (a + b) \quad [\mathrm{kJ}] \tag{1}$$

where:

q—energy expenditure for a separate operation during one shift, t—duration of operation [min],
a—energy expenditure of the selected body position [kJ/min],
b—energy expenditure of the selected range of muscle load [kJ/min].

Organization of work at a workstation:

$$20 \cdot (10.89 + 8.40) = 385.80\,\mathrm{kJ}$$

Remaining part of the calculation is performed by the scheme: [B].

Total energy expenditure as a fitter during one shift (8 h) [kJ]

$$Q_c = q_1 + q_2 + \cdots + q_n \quad [\mathrm{kJ}] \tag{2}$$

where: Q_c—total energy expenditure per position during one work shift, q_1, q_2, q_n—energy expenditure for a separate operation [kJ].

$$385.80 + 327.31 + 3619.00 + 613.20 + 98.01 + 37.80 = 5081.12\,\mathrm{kJ}$$

After analyzing the result, you can classify your duties as medium-heavy labor.

3.6 OWAS Method Analysis

Estimation of work nuisance is done though the photography of the working day, the assumed position of the body and its duration in a given position. The test consists in careful observation of the duties performed by the installer over several days. The results of the analysis should be averaged. Then we select one position of the body that has the greatest negative impact on the employee's musculoskeletal system. We use the isolated attitude to determine the level of static inconvenience. In addition, the external load has an impact on the final grade. Based on the combination of positions of individual parts of the body, we assign the position code, which is assigned to the appropriate grade of the position assessment. All indicators needed for reading can be found in tables designed by the Finnish Institute for Labor Protection.

3.7 *Body Position of the Assembler*

During tests, it can often be observed that the assemblers will assume these body positions: standing, standing inclined, sitting, walking. Table 4 shows list of the position this worker. During the analysis of the problem, the following should be taken into account: torso inclination angle and the frequency of lifting hands above the shoulder joint. Frequent change of body position is forced by the assembly of elements at different heights.

The graph (Fig. 1) shows the percentage share of a particular body position during one work shift. It can be seen that the assembler usually took an inclined standing position.

Table 4 List of positions taken by the assembler for 8 h working shift

LP	Activity description	Body position	Total time for 8 h [min]
1	Coming to work	Walking	5
2	Changing into work clothing	Walking	5
		Both arm work	
3	Preparing the workstation	Standing	8
		Both arm work	
4	Organising accessories	Standing	21
		Forearm work	
5	Bringing and putting back instruments	Walking	21
		Whole body work	
6	Drilling	Standing inclined	28
		Both arm work	
7	Installing side plates	Standing inclined	21
		Both arm work	
8	...	...	...
		...	
17	Assessing of completed works and signature on the form	Walking	21
18	Bathing and changing into personal clothing	Standing	10

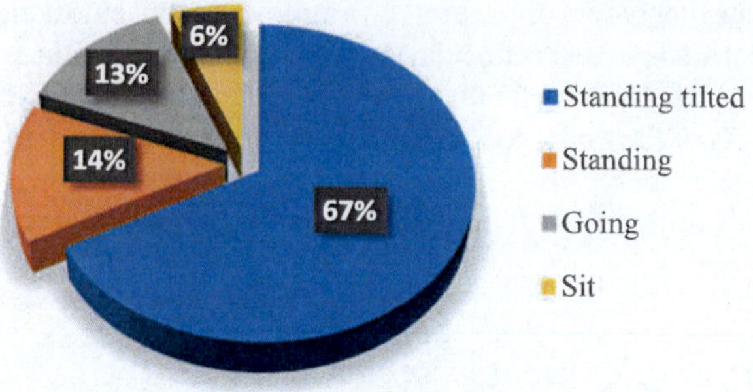

Fig. 1 Percentage share of a particular item for 8 h working shift

Using the tables developed by the Finnish Institute for Labor Protection, we can assign corresponding code for standing position inclined:

back bent forward		standing position with straight legs	
2	**1**	**2**	**1**
	both forearms below the shoulder joint		external load up to 10 kg

Then the position code has to be printed on the diagram and read the assigned category 2. The selected class does not require immediate position changes, but the employee's comfort should be improved in a short time. Frequent position of the figure in an upright standing position may have a negative effect on the musculoskeletal system.

After the interpretation of the final result, you can estimate static load on the medium level.

4 Conclusion

In the market economy, all enterprises strive to optimize production processes. One of the most important factors affecting production is timely deliveries. Many factories have dealt with problems related to the place and time of supply. As a result, employees of modern solutions in logistic engineering inside the factories started to consider the aspects of the quantity and location of the goods with greater attention. Ergonomics tools have been used for this purpose. I the article was used two methods: OWAS method because as a result of well-developed results you can get a lot of relevant information, eg the level of discomfort caused by static load.

Lehmann's tabular-chronometric method to assess energy expenditure. Thanks to accurate conclusions, it is possible to estimate the level of work severity.

On the basis of all written activities performed by employees, their duration, assumed body position and unit energy expenditure of a particular operation, the level of dynamic and static load was estimated. The energy expenditure of a fitter per one shift (8 h) is 5081 kJ. After analyzing the result, it is possible to assume the work load of the assembler at moderate level. The most heavy position is the standing position inclined. The employee stays in this position as much as 67% of the time of one work shift. After assigning the body position code, the category 2 can be read as a determinant of the static load on the medium level [5].

Based on analysis, it is possible to change, for example, the place of delivering the goods or rearranging the deployment of details on the trolley. It is worth noting that others researchers [7] extend these studies. Studies have shown that an isolated assessment of the working posture ignores the effects of interactions. Taking into account the effects of interactions leads to a greater accuracy of risk assessment results quantifying exposure to workload. It is very important for process management in the factory, which produced farm machinery.

References

1. Górska E (2007) Ergonomia: Projektowanie, Diagnoza, Eksperymenty, Oficyna Wydawnicza Politechniki Warszawskiej, Warszawa
2. Horst W (2011) Ergonomia z elementami bezpieczeństwa i ochrony zdrowia w pracy, Wyd. Politechniki Poznańskiej, Poznań
3. Koradecka D (1997) Bezpieczeństwo pracy i ergonomia tom 1 i 2, Centralny Instytut Ochrony Pracy, Warszawa
4. Aquino AJS, Batucal RB, Pangilinan ED (2021) Redesign of supermarket check-out counter using Ovako working posture analysis system (OWAS) and rapid entire body assessment (REBA). In: Proceedings of the 11th annual international conference on industrial engineering and operations management Singapore, 7–11 March
5. Wahyudi MA, Dania WAP, Silalahi RLR (2015) Work posture analysis of manual material handling using OWAS method. Agric Agric Sci Procedia 3:195–199
6. Fitri M, Laila W (2017) Kajian perbaikan postur kerja dengan metode OWAS (ovako working posture analysis system) (Studi Kasus di Pabrik Roti Cimpago Putih), Sekolah Tinggi Teknologi Industri Padang
7. Hellig T, Mertens A, Brandl C (2018) The interaction effect of working postures on muscle activity and subjective discomfort during static working postures and its correlation with OWAS. Int J Ind Ergon 68:25–33
8. Mahachandra M, Prastawa H, Susilo DY (2018) Working posture analysis of sweet whey powder handling at CV Cita Nasional warehouse using OVAKO Working Posture Analysis (OWAS). SHS Web Conf 49:02013
9. Seokhee Na DK, Chung MK (2020) Comparison of the Ovako working posture analysis system, rapid upper limb assessment, and rapid entire body assessment based on the maximum holding times. Int J Ind Ergon 77
10. Ming-Hsu W, Yi-Lang C, Wen-Ko C (2019) Using the OVAKO working posture analysis system in cleaning occupations. J Work 64(3):613–621
11. Lazar S, Klimecka-Tatar D, Obrecht M (2021) Sustainability, orientation and focus in logistics, and supply chains. Sustainability 13:3280. https://doi.org/10.3390/su13063280

Monitoring Agrometeorological Factors and Weight of the Bee Hive During Black Locust (*Robinia pseudoacacia*) Flowering

Atanas Z. Atanasov, Ivaylo S. Hristakov, Milan Koszel, and Artur Przywara

Abstract The study aimed at monitoring the influence of agrometeorological factors on the flow of nectar in the bee hive during the flowering of black locust (*Robinia pseudoacacia*). The study was conducted in 2021 growing season in the Northeaster Bulgaria in village Brestovica. Geographical location of the experimental region is 43°32′4.02″N and 25°45′14.10″E at an altitude of 223 m. The experimental apiary consisted of 122 bee colonies housed in Dadant-Blatt hives. The bees are of the species (*Apis mellifera macedonica*). During the study period of external factors are monitored, as outside air temperature (T_{air}), °C and air humidity (H_{air}) %, atmospheric pressure (A_p) hPa, soil moisture (S_m), % at 20 cm depth, soil temperature (S_t), °C at 10 cm depth, solar radiation (S_r) w m^{-2}, wind speed (W_s), m s^{-1}, precipitated rainfall (R), l m^{-2}. Also the air temperature inside (T_{in}) the hive was measured. The nectar flow is determined based on the changes in hive weight (Y). The real-time data were available via a web-based application Meteobot®.The results show that the 99% from the alteration of the Y is due to the factors days, time for reporting the indicators, H_{air}, T_{air}, S_m and $D_{p.}$ Changes in the weight of the hive due to the nectar secretion Y is highest at the outside air temperature $T_{air} \approx 27$ °C, air humidity $H_{air} \approx 48\%$ and soil moisture $S_m \approx 79\%$. It is insignificant the influence of the factors S_t; S_r; W_s; A_p and $T_{in.}$

Keywords Precision · Agrometeorological factors · Honey bee · Apiary · Robinia pseudoacacia · Productive potential

A. Z. Atanasov (✉) · I. S. Hristakov
Department of Agricultural Machinery, University of Ruse Angel Kanchev, Ruse, Bulgaria
e-mail: aatanasov@uni-ruse.bg

M. Koszel · A. Przywara
Department of Machinery Exploitation and Management of Production Processes, University of Life Sciences in Lublin, Lublin, Poland

S. Pascuzzi and F. Santoro (eds.), *Farm Machinery and Processes Management in Sustainable Agriculture*, Lecture Notes in Civil Engineering 289,
https://doi.org/10.1007/978-3-031-13090-8_25

1 Introduction

Black locust (*Robinia pseudoacacia*). is one of the most common invasive tree species in Europe [8]. It is widespread in Central European countries, such as Austria, Czech Republic, Germany, Hungary, Liechtenstein, Poland, Slovakia and others, where it displaces local tree species [12]. It occurs in countries with humid and cool climates, as well as in countries with dry and hot climates. In Bulgaria it occupies about 21, 2% of the total forest fund with a predominant age of trees from 11 to 15 years [3]. For the flat hilly areas of North-eastern Bulgaria, black locust is one of the most important main honeybees forage species in May. For some areas it is the first important main pasture for bees, and for others it is the second most important after oil rapeseed.

Black locust is a plant with high nectar secretion potential with high sugar content. According to some publications [4] for flowering duration of 5.5 days produces 1.6–3.7 mg of nectar per flower over 24 h with a sugar concentration of 34–67%. Some researchers [2, 13] prove that for Hungary and Poland the capacity to generate honey changes with the tree age. For the West Macedonia region [9] was found that the productive potential of trees aged 10 and over prevailing over 90% in an area is 112.6 nectar ha^{-1}, and for trees 6–10 years old, the productive potential is 86.8 nectar ha^{-1}. According to [5], the nectar content varies depending on the condition of the flowers (visited or unvisited). The same publication proves that bees are able to detect these differences, perform active choices preferentially selecting already-visited flowers: lower in nectar content but easier to manipulate. Although it has high productive potential, the yield of black locust honey for the flat hilly regions of Bulgaria varies from year to year. Nectar flow is very dependent on local weather conditions during some years the flowers yield little or no nectar. Even when the flow is good, the flowering period is short between 7 and 14 days, which requires the bee colonies to be optimally prepared. The study of climatic factors influencing foraging activity during the acacia flowering period is of great importance for obtaining higher honey yield. In modern beekeeping with the advent of information technology provides individual monitoring of biological units with the main purpose of controlling their development and productive potential. Many publications [6, 10, 11, 14] offer solutions for monitoring the environment and microclimate in the bee colony, through precise measurement and monitoring systems. Another researcher [7] using data on honeybee colony weights provided by electronic scales, investigated the effect of the immediate landscape on colony productivity.

Other researchers [1] have proposed an interesting solution for monitoring and analysing black locust flowering with unmanned aerial vehicles (UAVs). Unfortunately, the climatic diversity of the local regions and the environment diversity make these studies inapplicable for the individual regions in Bulgaria, which requires local experimental research.

The aim of our study is monitoring the influence of local agrometeorological factors on the flow of nectar in the bee hive during the flowering of black locust.

2 Materials and Methods

The study was conducted in 2021 growing season in the Northeaster Bulgaria in village Brestovica. Geographical location of the experimental region is 43°32′4.02″N and 25°45′14.10″E at an altitude of 223 m. In the region of great importance honey source plant for honeybee is black locust (*Robinia pseudoacacia*). The experimental apiary in Brestovica consisted of 122 bee colonies housed in Dadant-Blatt hives. The bees are of the species (*Apis mellifera macedonica).* Their strength was appropriate to the corresponding strength for successful pollination. There are in-colony factors and out-colony factors that can impact foraging activity and carrying capacity of honeybee. In our study a precision apiculture system (PAS) the measurement of the microclimate in the bee colonies and outside we used. The functional diagram of the PAS is shown on (Fig. 1).

During the study period of external factors are monitored, as outside air temperature (T_{air}), °C and air humidity (H_{air}) %, atmospheric pressure (A_p) hPa, soil moisture (S_m) at 20 cm depth, %, soil temperature (S_t) at 10 cm depth, °C, solar radiation (S_r) w m^{-2}, wind speed (W_s), m s^{-1}, precipitated rainfall (R), l m^{-2}. Also the air temperature inside (T_{in}), °C the hive was measured. The nectar flow is determined based on the changes in hive weight (Y), kg. The real-time data were available via a web-based application Meteobot® (Fig. 2). The experiment was performed on the days (d) of full flowering black locust, when the flow of nectar was highest. The time (t) for reporting the indicators of external and internal factors in the hive was made every 30 min at 9:00 a.m. to 20:00 p.m. To monitor the bee colony daily carrying capacity, the weight of the hive (Y) was measured. The recording of the change in the weight of the hive was carried out with an electronic beekeeping scales of TEHTRON-VAGA. Reporting was done of every 30 min at 9:00 a.m. to 20:00 p.m. The influence of weather conditions in the weight of the hive due to the nectar flow for the period of the blooming of the *Robinia pseudoacacia* was determinate by regression analysis.

3 Results and Discussions

The beginning of the flowering of the black locust for the region Brestovica started on 15.05.2021. Duration of flowering period was 16 days. We found that for the Brestovitsa region the average for the flowering period T_{air} is 18, 48 °C, H_{air} is 70, 74% and S_r is 212, 77 w m^{-2}. The amount of precipitation is 13, 5 l m^{-2}. The difference in the measured values of $T_{air,}$ $H_{air,}$ S_r and R determine the specific microclimate for the study region.

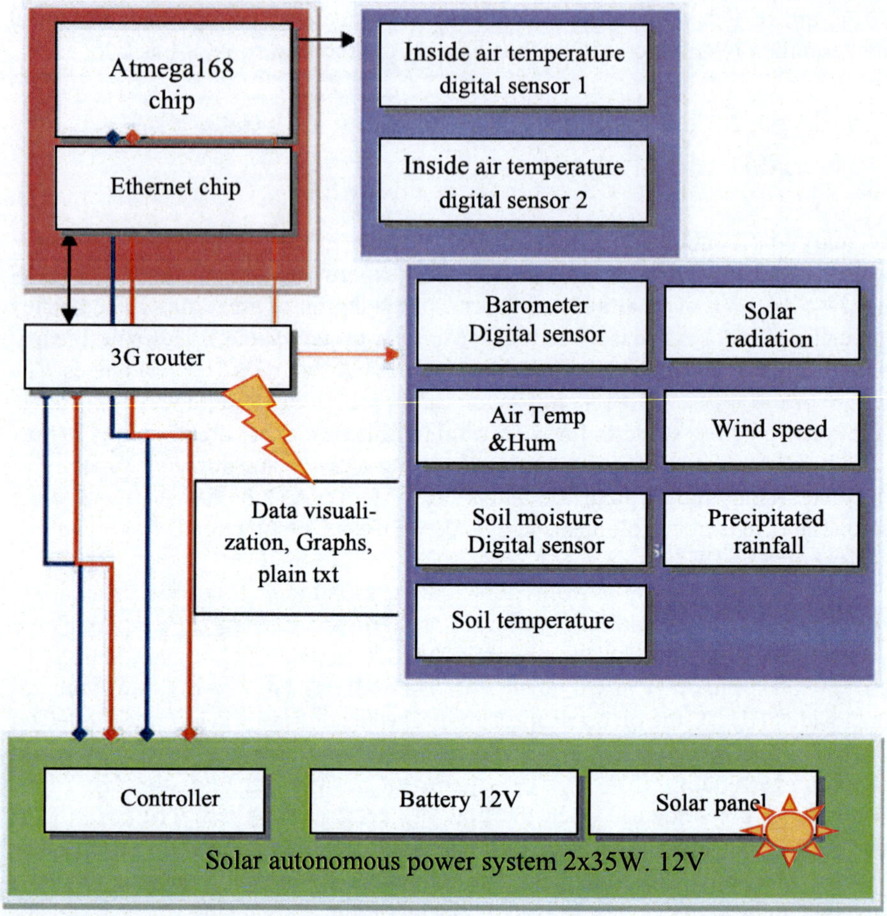

Fig. 1 Functional diagram of precision apiculture system (PAS)

For the beginning of flowering we marked the date when the first open flowers appear, the beginning of full flowering when 25% of the flowers are open and the end of full flowering when 75% of the flowers are overblown. For the period of our study we chose the days when the black locust was in full bloom in which we studied the hourly dynamics of nectar secretion during the flight hours of bees.

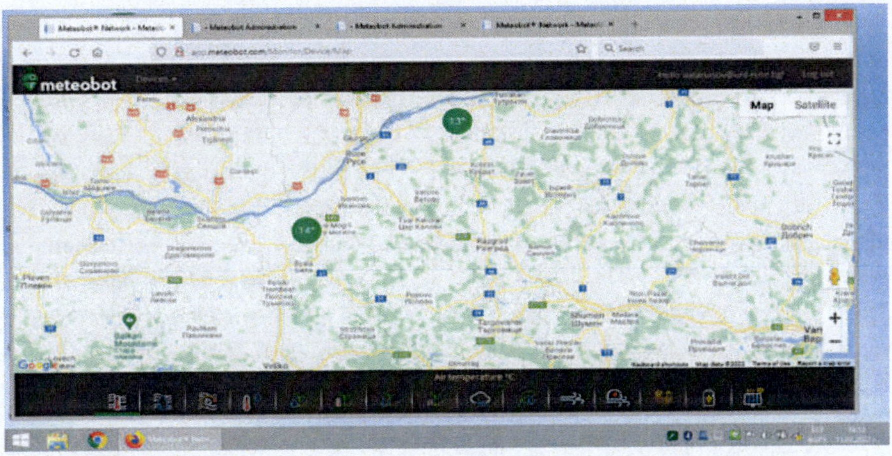

Fig. 2 Remote monitoring of changes in environmental factors in real time

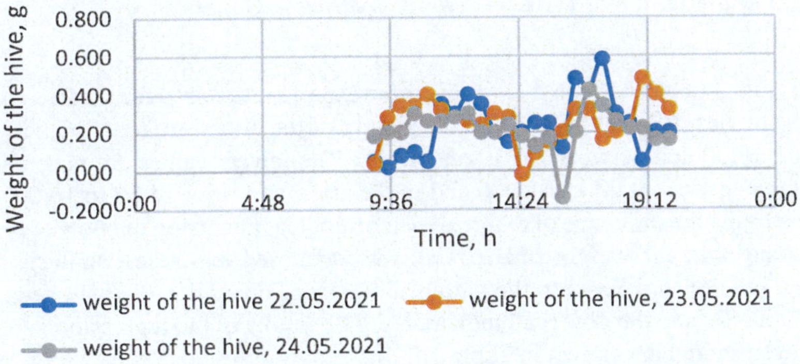

Fig. 3 Change in the weight of the hive for the measurement period in Brestovica

The inflow of nectar during the different hours of the day was measured as a difference in the change in the weight of the hive. On (Fig. 3) show the hourly dynamics of the change in the weight of the hive for the measurement period 22.05–24.05.2021 in the villages of Brestovica.

Table 1 Results of the regression analysis for the linear model for Brestovica

	Beta	Std. Err	B	Std. Err	t(71)	p-level
Intercept			−134,708	69,09,575	−1,94,959	0,055,680
d	**1,081,853**	**0,059,486**	**5,129**	**0,28,202**	**18,18,673**	**0,000,000**
t	**0,460,411**	**0,032,070**	**11,864**	**0,82,635**	**14,35,660**	**0,000,000**
Hair	**0,342,041**	**0,093,274**	**0,141**	**0,03,852**	**3,66,705**	**0,000,506**
Tair	**0,325,730**	**0,111,999**	**0,373**	**0,12,819**	**2,90,833**	**0,005,013**
Sm	**0,192,989**	**0,070,819**	**0,479**	**0,17,583**	**2,72,512**	**0,008,311**
St	−0,078,255	0,048,103	−0,162	0,09,988	−1,62,681	0,108,769
Sr	0,014,260	0,017,083	0,000	0,00,020	0,83,476	0,407,005
Ws	0,006,685	0,008,201	0,027	0,03,288	0,81,510	0,418,085
Ap	0,021,841	0,031,107	0,035	0,05,031	0,70,213	0,485,184
Dp	**−0,205,613**	**0,066,470**	**−0,401**	**0,12,948**	**−3,09,332**	**0,002,949**
Tin	0,005,911	0,008,297	0,070	0,09,867	0,71,245	0,478,819

Regression Summary for Dependent Variable: $R = 0, 99{,}832{,}589$; $R^2 = 0, 99{,}665{,}458$
Adjusted $R^2 = 0, 99{,}607{,}046$; $F\ (11, 63) = 1706, 2$; $p < 0, 05$

From (Fig. 3) it can be seen that for the three experimental days for the Brestovica region the highest values of change in the weight of the hive were reported between 11:30 and 14:00 and between 16:30 and 18:30. The lowest values are observed in the morning between 9:00 and 11:00 and in the afternoon from 14:30 to 16:00.

To determine the influence of external meteorological factors on the flow of nectar due to changes in the weight of the hive, we performed regression analysis. The functional relationship between the variables was considered d, t, T_{air}, H_{air}, A_p, S_m, S_t, S_r, W_s, R, T_{in} and the observed indicator Y. The results of the regression analysis for the linear model are shown in Table 1.

The results of the regression analysis for the linear model make obvious, that the coefficient of the distinctness $R^2 = 0{,}99$ or, that 99% from the alteration of the Y is due to the factors d, t, H_{air}, T_{air}, S_m and D_p is described from the linear model which looks like this:

$$Y = -134, 708 + 5, 129d + 11, 864t + 0, 141H_{air} + 0, 373T_{air} + 0, 479S_m - 0, 401D_p \quad (1)$$

The factor R is not included as it is a constant = 0 there is no precipitations. It is substantial the influence of d, t, H_{air}, T_{air}, S_m and D_p is proven, because the probability P-value is less than the level of significance 0, 05.

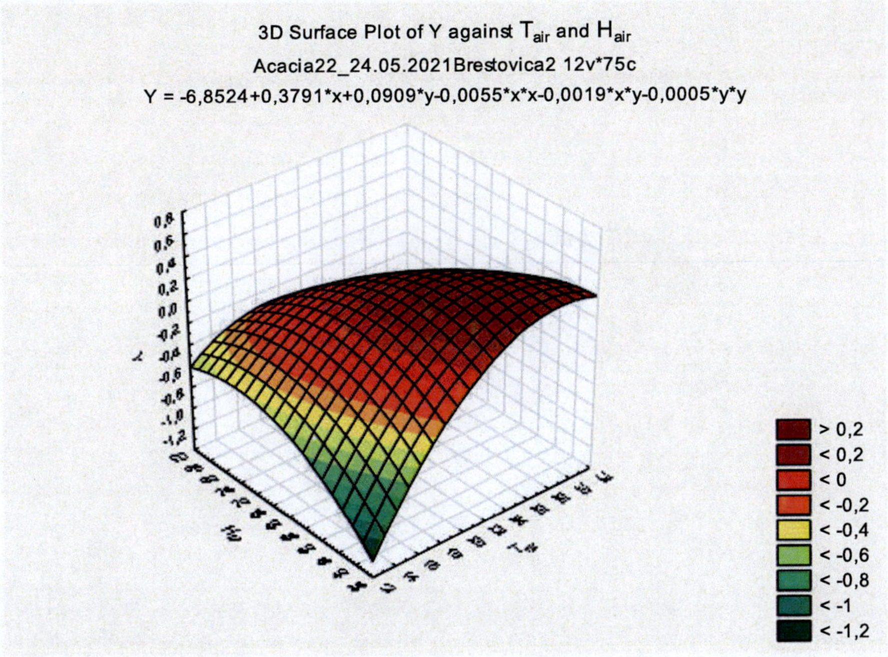

Fig. 4 The surface of the echo $Y = f\ (T_{air}, H_{air})$

It is insignificant the influence of the factors S_t; S_r; W_s; A_p and T_{in} because the probability P-value is greater than 0, 05. The factor d has the strongest influence, because for it the Student's criterion T = 18.18673 is the largest.

Reporting, the some main factors are T_{air}, H_{air}, and S_m there will be presented graphically the surface of the echo (Figs. 4, 5 and 6).

$$Y = f(T_{air} H_{air});\ Y = f(H_{air} S_m);\ Y = f(T_{air} S_m) \tag{2}$$

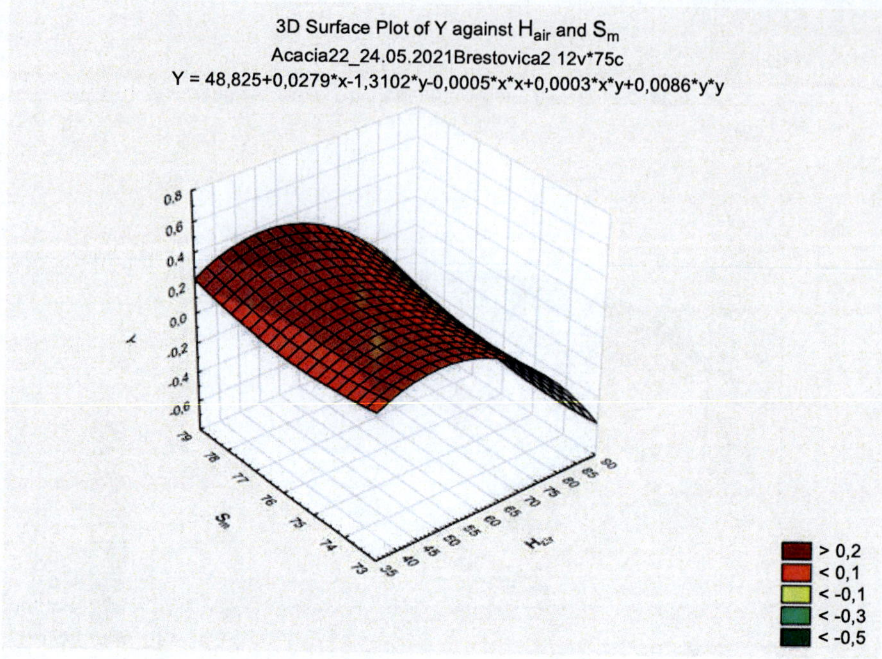

Fig. 5 The surface of the echo Y = f (H_{air}, S_m)

From (Figs. 4, 5 and 6) is seen, that the surface of the echo has a parabolic character and from it there could be difficultly defined the favorable values of T_{air}, H_{air}, and Sm into which the changes in the weight of the hive due to the nectar secretion Y has greater values. This question is better cleared out by means of the lines of the same echo, which are thus shown on (Figs. 7, 8 and 9).

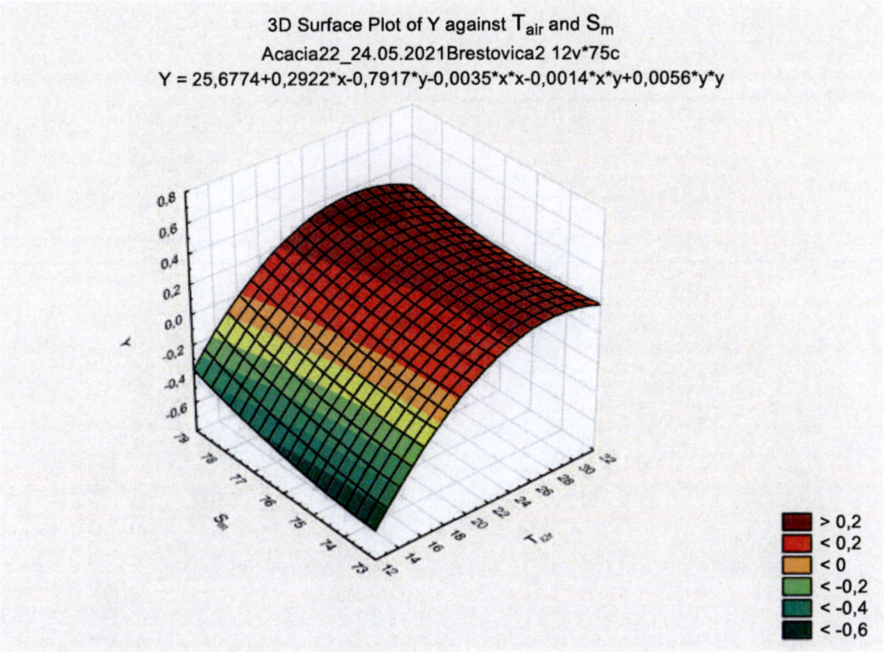

Fig. 6 The surface of the echo Y = f (T_{air}, S_m)

And the lines of similar echo f (T_{air}, H_{air}) = constant; f (H_{air}, S_m) = constant; f (T_{air}, S_m) = constant.

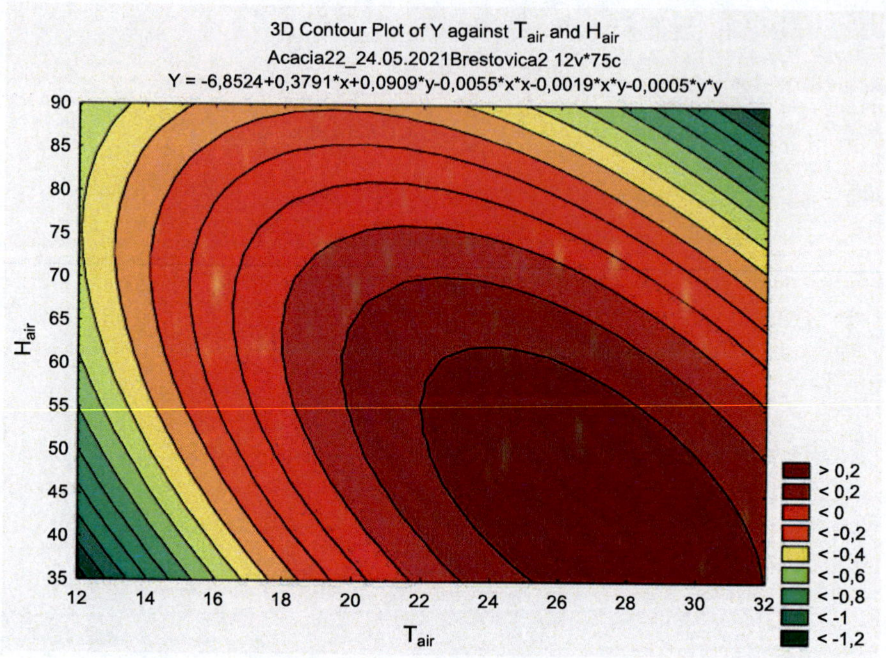

Fig. 7 A line of the same echo f (T_{air}, H_{air}) = constant

The lines of similar echo (Fig. 7) have an ellipse character and in the center of the most inner ellipse we will have the most favorable conditions we receive the Y greater than 0,2 g. These conditions are defined from $T_{air} \approx 27$ °C and $H_{air} \approx 48\%$.

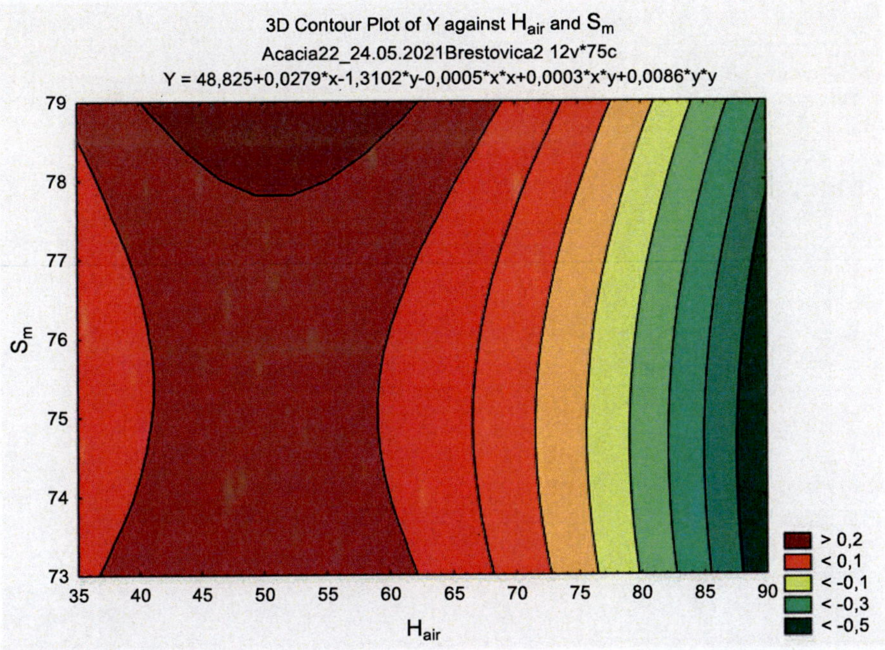

Fig. 8 A line of the same echo f (H_{air}, S_m) = constant

The lines of similar echo (Figs. 8 and 9).show that the best conditions for increasing the weight of the bee hive Y > 0.2 will be at $S_m \approx 79\%$, $H_{ai}r \approx 48\%$ and $T_{air} \approx 27$ °C.

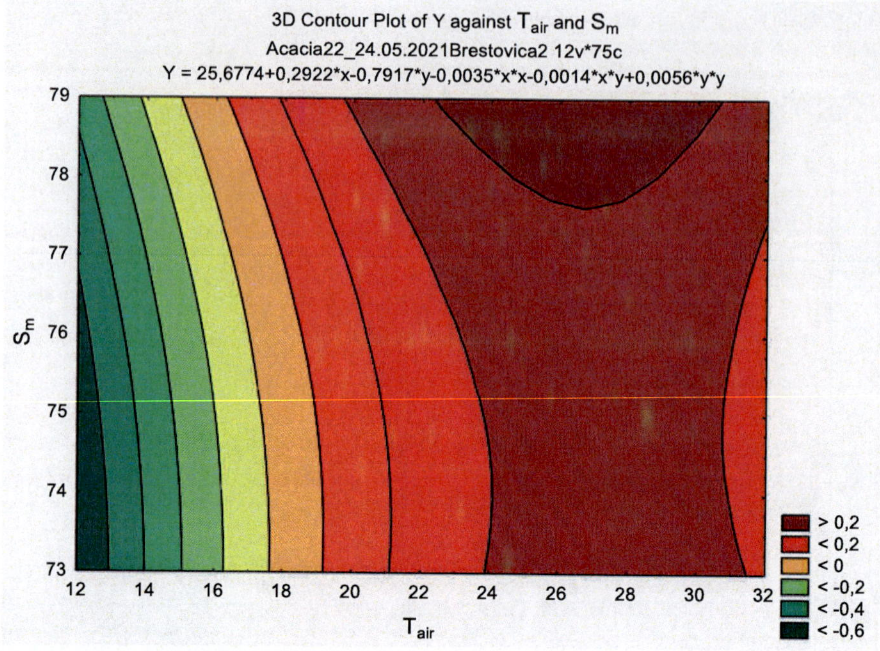

Fig. 9 A line of the same echo f (T_{air}, S_m) = constant

4 Conclusions

The influence local agrometeorological factors on the flow of nectar in the bee hive during the flowering of black locust (Robinia pseu-doacacia) is investigated. It is substantial the influence of d, t, H_{air}, T_{air}, S_m and D_p at $P < 0.05$.

The results show that the coefficient of the distinctness $R^2 = 0.99$ or, that 99% from the alteration of the Y is due to the factors d, t, H_{air}, T_{air}, S_m and $D_{p.}$ Changes in the weight of the hive due to the nectar secretion Y is highest at the outside air temperature $T_{air} \approx 27$ °C, air humidity $H_{air} \approx 48\%$ and soil moisture $S_m \approx 79\%$. It is insignificant the influence of the factors S_t; S_r; W_s; A_p and $T_{in.}$

Acknowledgements This research is supported by Bulgarian National Science Fund under Project KP-06-PN 46-7 "Design and research of fundamental technologies and methods for precision apiculture".

Polish society of agricultural engineering PTIR Lublin.

References

1. Carl C, Landgraf D, Van der Maaten-Theunissen M, Biber P, Pretzsch H (2017) Robinia pseudoacacia L. flower analyzed by using an unmanned aerial vehicle (UAV). Remote Sens 9:1091
2. Csiha I (2013) Black locust forest as bee-forage. In: Rédei K (ed) Black locust (*Robinia pseudoacacia* L.) growing in Hungary. ERTI, Sárvár
3. Enakiev Y, Dimov K (2020) Prerequisites for the use of white acacia (*Robinia pseudoacacia l.*) for energy needs. In: Conference ecology and agrotechnologies – fundamental science and practical realization, Sofia, Bulgaria, pp 258–264
4. Farkas A, Zajacz E (2007) Nectar production for the Hungarian honey industry. Eur J Plant Sci Biotechnol 1(2):125–151
5. Giovanetti M (2019) Foraging choices balanced between resource abundance and handling concerns: how the honeybee, Apis mellifera, select the flowers of Robinia pseudoacacia. Bull Entomol Res 109(3):316–324
6. Hong W, Xu B, Chi X, Cui X, Yan Y, Li T (2020) Long-term and extensive monitoring for bee colonies based on internet of things. IEEE Internet Things J 7(8):7148–7155
7. Lecocq A, Kryger P, Vejsnæs F, Bruun JA (2015) Weight watching and the effect of landscape on honeybee colony productivity: investigating the value of colony weight monitoring for the beekeeping industry. PLoS ONE 10(7):e0132473
8. Nicolescu VN, Rédei K, Mason WL (2020) Ecology, growth and management of black locust (*Robinia pseudoacacia L.*), a non-native species integrated into European forests. J For Res 31:1081–1101
9. Papadopoulou F, Tentsoglidou M, Pavloudakis F, Papadimopoulos N, Papadopoulos I (2018) Evaluation of honey producing potential of Robinia pseudacacia in reforested old lignite mines in West Macedonia. J Environ Sci Eng B 7:354–359
10. Sharif MZ, Di N, Liu F (2021) Monitoring honeybees (Apis spp.) (Hymenoptera: Apidae) in climate-smart agriculture: a review. Appl Entomol Zool
11. Stalidzans E, Zacepins A, Kviesis A, Brusbardis V, Meitalovs J, Paura L, Bulipopa N, Liepniece M (2017) Dynamics of weight change and temperature of *Apis mellifera* (Hymenoptera: Apidae) colonies in a wintering building with controlled temperature. J Econ Entomol 110(1):13–23
12. Vítková M, Müllerová J, Sádlo J, Pergl J, Pyšek P (2017) Black locust (*Robinia pseudoacacia*) beloved and despised: a story of an invasive tree in Central Europe. For Ecol Manag 384:287–302
13. Wojda T, Klisz M, Jastrzebowski A, Mionskowski M, Szyp-Borowska I, Szczygiel K (2015) The geographical distribution of the black locust (*Robinia pseudoacacia* L.) in Poland and its role on non-forest land. Papers Global Change IGBP 22(1):101–113
14. Zacepins A, Kviesis A, Komasilovs V, Muhammad FR (2020) Monitoring system for remote bee colony state detection. Baltic J Mod Comput 8(3):461–470

The Impact of Allelic State of *dhn1* and *rsp41* Genes on Grain Moisture Content of Maize Hybrids Within Marker Assisted Selection (MAS) for Drought Resistance

Larysa Prysiazhniuk, Yurii Honcharov, and Larysa Korol

Abstract This study was carried out in 2019 and 2020 in pilot plots of Research Institute of Agrarian Business (Vesele village, Dnipropetrovsk region, Ukraine). Four hundred sixteen maize hybrids were studied. The specific allele combinations of dhn1 and rsp41 genes in parents of maize hybrids were analyzed. According to obtained results the different combinations of favorable alleles were identified. All testers, which were used for specific combining ability (SCA) effects assessment, had various allele combinations of the *dhn1* and *rsp41* genes. Studied maize lines had different rates of drought tolerance according to allelic state of *dhn1* and *rsp41* genes. The estimation of SCA effects allowed to detect the optimal ratio of presence the favorable alleles and values of harvest grain moisture content in studied parents of maize of three-way cross hybrids. Thus, the low SCA effects of grain moisture content of testers were observed in A/G*A/A and A/G*A/G genotypes in 2019–2020. A/A*G/A*G/G, A/G*A/G*G/G, A/A*A/G*A/G, A/G*G/A*G/G and G/A*A/A*G/G of three-way crossed hybrids genotypes provided the low SCA effects during studied years. It was shown that allele combinations of the *dhn1* and *rsp41* genes. It was determined the greatest impact on SCA effects of grain moisture content both of testers and modified (three-way crossed) hybrids was allele combination—64 and 51% respectively.

Keywords Drought tolerance · Maize · Combining ability · MAS · Favorable alleles

L. Prysiazhniuk (✉) · L. Korol
Ukrainian Institute for Plant Variety Examination, Kyiv, Ukraine
e-mail: prysiazhniuk_l@ukr.net

Y. Honcharov
Research Institute of Agrarian Business, Dnipro, Ukraine

S. Pascuzzi and F. Santoro (eds.), *Farm Machinery and Processes Management in Sustainable Agriculture*, Lecture Notes in Civil Engineering 289,
https://doi.org/10.1007/978-3-031-13090-8_26

1 Introduction

Maize has a remarkable place among cereals and it is used as human food, animal feeding and industry [1]. It is estimated that 20–25% of the global maize area is affected by drought in any given year [2, 3]. Drought is a widespread abiotic stress that causes crop production losses in arid and semiarid areas worldwide, and global warming is predicted to further exacerbate drought frequency and impact [4]. Growing drought-resistant forms and hybrids is a way to overcome the effects of drought on reducing plant productivity. A drought-resistant hybrid of maize is said to be the hybrid ensuring 30% of its potential yield during flowering and physiological maturity period [5].

The one of important agronomical traits is also harvest grain moisture content. The dry matter in maize grains spoils easily due to the excessive grain moisture at harvest. High grain moisture contents at harvest necessitate grain drying prior to the transport and storage of ears and requires the extra expenses for gain the basis moisture content. Accordingly, low grain moisture at harvest has become the one of the major breeding aims [6–8].

The possibilities of molecular biology and plant genetics allow the use of knowledge about the structure of the genome and the genes expression to select lines with agronomic valued traits. Liu et al. [9] developed two functional CAPS (Cleaved Amplified Polymorphic Sequences) markers to identify allele states of *dhn1* and *rsp41* genes associated with drought resistance of maize [9, 10].

Li et al. [8] reported that GRMZM5G805627 and GRMZM2G137211 were identified as candidate genes underlying major QTLs for grain moisture in maize by combining genetic population analysis, transcriptomic profiling and gene editing. They concluded that the moisture content is controlled by a major QTL on chromosome 9, which may be associated with the regulation of CYR1-9 gene expression. Nevertheless, further experiments are needed to validate and understand the function of this genes [8]. Hence, nowadays maize breeders need some practical approach for selection promising maize lines and prediction of their contribution of better fitness-related traits to hybrids.

In our previous study it was reported that using two CAPS markers allowed to select maize lines which are drought tolerant [5, 11]. According to Liu et al. [9], the closest gene to SNP PZA01671.1 is related to *rsp41*, encoding an arginine/serine-rich splicing factor (RSP41). The RSP41 protein is a splicing factor that is part of the serine/arginine-rich protein (SR proteins) family of nuclear-localized RNA-binding proteins. Abiotic stress is known to affect alternative splicing of SR pre-mRNAs and to result in the translation of different protein isoforms with changed roles in the splicing of pre-mRNAs. The closest gene to PZA0355.1 is related to *dhn1*, encoding a dehydrin. Dehydrins (DHNs) are group 2 late embryogenesis abundant (LEA) proteins that appear in seeds late in development but also in leaves during drought, salt, or cold stress, or upon treatment with abscisic acid [9].

It is known that the value of an inbred line in the commercial production of hybrid maize is determined by two factors, the characteristic of the line itself with respect to yielding ability, pollen shedding, resistance to abiotic stress, etc. and the behavior of the line in hybrid combinations [12–14]. Combining ability may be considered as the potential of an individual inbred line to contribute better fitness-related traits to hybrid progeny. In maize breeding programs, knowledge of specific combining ability (SCA) of hybrid combinations as well as the identification and exploitation of heterotic groups, are crucial for successful hybrid production [15, 16]. This is because, information on the type of gene action controlling agronomical traits is key to identifying useful parents and hybrids, as well as designing appropriate strategies for breeding [17, 18]. According to the role of arginine/serine-rich splicing factor and dehydrins in water accumulation and water balance regulation in plant it can be practically value to estimate the impact of allele states of *dhn1* and *rsp41* genes in different maize genotype on SCA of harvest grain moisture content.

The aim of this research was to estimate the combining ability of grain moisture content maize genotypes and study the factors impact on SCA of maize hybrids with different allele combination of genes associated with drought resistance.

2 Materials and Methods

2.1 Plant Material

Four hundred sixteen maize hybrids were studied. These hybrids are modified hybrids of heterosis model Iodent × Lancaster germplasm (three-way cross hybrids). The effects of specific combining ability (SCA) were evaluated by the top-crossing method. To obtain test crosses there are used testers, which are sister sterile hybrids (Lancaster germplasm). All breeding materials were provided by Research Institute of Agrarian Business (Dnipro, Ukraine).

2.2 Laboratory Experiment

The laboratory studies were carried out in Laboratory of molecular genetic analysis, Ukrainian Institute of Plant Variety Examination (Kyiv, Ukraine) in 2019. Two CAPS markers dhnC397 and rspC1090 were used to identify the allele state of SNP A/G polymorphisms in the drought resistance genes *dhn1* and *rsp41* [5, 9]. To assess the impact of presence of favorable alleles on grain moisture content, genotypes both modified hybrids and testers with all combinations of detected alleles were used.

Table 1 The air temperature and amounts of precipitation during 2018–2019 maize growing cycle

Month	Air temperature, °C			Amounts of precipitation, mm		
	Daily average air temperature	2019	2020	Daily average amounts of precipitation	2019	2020
April	9.2	10.9	8.6	35.0	14.1	0.0
May	15.8	18.1	13.7	50.0	21.3	2.9
June	19.1	23.9	22.1	59.0	1.0	0.0
July	20.9	20.9	23.2	61.0	33.7	1.3
August	20.1	20.9	21.5	35.0	73.4	14.0
September	15.0	16.1	19.5	36.0	6.4	0.0

2.3 Field Experiment and Weather Conditions

The field experiment was undertaken in 2019–2020 on pilot plots of Research Institute of Agrarian Business (Vesele village, Dnipropetrovsk region, Ukraine). The field experiment was designed according to classical crosspollinated plant breeding methods [19]. The graine moisture content was estimated during harvesting using the plot combine (Wintersteiger, Germany).

The weather conditions rates of Research Institute of Agrarian Business in Vesele village (Sinelnykove district, Dnipro region) were provided by Sinelnykove weather station. The air temperature and amounts of precipitation during maize growing cycle in 2019–2020 are shown in Table 1.

2.4 Statistical Analysis

The coefficient of agrometeorological indicators from daily average amounts during studied years was computed using the equation below:

$$Dc = \frac{(X_i - \overline{X})}{\sigma}$$

where *Dc*—deviation coefficient; *Xi*—indicator of current weather; *X*—daily average amounts; σ—mean-square deviation.

The rate of deviation coefficients was determined according to scale: $Dc = 0$–1—close to normal conditions; $Dc = 1$–2—strong different conditions; $Dc > 2$—close to unique conditions [24].

The significant differences of studied indicators and the rate of factors impact on SCA of grain moisture content were determined by ANOVA using STATISTICA 12.0 software (trial version).

3 Results and Discussions

The polymorphism of CCAAAG(A) type by dhnC397 marker (*dhn1* gene), which is associated with drought-tolerance, is detected by presence only one band 164 bp after StyI restriction enzyme treatment. The presence of two segments 131 and 33 bp characterizes drought-susceptible genotypes with the polymorphism CCAAAG(G) type.

The CCAG/CCGG (SNP A/G) polymorphism by rspC1090 marker (*rsp41* gene) resulted in an amplification product change in a nearby HpaII restriction site. For the genotypes carrying the SNP CCGG(G) variant, PCR products could be digested with HpaII and produced 225 and 61 bp bands. This allele was associated with drought tolerance genotypes. Otherwise, only a 286 bp segment could be detected in the genotypes carrying the CCGG(A) variant due to the absence of the nearby HpaII site, which was associated with drought-susceptible genotypes [5, 9].

The maize lines, which are parents of studied hybrids, were analyzed by CAPS markers for presence of favorable alleles associated with drought-tolerance in our previous studies [5, 11].

For each type of allele combinations of testers SCA effects of grain moisture content was calculated (Table 2).

As results of analysis, the lowest SCA effect was in tester genotypes with favorable alleles of both *dhn1* and *rsp41* genes −2.6% in 2020. In 2019 the lowest SCA effect was obtained also for these genotypes——0.5%. It was shown that the low SCA effect was detected in tester genotypes with polymorphism CCGG(A) of *rsp41* gene in one of parent line in 2019 and 2020 −0.4 and −0.1% respectively. It was determined that significant differences were observed between tester genotypes with at least one favorable allele of both *dhn1* and *rsp41* genes during 2019–2020. It should be noted that the low SCA effect (−2.0%) was calculated for tester genotypes with unfavorable alleles by both of both *dhn1* and *rsp41* genes in 2020. Also, the low

Table 2 SCA effects of testers genotypes with different allele combinations 2019–2020

Allele combinations (*dhn1*/*rsp41* genes)		Average SCA effects, %	
		2019	2020
A/A	A/A	0.9	1.7
A/A	A/G	0.1	0.9
A/A	G/A	0.3	−0.5
A/G	A/A	−0.4	−0.1
A/G	A/G	−0.5	−2.6
A/G	G/A	−0.5	0.2
G/A	A/A	0.4	1.3
G/A	A/G	0.1	0.5
G/A	G/A	0.1	−2.0
$LSD_{0.05}$		0.16	

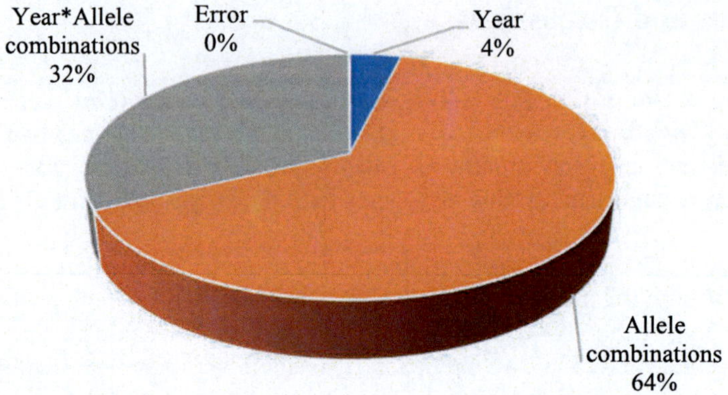

Fig. 1 The rate of factors impact (mean squares) on SCA of grain moisture content of maize tester genotypes during 2019–2020

SCA effect (−0.5) was detected in tester genotypes with unfavorable alleles of both *dhn1* and *rsp41* genes in one of parent line in 2020.

As results of rate calculation of factors impact, it was determined that the greatest influence on SCA effect of grain moisture content of tester genotypes has allele combination—64% (Fig. 1).

The Fig. 1 shows that year weather conditions have the lowest impact on SCA effect of grain moister content of tester genotypes—4%. However, it should be noted that interact of year weather conditions and allele combinations influences on studied indicator significantly—32%. It can be concluded that there is the high rate of interaction of year weather conditions with allele combinations impact expressed as the adaptation of tester genotypes with favorable alleles. It shows that tester genotypes with favorable alleles can adapt to different weather conditions. This can be proved by the low SCA effects of grain moisture content during studies years. Meanwhile, the major of tester genotypes with unfavorable alleles shown different SCA effects during studied years which indicants their low adaptation ability to different weather condition during maize growing season.

As results of SCA effects calculation for modified hybrids, it was determined that the lowest SCA effect was obtained for modified hybrids with follow allele combination: A/A*A/A*A/G—−0.7% in 2020 (Table 3). In this year the low SCA effect was also noted in modified hybrids with allele combination: G/A*A/A*A/G—−0.5%. However, in 2019 these modified hybrids shown positive SCA effects: 0.0 and 0.1% respectively. It can be noted, that the low SCA effects was calculated for modified hybrids with allele combination: A/A*G/A*G/G—−0.3% during studied years and for modified hybrids with allele combination: A/G*A/G*G/G—−0.2 and −0.1% in 2019–2020 respectively.

The modified hybrids with favorable alleles of *dhn1* and *rsp41* genes shown rather positive SCA effects during 2019–2020—0.0 and 0.1% respectively. The

Table 3 SCA of modified hybrids with different allele combinations 2019–2020

Allele combinations (*dhn1*/*rsp41* genes)			Average SCA, %	
			2019	2020
A/A	A/A	A/A	0.0	0.2
A/A	A/A	A/G	0.0	−0.7
A/A	A/A	G/G	−0.1	0.2
A/A	A/G	A/A	0.0	0.1
A/A	A/G	A/G	−0.1	−0.1
A/A	A/G	G/G	0.3	−0.1
A/A	G/A	A/A	0.0	−0.1
A/A	G/A	A/G	−0.2	0.2
A/A	G/A	G/G	−0.3	−0.3
A/G	A/A	A/A	0.0	−0.1
A/G	A/A	A/G	−0.1	0.3
A/G	A/A	G/G	0.1	0.1
A/G	A/G	A/A	0.0	−0.1
A/G	A/G	A/G	0.0	0.1
A/G	A/G	G/G	−0.2	−0.1
A/G	G/A	A/A	0.0	0.0
A/G	G/A	A/G	−0.1	0.2
A/G	G/A	G/G	−0.1	−0.2
G/A	A/A	A/A	0.0	0.0
G/A	A/A	A/G	0.1	−0.5
G/A	A/A	G/G	−0.2	−0.1
G/A	A/G	A/A	0.0	0.1
G/A	A/G	A/G	0.0	−0.1
G/A	A/G	G/G	−0.1	0.3
G/A	G/A	A/A	0.0	0.0
G/A	G/A	A/G	0.2	0.4
G/A	G/A	G/G	0.2	−0.1
$LSD_{0.05}$			0.03	

stable low SCA effect was noted in modified hybrids with allele combination: A/A*A/G*A/G—−0.1%. For modified hybrids with allele combinations A/G*G/A*G/G and G/A*A/A*G/G SCA effects were from −0.1 to −0.2% during studied years. It was observed that modified hybrids with the lowest SCA effect in 2020 had the positive SCA effects in 2019—from 0.0 to 0.1%.

Considering the weather conditions, the air temperature during August–September 2019 was close to normal temperature (Dc = 0–1). In comparison to 2019, in 2020 Dc was 2 which indicates strong different conditions. The deviation coefficients for amounts of precipitation during August–September 2019 were from 1 to −2. In 2020 during this period Dc was from −1 to −2. Therefore, according to obtained data, 2020 characterized of precipitation deficit which could influence on grain moisture content both tester genotypes and modified hybrids with different allele combinations.

The breeding potential per se in specific combinations (SCA) of maize lines derived from the material of temperate, subtropical and tropical origins and their overall performance in crosses (GCA) were investigated by Malik et al. [20]. According obtained results, the tropical material gave eminent GCA effects for days to pollen shedding, ear weight and grain moisture at harvest. Hence, it was concluded that tropical material contributed lateness in flowering which, ultimately, resulted in higher moisture at harvest [20].

Nasser et al. [17] examined GCA effects of the inbred lines and SCA of the crosses under combined drought and heat stress and well-watered environments. Authors assessed yield and stability of the derived hybrids under contrasting environments and investigated inter-relationships among traits of the hybrids under contrasting environments. The assessment of combining ability of studied maize genotypes the authors managed to select inbreds which could be used in hybrid combinations to develop outstanding drought and heat stress tolerant hybrids and synthetics [17]. The estimation of SCA effects allowed Erdal et al. [7] to select maize lines for earliness and lower grain moisture content. Simon et al. [1] assessed SCA of grain yield and other agronomic traits including the grain moisture content for maize germplasm. Significant SCA effects of grain yield, plant and ear height, plant and ear aspect, ears and plants harvested, ear rot, husk cover, moisture shown the possibility of studied maize lines for direct production as single cross hybrids or developed further as three-way cross hybrids [1]. Thus, when specific combinations are desired, especially in hybrid maize development, SCA effects could help in the selection parental material for hybridization.

To assess the rate of factors impact on SCA effects of grain moisture content of maize modified hybrids, the ANOVA was used. The results of assessment are shown on Fig. 2.

According to obtained results, it was determined that the greatest influence on SCA effect of grain moisture content of maize modified hybrids has allele combination—51%. Partitioning mean squares into its components revealed the low impact the year weather conditions impact on SCA effect of grain moisture content—1%. However, the interaction of allele combinations and year weather conditions influence on SCA effect significantly—48%. It is noticed that the maize modified hybrids which shown low SCA effects during 2019–2020, have not only favorable alleles.

The low SCA effects were calculated for modified hybrids with A/A and G/G alleles in one of hybrid parent (A/A*A/G*A/G and A/G*A/G*G/G). It may be explained both line genotypes characteristics, independently on allelic state of *dhn1* and *rsp41* genes, and three-way cross hybrids features. Three-way cross hybrids are hybrids of intensive type with high requirements cultivation technology and growing season weather conditions. Hence, the three-way cross hybrids are characterized of high plasticity of adaptation ability which expressed as higher than for testers impact

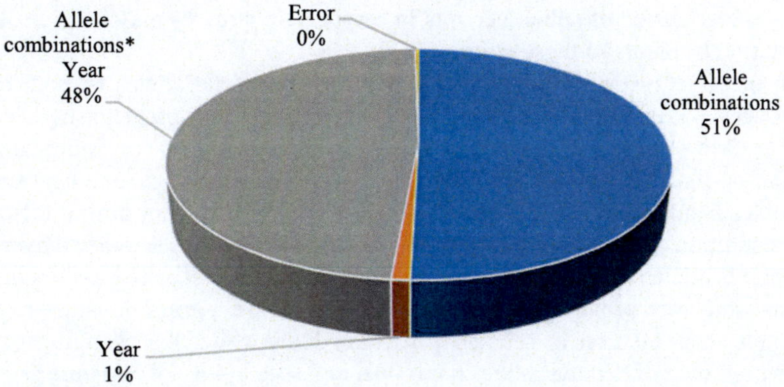

Fig. 2 The rate of factors impact (mean squares) on SCA of grain moisture content of maize modified hybrids during 2019–2020

on SCA effects of grain moisture content of the interaction of allele combinations and year weather conditions.

It was noticed that in parent maize lines, which demonstrated low SCA effects during studied years, the presence of favorable alleles of *rsp41* gene was observed (Table 3). Thus, the stable low SCA effects were obtained for modified hybrids with G/A alleles in one of hybrid parent (A/A*G/A*G/G; A/G*G/A*G/G; G/A*A/A*G/G). In our previous study the positive correlation was determined between the presence of a favorable allele of *dhn1* gene in maize lines and leaf surface temperature, a positive correlation was also found between the presence of a favorable allele of *rsp41* gene and the percentage of germinated seeds of maize lines in sucrose solution (r = 0.31) [11]. It can be considered that *dhn1* gene may be connected more with heat tolerance and *rsp41* gene responses for maize genotype tolerance to soil drought.

In higher plants dehydrins connected with *dhn1* gene are accumulated in seeds at later stages of embryogenesis and in vegetative tissues treated with ABA (abscisic acid) or subjected to stress factors (drought, salinity, and cold temperature). Usually, these proteins are not seen in premature embryos and young seedlings. It is interesting that appearance of the LEA (late embryogenesis abundant) proteins in embryonal tissue depends on content of endogenous ABA, whose level is usually increased during seed formation and especially before their maturation [21, 22].

The results which were obtained by Palusa et al. [23] clearly show that heat and cold strongly alter the alternative splicing (AS) of most SR genes. Hence, some of the heat- and cold-induced changes in gene expression may be caused by the AS of SR genes. It is reported that the intranuclear distribution of several SR proteins is regulated by phosphorylation and stresses such as heat and cold. Hence, some of the observed effects of heat and cold stresses on splicing could be the result of an altered distribution of SR protein in the nucleus. These results will have important implications in understanding the mechanisms of gene regulation under abiotic stresses. Altered ratios of splice variants in response to stresses may have a role in the adaptation of plants to these stresses [23].

It can be concluded that maize genotypes with favorable alleles of both *dhn1* and *rsp41* genes are tolerant for both types of drought air and soil. The single cross hybrids, as testers are in this study, could provide low harvest grain moisture content both under normal weather conditions and under combined drought and heat stress. On the other hand, the low SCA effects of grain moisture content during different weather conditions were obtained for three-way crossed hybrids (modified hybrids) with favorable alleles only of *rsp41* gene in parent maize lines. It can be assumed, that dehydrins which are provide heat tolerance may decrease a grain moisture release due to their accumulation in seed in particular. Meanwhile, the changing of the alternative splicing of SR genes does not influent on seed ability of moisture release. Thus, the most promising parent lines for three-way crossed hybrids which would be demonstrated both drought tolerance and low harvest grain moisture content under different weather conditions are genotypes with allele combination A/G and G/G. Moreover, the other parents of three-way crossed hybrids have to be contained at least one favorable allele of *dhn1* gene.

4 Conclusions

In the study, SCA effects both of tester and maize modified hybrids were found to be significant. It was determined that the lowest SCA effects of harvest grain moisture content were obtained for tester genotypes with favorable alleles of *dhn1* and *rsp41* genes during 2019–2020. By assessment of the rate of factors impact on SCA effects of grain moisture content it was found that the allele combination has the greatest influence on SCA effect of grain moisture content of tester genotypes—64%. It was determined that the stable low SCA effects were calculated for modified hybrids with different allele combinations of *dhn1* and *rsp41* genes. It was shown that these three-way crossed hybrids included both favorable and unfavorable alleles in one of their parents. It is assumed, that may be explained both genotypes of maize lines per se and specific of genes action which are associated with drought tolerance. The influence of allele combinations of modified hybrids on SCA effects of harvest grain moisture content was 51%, while the interaction of allele combinations and weather conditions of maize growing season impacted on 48%. According to obtained results of this study it is concluded that the promising parent lines for three-way crossed hybrids, which are drought tolerant and are characterized by low SCA effects of harvest grain moisture content, can include A/G and G/G alleles of *dhn1* and *rsp41* genes.

References

1. Simon Z, Kasozi LC, Patrick R, Abubaker M (2018) Gene action for grain yield and agronomic traits in selected maize inbred lines with resistance to Striga hermonthica in Uganda. J Food Secur 6(4):155–162. https://doi.org/10.12691/jfs-6-4-3
2. Arshad K, Hussain A, Aslam A, Azam A, Asad M, Bilal M (2021) Genetic analysis of drought resistance in maize using physiological, morphological and molecular markers. Asian J Biotechnol Genet Eng 4(2):1–10
3. Salami AE, Afolabi MS, Agbowuro GO (2019) Genetic studies of grain yield and other agronomic traits of low-n maize (Zea mays L.) using a diallel cross under nitrogen fertilizer levels. Int J Plant Breed Crop Sci 6(3):569–574
4. Hao Z, Li X, Xie C, Weng J, Li M, Zhang D, Liang X, Liu L, Liu S, Zhang S (2011) Identification of functional genetic variations underlying drought tolerance in maize using SNP markers. J Integr Plant Biol 53(8):641–652. https://doi.org/10.1111/j.1744-7909.2011.01051.x
5. Prysiazhniuk L, Honcharov Y, Piskova O (2018) Identification of drought-resistant maize lines by DNA markers. In: Spalevic V, Popovic V (eds) International GEA (Geo Eco-Eco Agro) Conference "Green Room Sessions 2018", pp 88–93
6. Kang MS, Zuber MS (1989) Combining ability for grain moisture, husk moisture, and maturity in maize with yellow and white endosperms. Crop Sci 29(3):689–692
7. Erdal Ş, Cengiz R, Öztürk A, Pamukçu M, Cengiz B, Dinçer C, Scott MP (2020) Mısırda Bazı Özelliklerin Genetik Analizi. J Inst Sci Technol 10(3):2153–2167. https://doi.org/10.21597/jist.660028
8. Li W, Yu Y, Wang L, Luo Y, Peng Y, Xu Y, Liu X, Wu S, Jian L, Xu J, Xiao Y, Yan J (2021) The genetic architecture of the dynamic changes in grain moisture in maize. Plant Biotechnol J 19(6):1195–1205. https://doi.org/10.1111/pbi.13541
9. Liu S, Hao Z, Weng J, Li M, Zhang D, Pan G, Zhang S, Li X (2015) Identification of two functional markers associated with drought resistance in maize. Mol Breed 35(1):1–10. https://doi.org/10.1007/s11032-015-0231-7
10. Ul Haq I, Ijaz S (eds) (2021) Sustainable winter fodder: production, challenges, and prospects, 1st edn. CRC Press, New York. https://doi.org/10.1201/9781003055365
11. Prysiazhniuk L, Honcharov Y, Shytikova Y, Chernii S, Hurska V, Tkachyk S (2020) The estimation of heat and drought tolerance maize lines using DNA markers. Agrofor 5(2):75–87. https://doi.org/10.7251/AGRENG2002077P
12. Fan XM, Chen HM, Tan J, Xu CX, Zhang YD, Luo LM, Huang YX, Kang MS (2008) Combining abilities for yield and yield components in maize. Maydica 53(1):39
13. Kanagarasu S, Nallathambi G, Ganesan KN (2010) Combining ability analysis for yield and its component traits in maize (Zea mays L.). Electron J Plant Breed 1(4):915–920
14. Matin MQI, Rasul MG, Islam AKMA, Mian MK, Ivy NA, Ahmed JU (2016) Combining ability and heterosis in maize (*Zea mays* L.). Am J BioSci 4(6):84–90. https://doi.org/10.11648/j.ajbio.20160406.12
15. Bhatnagar S, Betran FJ, Rooney LW (2004) Combining abilities of quality protein maize inbreds. Crop Sci 44(6):1997–2005
16. Gami RA, Chauhan RM, Patel MS, Patel JA, Solanki SD (2018) Line× Tester analysis for kernel yield and component traits in maize (*Zea mays* L.). Res Crops 19(3):465–471. https://doi.org/10.31830/2348-7542.2018.0001.17
17. Nasser LM, Badu-Apraku B, Gracen VE, Mafouasson HN (2020) Combining ability of early-maturing yellow maize inbreds under combined drought and heat stress and well-watered environments. Agronomy 10(10):1585. https://doi.org/10.3390/agronomy10101585
18. Kamara MM, Ghazy NA, Mansour E, Elsharkawy MM, Kheir A, Ibrahim KM (2021) Molecular genetic diversity and line× tester analysis for resistance to late wilt disease and grain yield in maize. Agronomy 11(5):898. https://doi.org/10.3390/agronomy11050898
19. Dospekhov BA (1985) Field experiment methodology (with the basics of statistical processing of research results, 5th edn. Agropromizdat, Moskow. (in Russian)

20. Malik SI, Malik HN, Minhas NM, Munir M (2004) General and specific combining ability studies in maize diallel crosses. Int J Agric Biol 6(5):856–859
21. Allagulova CR, Gimalov FR, Shakirova FM, Vakhitov VA (2003) The plant dehydrins: structure and putative functions. Biochem Mosc 68(9):945–951
22. Rorat T (2006) Plant dehydrins-tissue location, structure and function. Cell Mol Biol Lett 11(4):536–556. https://doi.org/10.2478/s11658-006-0044-0
23. Palusa SG, Ali GS, Reddy AS (2007) Alternative splicing of pre-mRNAs of Arabidopsis serine/arginine-rich proteins: regulation by hormones and stresses. Plant J 49(6):1091–1107. https://doi.org/10.1111/j.1365-313X.2006.03020.x
24. Yeremenko O, Kalenska S, Kalytka V, Malkina V (2017) Sun-flower productivity depending on agrometheorological conditions of the Southern steppe of Ukraine. Agrobiology, 2:123-130. (in Ukrainian)

Pollen Concentration of *Ailanthus Altissima* on the South Transdanubia Region in Hungary

Viktor J. Vojnich, Árpád Ferencz, Tamás Tóth, Károly Tempfli, and Donát Magyar

Abstract In areas where the tree of heaven (*Ailanthus altissima*) appears and multiplies, the original vegetation degrades and transforms. The invasive tree of heaven is also of great importance in urban environments, where it causes building damage, static problems and endangers utilities. *Ailanthus* pollen concentration was measured during the 3-year period 2018–2020 at three county capitals (Kaposvár, Szekszárd, Pécs) of the South Transdanubia region, Hungary (Somogy county, Tolna county and Baranya county), with a 7-day Hirst-type (Burkard) pollen trap. The extent of the prevalence of *A. altissima* can be deduced from its pollen concentration. For this purpose, multi-year pollen data are displayed on a result map in which areas characterized by different pollen concentrations are represented by colour codes. The mass appearance of *A. altissima* is a serious problem in almost all areas (national parks, forests, inner city zones and towns), where its control/eradication would cost millions of euros. Pollen monitoring provides information on the size of *A. altissima* stands and provide a basis for proposals and plans for measures to control this invasive tree species and mitigate the damage caused by it.

Keywords *Ailanthus altissima* (tree of heaven) · Invasive plant · South Transdanubia region (Hungary) · Pollen concentration · Pollen monitoring data

V. J. Vojnich (✉) · Á. Ferencz
Faculty of Agriculture, University of Szeged, Hódmezővásárhely 6800, Hungary
e-mail: vojnich.viktor@mgk.u-szeged.hu

T. Tóth · K. Tempfli
Faculty of Agricultural and Food Sciences, Széchenyi István University, Mosonmagyaróvár 9200, Hungary

D. Magyar
Laboratory Department, National Public Health Center, Budapest 1097, Hungary

S. Pascuzzi and F. Santoro (eds.), *Farm Machinery and Processes Management in Sustainable Agriculture*, Lecture Notes in Civil Engineering 289,
https://doi.org/10.1007/978-3-031-13090-8_27

1 Introduction

The tree of heaven (*Ailanthus altissima* (Mill.) Swingle) belongs to the family *Simaroubaceae*. The genus, numbering about 10 species, has a focal point of distribution in the front and back of India, as well as in the Far East. The basic variant of the species, *A. altissima* var. *altissima*, is native to China. The bark of *A. altissima* var. *tanakai*, which is widespread in Taiwan, is yellower than the base species and has shorter leaves. It is recognized by the reddish branches of *A. altissima* var. *sutchuenensis*. In addition to the tree of heaven, the downy tree of heaven (*A. giraldii* Dode) and the thorny tree of heaven (*A. vilmoriniana* Dode) are also among the temperate species of the genus [1–3].

Three of heaven is dioecious, with male and female flowers being borne on different individuals. The flowers open in loose endless buds, greenish-yellow, woolly on the inside, consisting of 5 or 6 cups, petals and fruit leaves. The numbers of stamens per flower is 5 + 5 or 6 + 6. The flowers produce a lot of nectar [4, 5].

The tree of heaven is native in areas along the lower reaches of the Yangtze River to north-eastern, central China (Hubei, Honan, Anhui, Jiangsu; Hunan, Jiangxi, and Zhejiang provinces) and Korea. The distribution this species is currently between 22°N and 43°N up to 1,500–1,800 m above sea level. Its worldwide spread began in the 1740s, when its seeds were transported to Paris on a land journey through Siberia (Russia). It was planted in London in 1751 as an ornamental tree. It came to North America in the 18th century, where it was first planted as an ornamental tree in northern cities. In 1856, it was reported [6] that the stock was already in a natural forest at the territory of South Tyrol (Austria). It was planted in Paris to replace the hybrid plane or London plane tree (*Platanus*) in 1875. *Ailanthus altissima* is widespread in most of the western hemisphere today. It is often planted in the subtropical and northern temperate zones, including throughout East Asia and Europe, as well as North America. Due to its favourable properties, it has spread to all inhabited continents [7–10].

The first data on the occurrence of *A. altissima* in Hungary is available from a documentation of planting experiments of this species in Villány in 1841–1843 [11, 12]. It has been established in the Great Plain since the middle of the 20th century, due to its stock-like, conscious settlements and spontaneous outbursts. Today, it occurs almost everywhere in the hilly, lowland areas of the country with a warmer climate. It is relatively rare in Western Transdanubia and in the higher regions of the central mountains [13, 14].

Fig. 1 *Ailanthus altissima* growing in Australia

The tree of heaven is considered to be an invasive species [15]. Nowadays, the problem of invasive species is gaining more and more attention. In areas where the *A. altissima* appears and multiplies, the original vegetation degrades and transforms (Fig. 1). The tree of heaven is also of great importance in urban environments, where it causes building damage, static problems and endangers utilities. In addition, it is worth mentioning that *A. altissima* pollen is an allergen [16].

The aim of our study is to investigate the pollen concentration of *Ailanthus altissima* (tree of heaven) in the South Transdanubia region of Hungary.

2 Materials and Methods

The pollen concentration of *Ailanthus altissima* was measured at the three county capitals (Kaposvár, Szekszárd, Pécs) of the South Transdanubia region (Somogy county, Tolna county, Baranya county) (Fig. 2) with the 7-day Hirst-type (Burkard) pollen trap (Fig. 3 and Fig. 4a and b), for the 3-year period 2018–2020.

The device collides the pollen content of the intake air onto a tape with a sticky surface (Melinex tape). The tape is divided into two-hour bands and the pollen grains stained with basic fuchsine are counted in two 0.5 mm bands 6–6 mm from the edges

Fig. 2 The South Transdanubia region (Somogy county, Tolna county and Baranya county) in Hungary

Fig. 3 Hirst-type (Burkard) pollen trap

at 400 × magnification. The results are given as a 24-h average, expressed in pollen grains/m^3 of air.

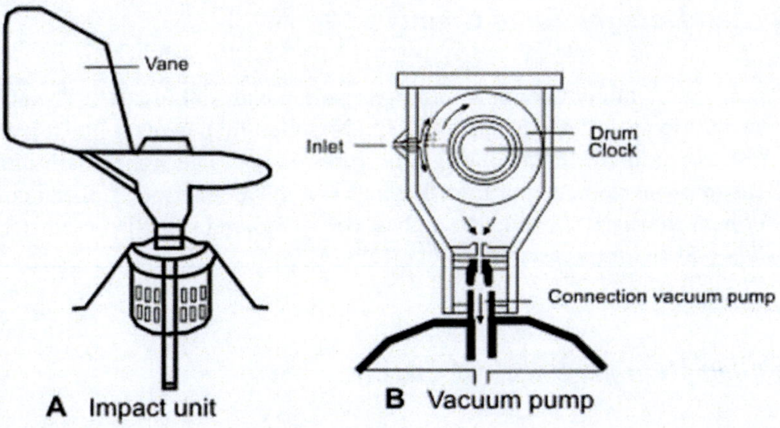

Fig. 4 **a** Impact unit in Hirst-type (Burkard) pollen trap. **b** Vacuum pump in Hirst-type (Burkard) pollen trap

The pollen season was defined as the day on which the sum of the daily average pollen concentration reaches 1% of the final amount and the end of the season when it reaches 99% [17].

3 Results and Discussions

3.1 Pollen Data for Somogy County

In Kaposvár, the pollen trap caught a total of 33 *A. altissima* pollen grains in 2018. The highest amount of pollen grains were counted on the 29th May (7 pollen/m^3). One year later, the pollen count of annual tree of heaven was 43, when the highest daily amount (11 pollens/m^3) was detected on June 15th. In 2020, 18 pollen grains were counted as an annual sum. The highest daily pollen count of 4 pollen m^3 was recorded in the city of Kaposvár on the 27th June 2020.

3.2 Pollen Data for Tolna County

In Szekszárd, the total number of *A. altissima* pollen was 140 in 2018. The highest daily concentration was detected on May 23rd (24 pollen/m^3). In the following year, a total of 190 *Ailanthus altissima* pollen grains were counted. The highest daily amount (65 pollens/m^3) was detected on June 8th. In 2020, the annual total pollen count of the tree of heaven was 109, while the highest daily amount (13 pollen grains/m^3) of *A. altissima* pollen was measured on June 15th.

3.3 Pollen Data for Baranya County

In Pécs, the total number of *Ailanthus* pollen was 198 in 2018. The highest daily concentration of pollen grains (39 pollen/m^3) was measured on May 19th. One year later, the total *A. altissima* pollen count was 161, while the highest daily concentration (58 pollen/m^3) was detected on June 7th. In 2020, we counted 121 *Ailanthus* pollen grains. The highest daily value (peak value) was recorded on June 11th (33 pollen/m^3) in Pécs.

The data on the daily pollen number of tree of heaven are shown in 2018 (Fig. 5), 2019 (Fig. 6) and 2020 (Fig. 7).

Fig. 5 Daily pollen concentrations of *Ailanthus altissima* (2018)

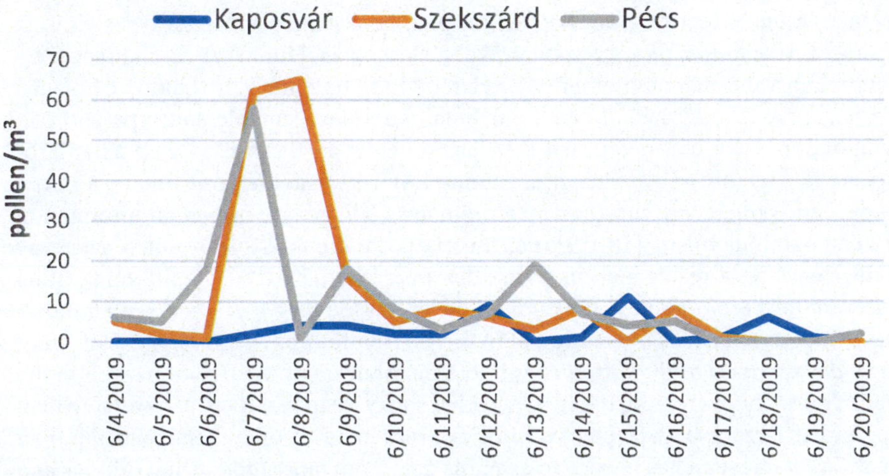

Fig. 6 Daily pollen concentrations of *Ailanthus altissima* (2019)

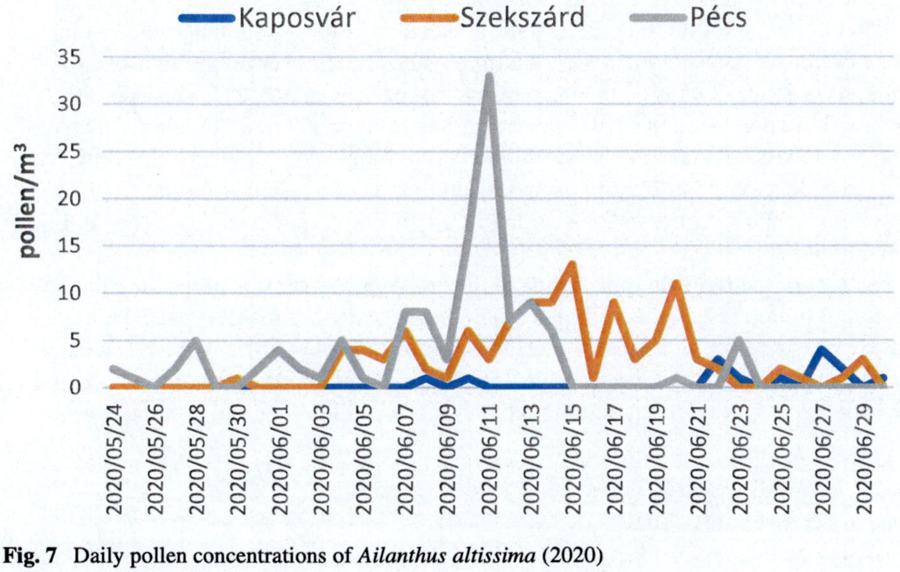

Fig. 7 Daily pollen concentrations of *Ailanthus altissima* (2020)

4 Conclusions

The highest total pollen count of *A. altissima* was measured in all three years in Baranya county (198 pollen in 2018; 161 pollen in 2019; 121 pollen in 2020), while the lowest value was measured in Somogy county in all three years (33 pollen in 2018; 43 pollen in 2019; 18 pollen in 2020). In Tolna county, the annual total pollen concentration varied between 109–190 pollen grains in the studied period.

Data on the extent of plant populations can be obtained in two ways: by direct survey of vegetation (e.g., National Weed Survey in Hungary) or indirectly by summarizing and mapping other data related to the potential distribution of plants. The latter group may include environmental variables (suitable soil type and pH, precipitation, etc.) that determine the plant's needs on the site as well as anthropogenic factors (crop structure, urbanization level, etc.), at the same time, data from atmospheric pollen concentration measurements also provide important information for anemophilous plants [18, 19]. For this purpose, the multi-year pollen data may be displayed on a result map in which the areas characterized by different pollen concentrations are represented by colour codes. Based on our studies, pollen monitoring data could also be used to study the distribution of *A. altissima.* Our work draws the attention to the differences in the distribution of the tree of heaven in the South Transdanubia region, based on which it can be seen that there can be more than a tenfold difference between the monitoring areas in terms of the total annual pollen count. Further research is needed to examine data from more monitoring stations and years that would allow for greater accuracy. In addition, we plan to include pollen data from other Hungarian cities, which can be explored on a national map to reveal the spatial differences in plant distribution.

The total annual pollen count (Table 1) of other pollen-producing woody plants (*Acer, Betula, Fraxinus, Moraceae, Platanus, Tilia* species) common in urban environments is generally higher than that of the tree of heaven [20, 21]. This is probably because these plants are mostly planted as urban ornamental trees and therefore have higher numbers of individuals and higher pollen emissions than *A. altissima.* As the tree of heaven is an invasive species, there is no so-called *Ailanthus altissima* containing species proposed for planting in urban green spaces in the List of Public Row Wood [22].

The size of the trees (height, radius of the crown projection) is proportional to the amount of pollen release according to some model calculations [23, 24]. The tree of heaven, like other noxious plants/weeds, is often removed before it reaches a larger size. This also results in lower pollen counts compared to other urban tree species, whose large specimens emit significant amounts of pollen (e.g. old sycamore trees).

Table 1 Total pollen number of plant species planted as urban ornamental trees in the South Transdanubia region (2019–2020)

Allergen species	City	Kaposvár		Szekszárd		Pécs	
	Year	2019	2020	2019	2020	2019	2020
Acer spp.		682	457	1549	1650	875	1208
Betula spp.		5360	3211	4800	5600	3455	2817
Fraxinus spp.		1291	1299	1956	3226	1718	2478
Moraceae		2090	---	5852	---	3419	---
Platanus spp.		10,465	2027	2067	857	10,381	4222
Tilia spp.		---	402	---	188	---	361

The situation of *A. altissima* is somewhat similar to that of *Broussonetia papyrifera*, which is also an invasive, anemophilous tree species in urban environments. However, their pollen production is not comparable because the pollen grains of *B. papyrifera* are pooled with other tree species having similar pollen grains (*Morus alba, Morus nigra*) as "*Moraceae* pollen" during aerobiological monitoring.

The mass appearance of *A. altissima* is a serious problem in almost all areas (national parks, forests, inner city zones and towns), where its control/eradication would cost millions of euros. Pollen data provide information on the size of *A. altissima* stands endangering the interiors, utilities and buildings of cities. Pollen monitoring provides a basis for proposals and plans for measures to control the invasive tree species and mitigate the damage caused by it.

Acknowledgements The authors would like to thank *Vivien Pál* for her work in producing pollen data from the National Public Health Center.

References

1. Hegi G (1924) Illustrierte Flora von Mittel-Europa. VII, München
2. Hu SY (1979) *Ailanthus*. Arnoldia 39(2):29–50
3. Udvardy L (2004) Bálványfa (*Ailanthus altissima* [Mill.] swingle). In: Mihály B, Botta-Dukát Z (szerk.) Biológiai inváziók Magyarországon. Özönnövények, A KvVM Természetvédelmi Hivatalának Tanulmánykötetei 9, Természet BÚVÁR, Budapest
4. Barkman JJ (1988) New system of plant growth forms and pheno-logical plant types. In: Werger MJA, Aart PJM, van der During HJ, Verhoeven JTA (eds) Plant form and vegetation structure, pp 9–44. SPB Academic Publishing, The Hague
5. Bartha D (1994) Magyarország faóriásai és famatuzsálemei. Erdészet-történeti Közlemények 15:242
6. Ascherson P, Graebner P (1914) Synopsis der Mitteleuropäischer Flora. Leipzig-Berlin, V/1, pp 220–370
7. Good R (1974) The geography of the flowering plants. Longman, London
8. Böcker R, Kowarik I (1982) Der Götterbaum (*Ailanthus altissima*) in Berlin (West). Berlin 26:4–9
9. Kowarik I (1983) Zur Einbürgerung und zum pflanzengeographischen Verhalten des Götterbaumes [*Ailanthus altissima* (MILL.) SWINGLE] in französischen Mittelmeergebiet (Bas-Languedoc). Phytocoenologia 11:389–405
10. Kowarik I, Böcker R (1984) Zur Verbreitung, Vergesellschaftung und Einbürgerung des Götterbaumes [*Ailanthus altissima* (MILL.) SWINGLE] in Mitteleuropa. Tuexenia 4:9–29
11. Bartosságh J (1841) Beobachtungen und Erfahrungen über den Götterbaum (*Ailanthus glandulosa* L.) Ofen, III
12. Bartosságh J (1843) Folytatólagos értesítés a bálványfa (*Ailanthus glandulosa*, Götterbaum) terjedése körül. Magyar Gazda 3:298–300
13. Soó R, Jávorka S (1951) A magyar növényvilág kézikönyve I–II. Akadémiai Kiadó, Budapest
14. Kovács M, Priszter Sz (1974) A flóra és vegetáció változása Magyarországon az utolsó száz évben. Botanikai Közlemények 61:185–196
15. Landenberger RE, Kota NL, McGraw JB (2007) Seed dispersal of the non-native invasive tree *Ailanthus altissima* into contrasting environments. Plant Ecol 192(1):55–70
16. Mousavi F, Majd A, Shahali Y, Ghahremaninejad F, Shoormasti RS, Pourpak Z (2017) Immunoproteomics of tree of heaven (*Ailanthus atltissima*) pollen allergens. J Proteomics 154:94–101

17. Ziska LH, Makra L, Harry SK, Bruffaerts N, Hendrickx M, Coates F, Saarto A, Thibaudon M, Oliver G, Damialis A, Charalampopoulos A, Vokou D, Heiđmarsson S, Guđjohnsen E, Bonini M, Oh J-W, Sullivan K, Ford L, Brooks GD, Myszkowska D, Severova E, Gehrig R, Ramón GD, Beggs PJ, Knowlton K, Crimmins AR (2019) Temperature-related changes in airborne allergenic pollen abundance and seasonality across the northern hemisphere: a retrospective data analysis. Lancet Planet Health 3(3):e124–e131. https://doi.org/10.1016/S2542-5196(19)30015-4
18. Thibaudon M, Šikoparija B, Oliver G, Smith M, Skjøth CA (2014) Ragweed pollen source inventory for France-the second largest centre of Ambrosia in Europe. Atmos Environ 83:62–71
19. McInnes RN, Hemming D, Burgess P, Lyndsay D, Osborne NJ, Skjøth CA, Vardoulakis S (2017) Mapping allergenic pollen vegetation in UK to study environmental exposure and human health. Sci Total Environ 599:483–499
20. Udvardy O, Kajtor-Apatini D, Magyar D, Szigeti T (2020) A magyarországi Aerobiológiai Hálózat tájékoztatója 2019. Nemzeti Népegészségügyi Központ, Budapest
21. Kajtor-Apatini D, Magyar D, Udvardy O, Bócsi E, Pál V, Szigeti T (2021) A magyarországi Aerobiológiai Hálózat tájékoztatója 2020. Nemzeti Népegészségügyi Központ, Budapest
22. Szabó K (2019) Közterületi Sorfák Jegyzéke. Magyar Díszkertészek Szövetsége, Budapest
23. Cariñanos P, Casares-Porcel M, Quesada-Rubio JM (2014) Estimating the allergenic potential in urban green zones: a case study in Granada. Landsc Urban Plan 123:134–144
24. Magyar D, Páldy A, Szigeti T, Szilágyi A, Orlóci L (2020) A potenciális allergenitás felhasználási lehetősége a zöldterületek minősítésében és az allergén terhelés szabályozásában. Egeszsegtudomany 64(4):57–80

Social Networks for Sustainable Development in Greece: The Case of Messenia

Athanassios Papageorgiou

Abstract The deadlocks encountered in the agricultural sector of an increasing number of countries, over the last decades, concerning production cost, products' distribution as well as "extra-financial" social and environmental burdens, have strengthened the idea that conventional agriculture has reached its limits. Integrated and organic farming are a common application method of sustainable agriculture, but in the case of Greece its results has been limited. In many cases, farmers have been reluctant to apply sustainable agriculture for an extended period of time. Usually, the transition towards SA is done gradually on only some parts of the farms and it's difficult to have a clear image of the cost-efficiency between sustainable and conventional agriculture, making the farmer's decision to fall back to conventional agriculture easier. The contribution of the social networks for sustainable development has been highlighted from many researchers; however, the level of their impact has been fluctuating. This article analyses the factors affecting the farmers' decisions concerning their cultural practices, as well as the factors related to the attractiveness and the contribution of social networks towards sustainable development.

Keywords Conventional agriculture · Sustainable development · Farmers' networks · Social attitude · Greece

1 Introduction

The increased competition observed, in the global agri-food system and the problems encountered in the distribution of many products due to market saturation and malfunctions regarding their quality, have strengthened the idea that conventional agriculture (CA) has reached its limits. For regions where the agricultural sector has seen a noticeable growth, including many countries of the European Union (EU), the crisis caused by overproduction and the drop in prices, leads to a cultural crisis, the loss of the possibility of product differentiation, due to the homogenization

A. Papageorgiou (✉)
Department of Agronomy, University of the Peloponnese, Antikalamos, 24100 Kalamata, Greece
e-mail: a.papageorgiou@uop.gr

S. Pascuzzi and F. Santoro (eds.), *Farm Machinery and Processes Management in Sustainable Agriculture*, Lecture Notes in Civil Engineering 289,
https://doi.org/10.1007/978-3-031-13090-8_28

of the productive system. Agricultural income is decreasing, since it is more and more correlated to the fluctuations of a highly specialized market due to the limited product variety and based on European funding. Furthermore, CA practices incorporate "extra-financial" dangers related to the rural society as many farmers have been pushed out of the sector and to the natural resources [1, 2].

The environmental cost of the high-input CA led to concern and scepticism among both the residents of rural areas and the agricultural leaders alike. Word Bank and United Nations experts were among the first to systematically analyse the environmental impact of the applied development model 40 years ago [3, 4], while at 1992, the reform of the Common Agricultural Policy, constituted the first challenge of the productivist cultural practices, highlighting the effects it had on the environment [5], and the undesirable excesses in production. Agricultural activity is considered to be an element of a system, which consists of the economic, the social and the environmental conditions of each region comprising and functioning as a whole [6], an idea that is also among the core ideas of sustainable agriculture (SA).

Furthermore, the efficiency and the profitability of the agricultural activity is not only measured using indexes such as productivity per hectare or production cost, but also production value. The latter can be twofold, quantitative and qualitative (i.e., organoleptic, dietary and even symbolic) and even in cases where the production is relatively low, as many times it is the case for alternative agriculture (AA) like integrated (IF) or organic farming (OF), revenues can be balanced via a higher price. The goal of an AA is to ensure the survival of the agricultural holding via decreasing the production cost, utilizing the whole of the available resources, and/or via the income increase by expanding the activity range, including processing, direct trade, or other services in or out of the farm.

Although in Greece there were multiple attempts to apply the SA, via IF or OF, its impact was limited because farmers' majority had an opportunistic approach. Their interest was dependent on the related funding leading many times to them going back to the CA and to the dismantling of networks already put in place, when the funding was over.

The importance of networks in the distribution and management of knowledge [7, 8] and in empowering pro-environmental agriculture has been stated by many researchers. Hassanein and Kloppenburg (1995) mention the contribution of these networks on "farmer-generated knowledge, holistic and ecological thinking, and practical research" [9] for alternative agriculture, enabling the management of the production in a sustainable way. Others refer to the existence of "socio-technical niches" that achieve a spatial valorisation through the commitment of the key actors in preserving cultural, ecological and environmental integrity and in finding ways to create economic benefits for the local community [10, 11].

The aim of this study is to analyse the factors that play a role in developing and maintaining sustainable agriculture in Greece and the networks that apply it. The area of the study is the prefecture of Messenia, in south-west of Greece, where the agricultural sector plays a central role in the local economy and there are individual integrated and organic farmers, IF greenhouses and SA networks.

The paper consists of six sections. The theoretical framework approaching the process towards sustainable development and networking. Next a short section on methodology and the case description. Results are discussed in the fourth section and are structured in two sub-sections, data analysis and the factors affecting networks and sustainability resilience. Section five is centred on the discussion, thus leading to the conclusions.

2 Theoretical Framework

Over the years, farms with the greatest exposure to intensification began to face numerous problems. The application of CA increased the average crop yield per hectare, but the widespread use of agricultural machinery and the increased inputs, often resulted in a disproportionate increase in production costs. Many farmers are greatly in debt, partly because of heavy investments in specialized machinery and other equipment. The repayment of the debts including interest introduces the need to adopt new techniques and to develop their infrastructure continuously, but every time with an unchanged income [12, 13]. The same farms are associated with declining soil productivity, deteriorating environmental quality, reduced profitability and threats to human and animal health.

SA, by respecting biodiversity, uses local populations, well-adapted to the local habitat, resistant to parasites and diseases, offering a variety of ecological services including recycling of nutrients, regulation of microclimate and local hydrological processes, suppression of undesirable organisms, detoxification of noxious chemicals, etc. [14, 15], and at the same time ensures cost reduction, by using low input cultural practices. These production systems contribute to the higher authenticity and quality of the products, and to the increase of their competitiveness [16].

However, as the Brundtland Report (1987, 15) consider "sustainable development is not a fixed state of harmony, but rather a process of change in which [...] painful choices have to be made". In relation to the agricultural sector, sustainable development implies a continuously ongoing dynamic process [17, 18] and the transition from a CA to a SA agriculture presumes the adoption of practices and the restructuring of the landscape, with a higher degree of complexity than the one encountered in the CA, since it takes into account the specific characteristics of each region individually. The more these characteristics change the higher the risk becomes when applying it. Thus, the need to aid the farmers during the transition period to adapt to the changes as well as the importance of networks for interaction, sharing, exchange of experience and learning becomes clear [19–22].

According to Moschitz et al. (2015) "a network consists of individuals and organizations and the relations between them". However, during the transition period, from the CA to the SA, establishing and especially maintaining these networks can encounter multiple hurdles. The hurdles are mainly related to the complexity of the transition in both the technical and collective action levels since everyone has his/her own perception of resolving the emerging issues. There is a social transformation that takes place making the whole process even more complex [23]. Pretty and Ward (2001), when referring to the social capital whose "central aspects are identified: relations of trust; reciprocity and exchanges; common rules, norms and sanctions; connectedness, networks and groups" mention that "as it lowers the costs of working together, social capital facilitates cooperation. People have the confidence to invest in collective activities" [24]. Once the group acts as a unified front, its members enjoy being a part of the group, working for a common goal, lowering significantly the transactional and organizational costs, and increasing their effectiveness [25].

3 Materials and Methods

3.1 *Methodology*

Taking for granted that the challenge towards an alternative, locally oriented, SA is not taken up individually, in many regions farmers organize themselves in small or larger entities, such as formal or informal groups, cooperatives or networks.

The study takes place in the prefecture of Messenia, a region in the south-west of Peloponnese, and includes individual integrated and organic farmers, IF greenhouses, SA's networks, producers that have distanced themselves from the groups, and existing networks. It is not restricted to formal groups, since by including from small, maybe informal groups, in rural communities to larger networks with far broader impact, we present the full spectrum of awareness about SA in the region.

The selection of the criteria for the sample delimitation is related to the application (or lack of) of sustainable farming practices based on the fact that SA can be multidimensional according to its objectives such as: the production of goods and services that respect the environment (sustainability related to the practices and systems used), the water management, the management of the soil capital and the help to avoid its erosion, the improvement of the climate, the protection from the natural disasters, the conservation of biodiversity and the preservation of the natural landscape [26].

Taking into account the specific characteristics presented in the study's region, the following tools are used:

- Sustainable practices: examining the protocols and the way they are applied.
- Sustainable materials: examining the propagating material, the animal's breeds used in the process and the way they affect biodiversity.
- Sustainability dissemination: examining the dynamic with which SA is diffused.

Relative to the first two criteria, the three following combinations for the sample delimitation are encountered:

1. Sustainable practices on hybrid varieties.
2. Sustainable practices on traditional varieties - local populations.
3. Conventional practices on traditional varieties - local populations.

The research questionnaire consists of four parts: a) characteristics of the questionee (farmers' network or individual) and their professional background, b) production and distribution of products, c) sustainability in farmers' decision making and d) experience from participating in a sustainable development network.

Due to the lack of regional data related to the evolution of SA in the Greek Ministry of Agriculture, since there is not a complete record of the farmers engaged in it, the sampling process was purposive rather than random. This point constitutes a gap in the current study since it is not clear if the producers are also sensible in subjects of biodiversity. There is the sense that in the region individuals and informal groups of farmers are engaged in maintaining local populations not only for personal use but also for marketable crop and livestock production.

3.2 Case Description

In Greece we encounter two types of SA, with varying degrees of strict application, IF and OF.

The first attempts towards an alternative agriculture can be identified in the early 80's. It coincides with the arrival of foreigners, mainly from the north of Europe. They introduced new agricultural practices, mostly related to OF for auto-consumption, but at the same time they marked the beginning of a different approach on the agricultural productive process. In Greece, at that time, there was no interest in OF, since the CA was not intensified, as much as it was in the rest of Europe. The delayed adoption of productivist agricultural practices in Greece ended up being an advantage, at a later stage, for the transition from conventional to sustainable agriculture [27].

Alternative agriculture started evolving towards its industrial form in 1982, thanks to the interest of a Holland industry for organic raisin. In 1988, Friedrich Blauel, an Austrian olive oil importer, coordinated a project of organic olive oil production in the region of Mani, a semi-mountainous region of Messenia. This led to the creation of the first organic farmers' cooperative in Greece, but this network doesn't exist anymore. This represented the first wide scale project of OF in the country, before the Council Regulation (EEC) No 2092/91, in which a German protocol was applied.

The area of study is the Prefecture of Messenia, a region without large valleys, a factor that justifies its lack of agricultural development. It represents 3% of the agricultural area of the country and 4% of agricultural holdings, from which 3% engages with OF. Agricultural area under organic practices corresponds to 4.1% of the total area, 89.9% of which is on olive growing, while for Greece as a whole this culture corresponds to 47.8% of the agricultural area under OF. The rural exodus, at least early on, contributed to a limited intensification (e.g.: limited use of chemicals) of the production process in relation to the rest of Greece, and to the preservation of local populations, centred on perennial species such as olive trees (representing 87.6% of the agricultural area of the prefecture), fig trees, prickly pears, artichokes, aromatic plants and vines for raisin production, although the latter is almost extinct.

The empirical material of the study was collected in 2019 from nine out of thirty-one municipalities of Messenia through twenty-four personal interviews in the interviewee's workplace, based on a semi-structured questionnaire. Four were with Public Service executives and among them three were related to IF and OF and one to biodiversity, the local coordinator for the European program "On farm conservation of rare and endangered local animal breads". Two were with representatives of certification organizations for the agricultural practices and two with individuals active in educational programs around biodiversity, who have also created local seeds banks.

Concerning the production process, seven interviews were made with representatives/directors of farmers' networks. Five of these interviews were with the directors of agricultural cooperatives being almost exclusively occupied with olive oil production, one with an executive of "SYKIKI" cooperative of dried figs, and one with the coordinator of a producers' group of raisins. Also, two interviews were with representatives of informal groups of farmers being in transition from CA to OF, one on olive oil and another on aromatic herbs. One interview was made with a former member of a cooperative, one with a farmer who decided to leave an informal group to follow his personal production protocol and one with a tomato greenhouse manager. Furthermore, about the preservation of traditional varieties using pro-environmental practices three interviews were made, one with a producer of dried figs, one with a producer of prickly pears and one with a producer of artichokes (Table 1).

Finally, relatively to the diffusion of biodiversity one interview was made with a former member of an urban garden.

4 Results

4.1 Data Analysis

Five out of the seven interviewed agricultural cooperatives are almost occupied with the production and the distribution of olive oil and only one farmer uses the network for the distribution of table olives. Apart from one cooperative, which has as active members all the members of the community, the rest represent a more or less significant part of olive oil producers.

Farmers belonging in a cooperative can be divided into two main subgroups: simple members and those who have created integrated management organisations (IMO), i.e. those who are applying SA practices, following the IF standard AGRO 2.[1] A small percentage of the farmers are involved in OF.

All the aforementioned cooperatives are part of the "Olive growing farm organizations" machinery supply operational program, and although their production,

[1] The Integrated Management System is a system of organizing an agricultural holding which includes, inter alia, Good Agricultural Practice, Workers' Safety and Health, Product Safety, Traceability and Environmental Action.

Table 1 Basic data of the Messenian agriculture and the selected sample

	Total	Sample population					
		Farmers' network			Individuals		
		All members	IF	OF	Total	IF	OF
Olive growing							
Holdings	26,291	712	394	55	2		1
UAA (Ha)	85,632.00	2,631.00	1,219.40	166.00	11.00		4.00
Raisins							
Holdings	1,619	250	45				
UAA (Ha)	1,525.70	260.00	70.00				
Dried figs							
Holdings	291	270	124	7	1		
UAA (Ha)	766,14	715.40	326.84	15.63	1.20		
Hydroponic vegetables							
Holdings	24				1	1	
UAA (Ha)	28.30				4.00	4.00	
Artichokes							
Holdings	17				1	1	
UAA (Ha)	38.00				7.80	7.80	
Prickly pears							
Holdings	2				1		
UAA (Ha)	2.02				1.15		

Notes IF (Integrated Farming), OF (Organic Farming)

volume wise, is not substantial, they have managed to supply their products on the domestic and international market.

The raisins cooperative was founded in 2005. Part of the cooperative has created an IMO and applies AGRO 2.1 and AGRO 2.2 protocols. However, due to the lack of motives for sustaining this cultivation the cooperative is shrinking since the older members have left and the younger members are switching towards olive growing.

The most established farmers' network in the region is "SYKIKI" dried figs cooperative, providing consulting services to the farmers throughout the production process, on the preparation of the final product and the market offering. It was founded in 1923 and nowadays it consists of around 500 members (not only Messenia's), 124 of them are part of an IMO following AGRO 2.1 and AGRO 2.2 protocols. The region has only one local variety of figs and each year the cooperative, having designated the special farms where the mother plants are growing, receives the propagating material, reproduces and distributes the totality of the trees for the new plantations of the region, maintaining in this way this local population. Out of the total production around 30% represents IF and 5% OF.

Two informal farmer groups are in transition from conventional to OF. The first one consists of six people. Apart from olive oil, it also produces table olives, dried grapes, and one of the members is producing aromatic herbs. They have joined the Interreg Mediterranean project ARISTOIL,[2] a project co-financed by the European Regional Development Fund. The second group consists of three members, occupied with the cultivation of local mountain tea. The production takes place on a Taïgetos's plateau at an altitude of over 1700 m. They have joined two research programs, one via creating a cluster with four other companies and the Department of Pharmacology of the University of Athens and the second program is called UBQ2 "remote crop monitoring infrastructure program" (PA 2014–2020). Two out of the three members used to work in a different field, but, as it was also the case for the members of the first group, they decided to leave the city centre and return to their place of origin due to the economic crisis.

The tomato greenhouse belongs to a single-person and it was founded in 2012, with the contribution of a European funding program. It uses a hydroponic production process on improved varieties and applies an integrated management system, including a system to recycle its liquid waste. Since its founding it has been on a development path and in 2017 has expanded its facilities. This greenhouse constitutes an agricultural enterprise engaged in developing a self-assessment protocol for their particular operations [15].

The sample also includes a producer, who for four years used to participate in an IMO, but two years ago selected to differentiate him to apply OF. He follows a personally defined protocol which has as its goal the least possible human interventions in the reproductive process of the ecosystem. He has joined the ARISTOIL project. He is young and although he used to work in a different sector, he chose, due to the economic crisis, to make use of the family holdings.

Organic livestock production is almost non-existent. The production of organic meat is very restricted due to both the lack of demand and the lack of certified slaughterhouses. The limited demand is covered by individual butchers. The main interest for organic livestock production is focused on milk production for dairy products and mainly sheep and goat milk for the production of Feta cheese.

Concerning the preservation of endangered species and breeds animal genetic resources, the data collected following the Council Regulation (EEC) No 1257/1999 "on farm conservation of rare and endangered local animal breeds" show that in the period 1988–2012 two species, sheep and equine, have been recorded. The population of sheep has a significant decrease, while the population of equine remains steady. Since 2013, there is no official data related to the preservation of indigenous species because no breeder meets the requirements needed to join this program since the pastures were not certified.

[2] Goal of this project is the "reinforcement of Mediterranean olive oil sector competitiveness through development and application of innovative production and quality control methodologies related to olive oil health protecting properties" [28].

4.2 Sustainable Agriculture and Networks Resilience

The first wide application of SA, via OF, in Greece started after the Council Regulation (EEC) No 2029/91 and especially after 1995 with the inflow of European funding. At that time, the first certification organizations were founded.

In the early 2000's, the IF programs started with the creation of new certification organizations. It's via these organizations and their representatives in rural areas that IMOs began to be formed as sub-groups inside of existing cooperatives.

Apart from the dissemination of information about the new programs, cooperatives are also places where farmers are gathered and thus allowing the wider spread of AA in the country. As Nelson et al. (2014) mention, in the case of OF, representatives from the certification organizations acted as "bridges" by connecting the farmer to a new sub-network or source of information [29].

This period marks the beginning of an endless back and forth for the SA in Greece, represented by the continuous vacillation from the farmers' side regarding the cultural practices, meaning their engagement in SA, and their decision to work either as individuals or as a part of a group.

Networks are organized around a marketable product, possible variations and its by-products. Much attention is dedicated to the products' concept and development, the production's rules and procedures, and to the transfer of knowledge. In addition, they are occupied with building distribution networks, in domestic and international markets, having as a direct effect, the improvement of the managerial skill of its members.

Inside the networks, three sub-groups can be identified, simple members representing the majority, the IMOs and those who are occupied with OF. However, due to the fluctuating interest of the farmers to participate in these networks, there is a continuous shift regarding the active members of both the total network and their internal subgroups. As shown in the following (Fig. 1) the structure of these networks are not set by strictly set sub-teams but rather depend on the approach the members follow regarding the SA. They may transition from the CA to the SA (integrated or organic) or follow the opposite route. This fluctuation depends on multiple factors, such as the existence of financial incentive and the internal organization of the networks. Regarding the members in the sub-groups, apart from the aforementioned factors, their participation also depends on the existence of certification agencies, infrastructure for the certification procedure, seeds' banks, and the support received from the Ministry of Agriculture. The last factor seems to be more important for those who don't have a prior farming experience but come from a different profession. All these factors play a role in the decision of the farmers relative to the cultural practices, and their participation in the networks.

Relatively to the dynamic of the SA networks, although at first there is excitement about the opportunities the participation in these networks may offer, the knowledge diffusion and the ties created among its member, after the first funding phase, lasting usually 3–5 years, the expectation start to tone done and the networks' shrinking begins. Farmers' attitude depends on the existence of positive outcome which is

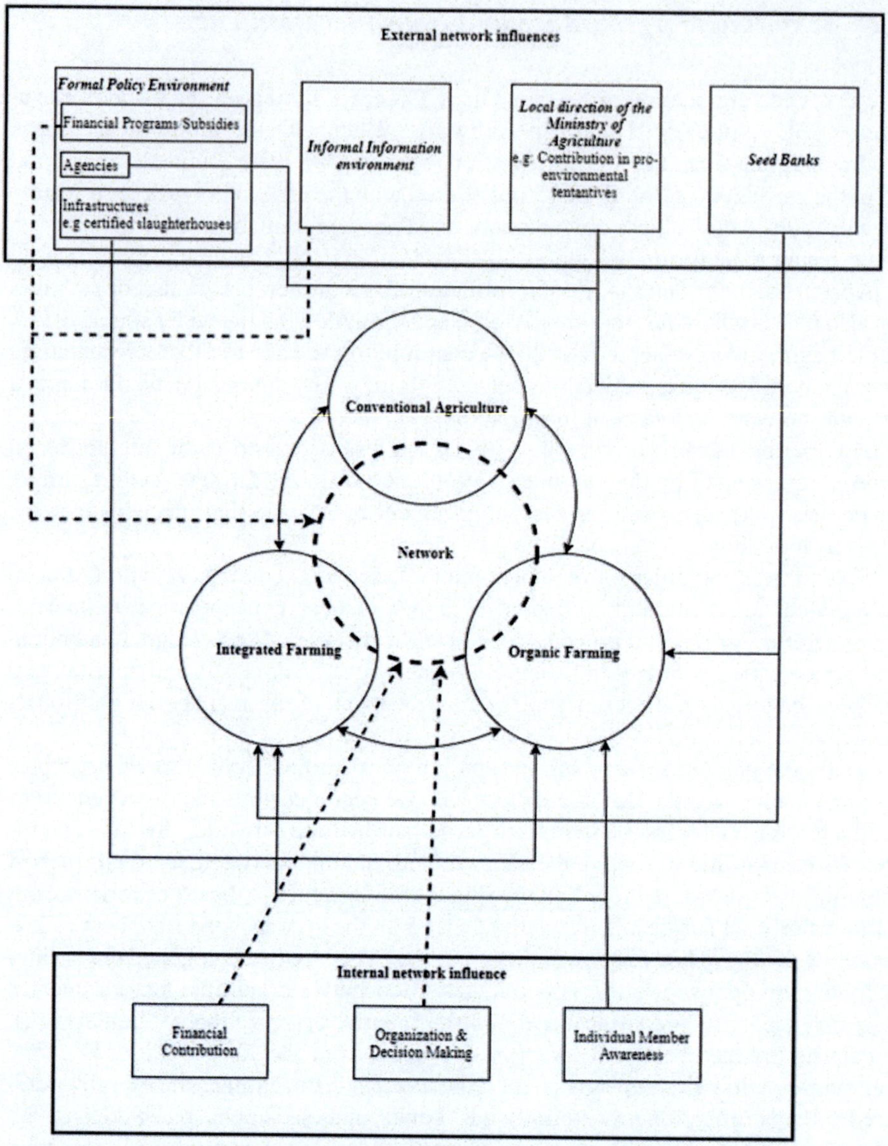

Fig. 1 Internal and external factors affecting networks and pro-environmental practices

received as a reward from the repeated sustainable actions [30, 31], as well as on whether they tend to think about the short term and/or have difficulties in estimating the benefits of long-term solutions around IF and especially OF [32]. Furthermore, the ties among the members tend to grow stronger as we observe simple members, members of IMOs and members of OF [29]. However, these ties are exposed to factors that can diminish their strength and as a result affect the membership of the networks. As Portes (1998) notes about social capital, these factors are about the feeling of exclusion of outsiders, excess claims on group members, restrictions on individual freedoms and downward levelling norms [33]. The two last factors play a major role especially on new farmers, as the ones observed in the sample of this research. The departure of one farmer of the sample is related to his decision to apply a strict protocol of pro-environmental farming which validates the premise that "sustainable behaviours change if people are made aware of the severity of the problem if they know what to do or if they realize that it is in their best interest to do so" [34].

Furthermore, organizational problems can occur when there is a feeling of inequality among the members, and/or there are farmers giving the impression of downgrading the efforts of other members, having as consequence the loss of trust. Also, problems can emerge when there are "self-regarding" preferences as "people who free-ride on the contributions of others" [32].

As a conclusion, all these problems lead to the shrinking of the networks, independently of the positive attitude regarding their operation, and usually the farmers that choose to lead act independently, capitalizing on funding or other tools, different from the ones used by the cooperatives.

5 Discussion

Apart from the "extra-financial" impact to the society and the environment, the financial impact of the SA towards improving farmers' income is twofold: the reduction of the production cost and the product differentiation depending on the applied protocols, leading to a potential increase in pricing. A producer from the sample managed to increase the price of his virgin olive oil by two and a half fold in relation to the conventionally produced one. The SA's networks diffuse the knowledge on production practices, on funding tools, facilitating the inclusion of farmers in programs that otherwise would be difficult to join, such as the "Olive growing farm organizations" machinery supply operational program, and on projects for the treatment of the remnants of the production cycle (e.g. compost). Finally, networks help via mitigating the perception of risk, mainly during the transition towards this new approach.

However, when farmers don't have a stable approach regarding the cultural practices, swaying between different systems, they can be overwhelmed by the created confusion. In many cases, the conventional practices aren't abandoned. The transition towards SA is usually done gradually on only some parts of the farms. With this fractured productive process, it's difficult to have a clear image of the cost-efficiency between SA and CA, making the farmer's decision to fall back to CA easier. A lack of social resilience is apparent, regarding the shift towards the sustainable approach of the agricultural activity and the ties created in the networks between the farmers are usually weak, thus not providing a motive to continue working as a group for addressing problems arising [35], leading them to quit the network and usually go back to CA.

The venture has a level of perpetuity when it is linked with specific markets such as the organic products and when the farmers manage to see economic benefits in a short period. When the incentive for SA is exclusively for absorbing the funding and this venture has strictly opportunistic origins, the farmers lead back to CA practices when the funding is finished.

Regarding the attractiveness of the cooperatives, it depends on their necessity for conditioning and distributing the final products. When this service is provided by other companies, the cooperative can act as a bridge for the farmers in this new productive process but after some time, when their necessity is less evident, it's abandoned. This happened in the case of the first cooperative of organic olive oil in Mani.

The attractiveness of the SA in the region of Messenia is limited. The number of participants in the IMOs depends on whether funding is provided. Once the program is ended, producers fall back to CA. According to Koutsouris (1998) "evolution and development can be conceived as a process of 'creative destruction' when new innovations in effect lead to the destruction of established practices" [36]. However, the majority of the farmers select to avoid the hurdle, and prefer to have the best possible financial result while putting in the least possible effort, mainly related to the workload. The casualty of this mentality was OF, which although, when started, had a militant position it ended up being the rear-guard, marginalized and without a viable future.

The exception to this paradigm, are people that understand the limited effectiveness of the CA and follow alternative farming and they are willing to collaborate and create groups to facilitate the transition. Among the interviewees there is a strong impression that there will be a rise in the number of the IMOs' members, although at the same time it seems that there is a lack of social sensibility, of what the economic theory of natural resources highlights as cultural capital, which in turn justifies the great diversity of sustainable development applications in Greece [37].

6 Conclusion

The diversity of the Greek rural countryside allows for the multidimensional development of the agricultural sector, and the product differentiation. The application of sustainable protocols taking into account the regional and local conditions can enhance the economic, social and environmental capital over time, offering even more diverse capabilities for the future. However, the vast majority of the farmers had an opportunistic approach. Although the pre-existing agricultural system, at least up until 1980's, was not intensified and the transition towards SA could be smooth, there was not an interest to move towards this direction.

The long-lasting economic crisis that brought drop to the farmers' income and the degrading of the standard of living of the society demonstrates the need for change. During the last decades, farmers' interest in joining groups and networks that could assist them in establishing a different sustainable production system is increasing.

They are new farmers, experienced innovative farmers, or farmers' offsprings who returned to their country of origin since they could not find work in the city centres.

Investing in SA is not done exclusively by farmers. There are also individuals that understand the arising business opportunities and are interested in the future development of their region. They follow production protocols taking advantage of the market's concerns, and generate revenue in the process, having already understood that the countryside is not a short-term source of income that can produce for an infinite amount of time, but it is an evolving ecosystem affected by everyone.

Finally, raising awareness on an individual level is not the most important aspect, since some will understand the importance of adapting a pro-environmental approach. The goal needs to be the shift of the general public's mentality and social networks can play a vital role in this process.

Acknowledgements The author wishes to thank the Department of organic farming of the Direction of Agricultural Ministry in Kalamata, the Director of the Agricultural Ministry Office in Trifylia and the Director of Olive, Fruit and Vegetables Institute of Kalamata for their contribution in finalizing the sample.

References

1. Briel B, Vilain L (1999) Vers l'agriculture durable. Éditions Educagri, Paris
2. Latouche S (2010) Le pari de la décroissance. Éditions Pluriel, Paris
3. Austin JE (1981) Agroindustrial project analysis. In: EDI series in economic development, world bank. The Johns Hopkins University Press, Baltimore
4. Brundtland G (1987) Report of the World Commission on Environment and Development: Our Common Future. United Nations General Assembly document A/42/427. Oslo
5. Andersen PP, Lorch RP (1994) Alleviating poverty, identifying agriculture and effectively managing natural resources. In: Food, Agriculture and the Environment. Discussion Paper l. International Food Policy Research Institute, Washington
6. Biology on line Homepage. https://www.biologyonline.com/dictionary/system. Accessed 04 Apr 2022
7. Goulet F (2013) Narratives of experience and production of knowledge within farmers' groups. J Rural Stud 32:439–447
8. Núñez-Espinoza JF, Figueroa-Rodríguez OL, Jiménez-Sánchez L (2014) Elementos para analizar redes sociales para el desarrollo rural en México: El caso RENDRUS [Element available to analyse social networks for rural development in Mexico: The RENDRUS case]. Agricultura Sociedad y Desarrollo 11(1):1–24
9. Hassanein N, Kloppenburg JR Jr (1995) Where the grass grows again: knowledge exchange in the sustainable agriculture movement. Rural Sociol 4:721–740
10. Marsden T, Smith E (2005) Ecological entrepreneurship: sustainable development in local communities through quality food production and local branding. Geoforum 36(4):440–451
11. Vlachos G, Karanikolas P, Koutsouris A (2017) Integrated farming in Greece: a transition-to-sustainability perspective. Int J Agric Resour Gov Ecol 13(1):43–59
12. Latouche S (2006) Pour ou contre la croissance économique? In: Guibert B, Latouche S (eds) Antiproductivisme, altermondialisme, décroissance, pp. 71–89. Éditions Parangon, Lyon
13. Petit M (2011) Pour une agriculture mondiale productive et durable. Éditions Quae, Paris

14. Karanikolas P, Bebeli P-J, Thanopoulos R (2018) Farm economic sustainability and agrobiodiversity: identifying viable farming alternatives during the economic crisis in Greece. J Environ Econ Policy 7(1):69–84
15. Koohafkan P, Altieri M, Gimenez EH (2012) Green agriculture: foundations for biodiverse, resilient and productive agricultural systems. Int J Agric Sustain 10(1):61–75
16. Papageorgiou A (2012) Sustainable development as a solution to agricultural competitiveness: the case of Greece in the era of European integration. J Central Eur Agric 13(3):458–474
17. Bidou D (2015) Le développement durable, une affaire d'entrepreneurs. Éditions PC, Paris
18. Kroma M (2006) Organic farmer networks: facilitating learning and innovation for sustainable agriculture. J Sustain Agric 28:5–28
19. Esparcia J (2014) Innovation and networks in rural areas: an analysis from European innovative projects. J Rural Stud 34:1–14
20. Milestad R, Darnhofer I (2003) Building farm resilience: the prospects and challenges of organic farming. J Sustain Agric 22(3):81–97
21. Moschit H, Roep D, Brunori G, Tisenkopfs T (2015) Learning and innovation networks for sustainable agriculture: processes of co-evolution, joint reflection and facilitation. J Agric Educ Extens 21(1):1–11
22. von Wiren-Lehr S (2001) Sustainability in agriculture - an evaluation of principal goal-oriented concepts to close the gap between theory and practice. Agr Ecosyst Environ 84(2):115–129
23. Carolan MS (2006) Social change and the adoption and adaptation of knowledge claims: whose truth do you trust in regard to sustainable agriculture? Agric Hum Values 23:325–339
24. Pretty J, Ward H (2001) social capital and the environment. World Dev 29(2):209–227
25. Servigne P, Chapelle G (2017) L'ENTRAIDE. L'autre loi de la jungle. Éditions Les liens que libèrent, Paris
26. Griffon M (2010) Pour des agricultures écologiquement intensives. Éditions de l'aube, Aix-en-Provence
27. Beopoulos N (1997) Environment and economic development in rural areas: the contribution of organic farming. Social Res Rev 92–93:183–203 [in Greek]
28. Aristoil Homepage. https://aristoleo.com/el/aristoil-interreg-med-programme-gr/. Accessed 04 Apr 2022
29. Nelson KC, Brummel RF, Jordan N, Manson S (2014) Social networks in complex human and natural systems: the case of rotational grazing, weak ties, and eastern US dairy landscapes. Agric Hum Values 31:245–259
30. Hosseinzade M, Jafari H, Ahmadian MA (2018) Rural entrepreneurship and sustainable development towards environmental sustainability (Central Bardaskan City area). Ukrainian J Ecol 8(2):235–245
31. Ouellette J-A, Wood W (1998) Habit and intention in everyday life: the multiple processes by which past behavior predicts future behavior. Psychol Bull 124(1):54–74
32. Fehr-Duda H, Fehr E (2016) sustainability: game human nature. Nature 530(7591):413–415
33. Portes A (1998) Social capital: its origins and applications in modern sociology. Ann Rev Sociol 24:1–24
34. Ro M, Brauer M, Kuntz K, Shukla R, Bensch I (2017) Making cool choices for sustainability: testing the effectiveness of a game-based approach to promoting pro-environmental behaviors. J Environ Psychol 53:20–30
35. Magis K (2010) Community resilience: an indicator of social sustainability. Soc Nat Resour 23(5):401–416
36. Koutsouris A (1998) Networking for a sustainable future: the case of development agencies. In: Doppler W, Koutsouris A (eds) Rural and farming systems analyses: environmental perspectives. Margraf Verlag, Weikersheim, Germany, pp 114–125
37. Burgenmeier B (2005) Économie du développement durable, 2éd. Éditions De Boeck Université, Bruxelles

The Prospects of Using Vegetable Oil in the Building Industry

Magdalena Kachel and Anna Krawczuk

Abstract The use of vegetable oil-based products is due to their poor technical performance, low prices of mineral oil-based products, and lack of awareness of environmental consequences. Our research has shown that better wetting parameters characterize the applied adhesive based on vegetable oil, as the angle θ between the surface of the wetted wood and the tangent to the droplet placed on this material is lower (11.9°) compared to the current market mineral oil-based fluid (28.6°).

Keywords Vegetable oil · Free fatty acids (FFA) · Contact angle · Adhesive liquids · Biodegradability

BOD	Biochemical Oxygen Demand
COD	Chemical Oxygen Demand
AN	Acid Number [mg KOH/g]
ThOD	Theoretical Oxygen Demand
FFA's	Free Fatty Acids [%]

1 Introduction

The construction industry is committed to using greener, cleaner and sustainable products. Due to the resulting environmental benefits, the resulting preparations based on ecological ingredients are increasingly used. Products based on mineral oils and the emission of volatile organic compounds produced by them in times of global warming and the formation of photochemical "smog" are serious problems.

M. Kachel (✉) · A. Krawczuk
Department of Machinery Exploitation and Management of Production Processes, Faculty of Production Engineering, University of Life Sciences in Lublin, 28 Głęboka Street, 20-612 Lublin, Poland
e-mail: magdalena.kachel@up.lublin.pl

S. Pascuzzi and F. Santoro (eds.), *Farm Machinery and Processes Management in Sustainable Agriculture*, Lecture Notes in Civil Engineering 289,
https://doi.org/10.1007/978-3-031-13090-8_29

One of the current priorities of the EU is to achieve 50% recycling by 2025. Therefore, work on vegetable oil-based release agents has been going on for many years [1]. However, the market application of these products is still limited due to poor technical performance, low prices of mineral oil-based products, lack of awareness of environmental consequences, and many other reasons, the market share of this type of oil is still small. Almost any substance that eliminates or reduces adhesion can be used as a release agent, as long as it is not harmful to concrete or formwork [2]. The various variants of separating agents based on vegetable oil include pure vegetable oils, fatty acid esters (ester oils), and emulsions of fatty acid esters or pure vegetable oils in water (emulsions) [3], usually modified with additives to improve their performance or application.

Because hardened concrete adheres firmly to the mold and causes undesirable effects during demolition, the surface of the concrete molding is damaged, or the mold (formwork) is damaged. Any damage that occurs requires repair due to the required quality of the structure and the expected visual parameters. To avoid complete disqualification and demolition works of elements or their possible reconstruction, it is advisable to use separating (adhesive) means, ensuring minimization of the adhesive forces between the materials in contact. We are dealing with a shortage of national architectural regulations and standards regarding concrete or wood in the current situation. The available specifications consider the provisions and requirements for surface quality contained in the German guidelines [4]. In the standard PN-EN 13,670 [5] – Execution of concrete structures in Sect. 5.2.2, where basic requirements for anti-adhesive agents are formulated.

New molding oils, according to the applicable normative regulations, should contain biodegradable components, including substances of plant origin (separating agents based on mineral oils and separating agents based on vegetable oils). These oils should also have appropriate properties. The current parameters of vegetable oils tell us about a high FFA content up to about 4%, AN content below 5 mg KOH/g, and water up to about 3%.

The research aimed to analyze two adhesives available on the domestic market. Rapeseed oil was analyzed for the content of FFA, AN, phosphorus (P), and water (H_2O) and then subjected to the transesterification process. After trans-esterification, the oil was analyzed for FFA, AN, P, and H_2O content. The agents were analyzed for aerobic biodegradability, and the surface tension and the contact angle were measured during the application of these agents to the wood.

2 Materials and Methods

2.1 Research Subject

Vegetable oil in the form of industrial rapeseed oil was trans-esterified, and then an adhesive was obtained. Two adhesive compounds were used for comparative purposes in the next step and the assumed research methods. The first was a liquid-based vegetable oil obtained from rapeseed oil (industrial). The second was an adhesive fluid based on mineral oil on the domestic market. Both substances were analyzed for biodegradability. Then, measurements of the surface tension of the fluids mentioned above were carried out, and after the droplets were applied to the wood sample, the contact angle measurement was carried out.

2.2 *Analysis of Rapeseed Oil Parameters*

The content of AN/FFA was determined based on the titration method, phosphorus content following the PN-88/A-86930 standard [5], while the water content was tested using the Karl Fischer method on the TitroLine 7500 KF trace apparatus by DONSERV.

2.3 *Biodegradability Testing*

The study of aerobic biodegradability in the water environment of the agent based on vegetable (rapeseed) oil and the agent based on mineral oil was carried out using manometric respirometry. This method consists in incubating in aerobic (aerobic) conditions, with constant stirring, at a constant temperature (22 ± 1 °C), for 28 days of test material with a known concentration (usually 100 mg/l, giving at least 50–100 mg ThOD/l substance, which is the only source of organic carbon), an appropriate amount of inoculum (inoculum) – up to 30 mg DM/l and a mineral nutrient solution in tightly closed respirometer vessels. The activated sludge was collected from the aerated reservoir of the "Czajka" Sewage Treatment Plant in Warsaw, which treats mainly domestic sewage.

Biodegradation is determined by the automatic determination of oxygen consumption and is directly expressed in mg/l BOD. The amount of oxygen taken up by the sludge microorganisms during the biodegradation of the test material (corrected for a parallel blank) is expressed as a percentage of the ThOD calculated from the chemical formula or COD.

During the study, the following were in accordance with Annexes 1 and 3 to SPO/BS/01/b (issue 5) and PN ISO 9408: 2005 [6].

2.4 *Measurements of Surface Tension and Contact Angle*

Measurements of the surface tension and the contact angle were carried out on the KRÜSS DSA30 measuring device. The sagging drop method was used to measure the surface tension. Based on the image of a liquid hanging drop, the drop shape analysis was carried out, because of which the surface tension was determined. Dosing of the adhesive was done manually with a syringe by turning the adjusting screw, which was placed in the syringe barrel instead of the plunger. After turning the regulating screw through the angle of 135°, a drop of liquid with a volume of 15 μl flowed from the needle. Knowing the diameter of the dosing needle, which was 1.828 mm, and the density of the analyzed sample, the surface tensions of the liquid were determined. Thirty measurements of surface tension were performed.

Measurements of the contact angle of the adhesive liquid were carried out on wood samples (aspen). Using a NICHIRYO Le-20 automatic pipette, a measuring drop of 4 μl volume was applied to the material's surface. Based on the digital analysis of the applied droplet image, the angle between the wood surface and the tangent to the measuring drop at the point of contact of these phases was determined. Ten contact angle measurements were made on both the left and right sides of the applied drop of adhesive.

3 Results

3.1 Vegetable Oil Analysis

The table below (Table 1) presents the basic parameters of rapeseed oil used for the analysis. The carried out transesterification process and the processes involved allowed for the reduction of the considered parameters. The content of acid number, free fatty acids, phosphorus, and water decreased compared to the control sample. The obtained values were lower and amounted to 18.60, 18.60, 3.70, and 2.5%, respectively.

Figures 1 and 2 show the biodegradability results for the analyzed adhesive liquids. The tested material is susceptible to aerobic biodegradation in the aquatic environment.

A 28-day study of the final aerobic biodegradability of the fluid on rapeseed oil baize by manometric respirometry following the OECD 301F [7] guidelines was carried out at a temperature of 22 ± 1 °C. On the 28th day of the test, the aerobic

Table 1 Properties of the tested vegetable oil

Parameters studied	Determination of AN/FFA [mg KOH/g/%]	Phosphorus content (P) [mg/kg]	Water content [%]
Initial parameters – control	4.30/2.15	77.50	1.283
Oil parameters after trans-esterification	0.80/0.40	2.87	0.032

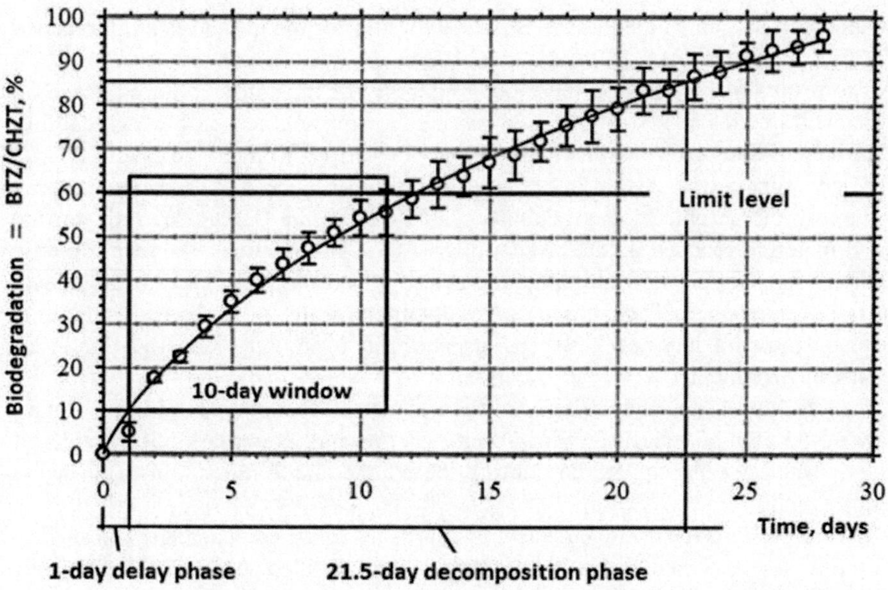

Fig. 1 The slope of the biodegradation curves of the test material and reference material based on vegetable oil

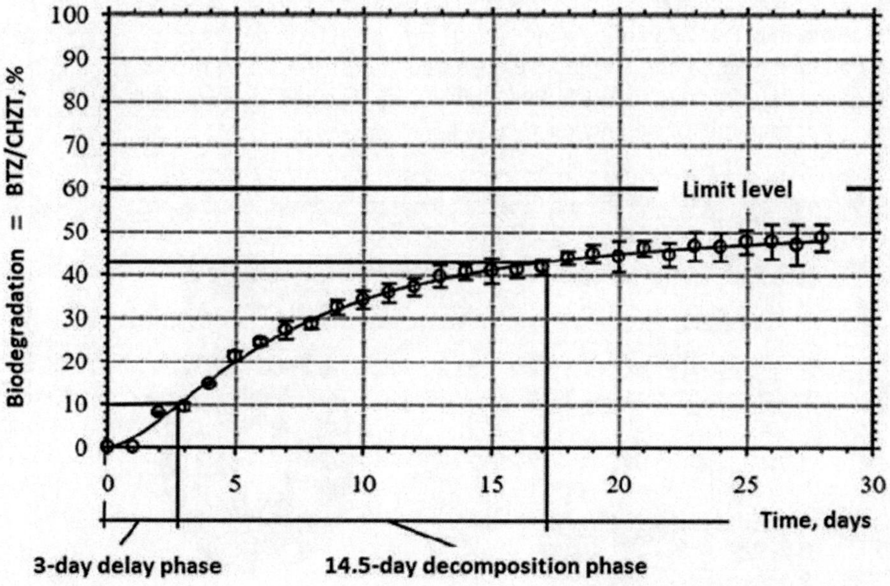

Fig. 2 The slope of the biodegradation curves of the test material and the reference material of a commercially available agent

biodegradability, calculated as the BOD/COD ratio for the tested material, reached 95.1%.

The results of the tests on the biodegradability of the adhesive liquid based on mineral oil are presented in Fig. 2. A 28-day study of the final aerobic biodegradability of the preparation based on mineral oil by manometric respirometry, following OECD 301F guidelines, was carried out at a temperature of 22 ± 1 °C. On the 28th day of the test, the aerobic biodegradability, calculated as the BOD/COD ratio for the tested material, reached 47.8%. According to OECD guidelines, the respirometric method's threshold level for high biodegradability is 60% COD. The threshold level has not been reached. Thus, based on the obtained results, it can be concluded that the tested material does not show high susceptibility to aerobic biodegradation in the aquatic environment.

Figures 3 and 4 show the results of surface tension measurements and an exemplary view of the applied measuring drop. Figure 5 presents exemplary photographs of drops of adhesive liquids applied during the contact angle measurement on wood samples.

Table 2 presents the average measurement results of the contact angle and surface tension of the analyzed anti-adhesive liquids and distilled water as the reference liquid.

Similar values of surface tension characterize the analyzed anti-adhesive liquids. The surface tension of the vegetable oil-based adhesive is 27.57 mN/m, while the mineral oil-based adhesive is 26.3 mN/m. The surface tension of these fluids is about 60% lower than that of distilled water (72.01 mN/m). The lowest contact angle value was obtained by applying a vegetable oil-based adhesive of 11.9° to a wood sample. The contact angle value is nearly 80% lower than distilled water (54.9°) and about 60% lower than that of the market release agent (28.6°).

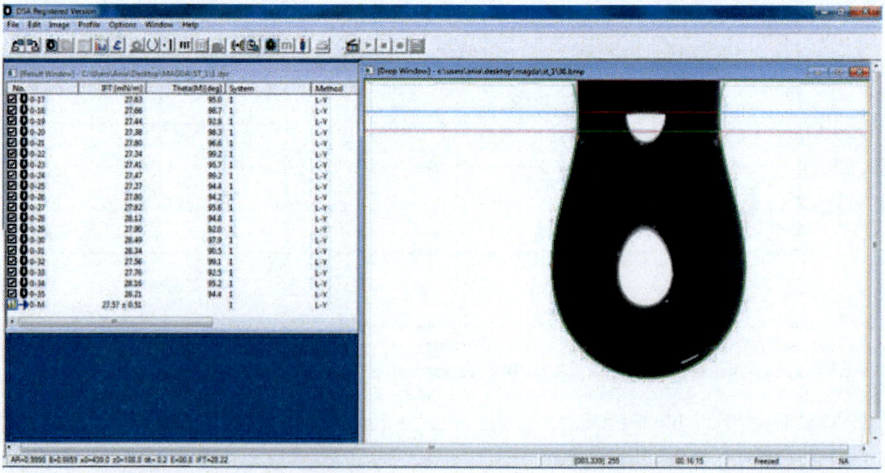

Fig. 3 View of the Drop Shape Analysis program window with a photo of a drop of a vegetable oil-based adhesive liquid during surface tension measurements

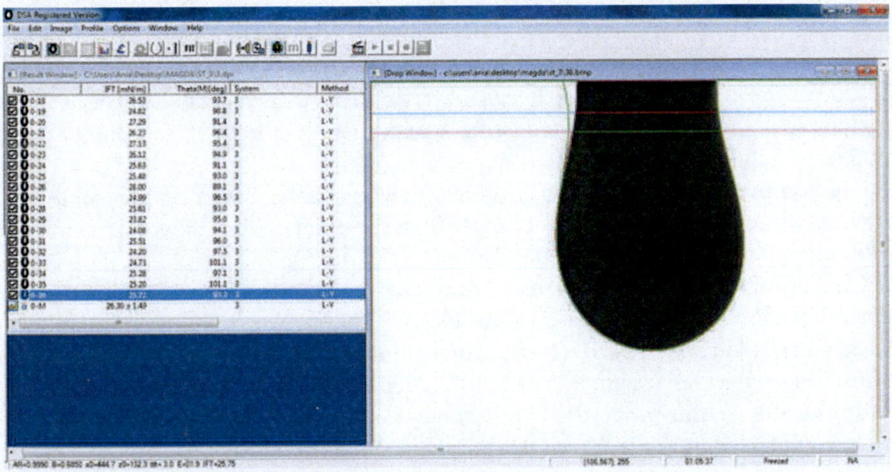

Fig. 4 View of the Drop Shape Analysis program window with a photo of a drop of a mineral oil-based adhesive liquid during surface tension measurements

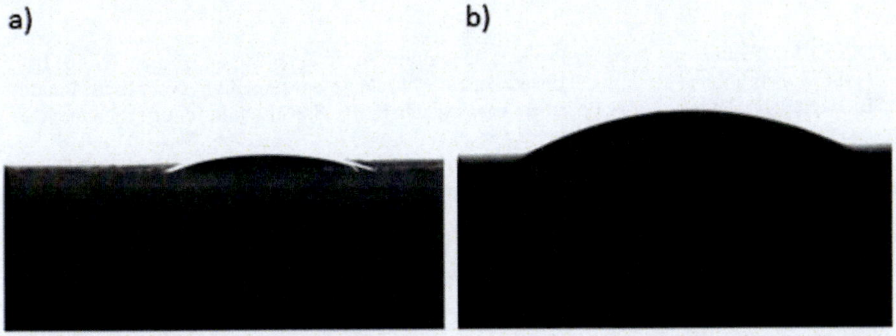

Fig. 5 An example photo of a drop of adhesive liquid: **a** based on vegetable oil and **b** available on the market, applied on wood (aspen) during the measurement of the contact angle

Table 2 Average values of the measurement results of the contact angle and surface tension of the analyzed anti-adhesive liquids

Wood	Liquid	Contact angle [°]	Surface tension [mN/m]
Aspen	distilled water	54.9	72.01
	AD – vegetable oil	11.9	27.57
	AD – commercial	28.6	26.30

4 Conclusions

Based on the obtained results, it can be concluded that vegetable oil as a base compound in the production of industrial agents ensures at least 95% certainty of its biodegradability.

The use of adhesive products for concrete/wood molds based on vegetable oils provides a good/comparable contact angle from the point of view of wood treatments with mineral oil-based fluids on the market.

Similar values of the surface tension characterize the tested adhesives. The surface tension of the vegetable oil-based adhesive is 27.57 mN/m, while the mineral oil-based adhesive is 26.3 mN/m. Both analyzed fluids show good wood wetting properties. The contact angles are less than 90°. However, the mineral oil-based adhesion exhibits better wetting properties (11.9° contact angle) than a commercially available mineral oil-based fluid (28.6° contact angle).

References

1. DG XIII/D Technology Transfer Projects, Definition Phase Final Report, Project IN 10110 D (1996) Substitution of Mineral Oil Based Concrete Mould Release Agents by Non-Toxic, Readily Biodegradable Vegetable Oil Based Release Agents in the Construction Industry (Sumovera), Amsterdam
2. Barnat-Hunek D, Szafraniec M (2021) Influence of biodegradable release oils on the physical and mechanical properties of light-colored architectural concrete. Materials 14:4630. https://doi.org/10.3390/ma14164630
3. Barnat-Hunek D, Grzegorczyk M, Łagód G (2019) Influence of temperature difference on thermal conductivity of lightweight mortars with waste aggregate. AIP Conf Proc 2170:020003
4. Sichtbeton M (2004) Planung, Ausschreibung, Vertragsgestaltung, Ausführung und Abnahme, BDZ/DBV; Merkblatt Sichtbeton: Berlin, Germany
5. PN-88/A-86930 / PN-ISO 10540–1:2005. Edible vegetable fats - Test methods - Determination of phosphorus content
6. PN ISO 9408:2005. Jakość wody -- Determination of the total aerobic biodegradation of organic compounds in the aquatic environment by determining the oxygen demand in a closed respirometer
7. OECD Guidelines for the Testing of Chemicals, Section 3 - Environmental fate and behaviour, Ready Biodegradability, 301F Manometric Respirometry Test, przyjęte. Accesse 17 July 1992

The Problem of Waste on the Food Market

Agnieszka Dudziak and Monika Stoma

Abstract The process of wasting food is an increasingly serious global problem, moreover, it has many negative consequences for society and the environment. Non-renewable resources are used and greenhouse gases are emitted, which are responsible for the heating of our planet. Food production uses huge amounts of water, energy and fuel, and food that is disposed of in landfills undergoes putrefactive processes, during which the very harmful compound methane is produced. There are many concepts that encourage society to counteract this phenomenon, such as the concept of zero waste or less waste, but first of all, there is need to change the way of thinking for these concepts to exist and contribute to actual, long-term changes. Many consumers have already changed their approach to the problem of food waste, through increasing awareness of human participation in upsetting the economy of the circulation of products on the market. This is a positive symptom, but it cannot be said that a lot has been done on this topic, because much is still ahead of us.

The aim of this paper is to present the problem of food wastage among consumers, the implementation of which was carried out with the use of own research, showing how consumers perceive the problem and whether they are familiar with the concept of zero waste.

Keywords Food · Waste · Consumer behavior · Zero waste

1 Introduction

The problem of wasting food on the food market has been present in the society for years. The processing sector is responsible for the generation of a large amount of food losses, the number of which depends on the advancement of techniques and the type of raw material processed. Food losses and waste occur at all stages

A. Dudziak · M. Stoma (✉)
Department of Power Engineering and Transportation, Subdepartment of Logistic and Business Management, Faculty of Production Engineering, University of Life Sciences in Lublin, 20-612 Lublin, Poland
e-mail: monika.stoma@up.lublin.pl

S. Pascuzzi and F. Santoro (eds.), *Farm Machinery and Processes Management in Sustainable Agriculture*, Lecture Notes in Civil Engineering 289,
https://doi.org/10.1007/978-3-031-13090-8_30

of the product manufacturing chain (cultivation / breeding - processing - storage - sale - consumption). The causes of large losses include, among others, improperly selected machines, incompetent workforce or disloyal suppliers of raw materials. The meat and dairy sectors are responsible for the consumption of large deposits and the emission of greenhouse gases [1].

In Poland, about 235 kg of food is wasted per year per capita and our country is in the seventh position among other European countries in terms of food waste. Irrational purchasing behavior of consumers often wastes food. Moreover, it is assumed that approximately 1.3 billion tons of food is wasted during deliveries. On the other hand, around 680 million people in the world suffer from hunger [2]. The irony, then, is the simultaneous growing amount of food wasted and the number of starving, malnourished people. It should be added that the outbreak of the Covid-19 pandemic had a negative impact on the increase in food prices, and the ongoing war in Ukraine will certainly further aggravate this problem in the near future.

It is worth mentioning that in recent years there are more and more concepts and ideas that place particular emphasis on the care for the use of products as much as possible. One of such concepts is the concept of "zero waste" [3, 4]. It treats waste as a substrate to produce another product. The definition adopted by the Zero Waste International Alliance is: "Zero Waste is the protection of all resources by responsible production, consumption, reuse and recovery of all products, packaging and materials, without burning them, and without discharges to land, water or air that endanger the environment or human health". Whereas, functioning according to the idea of "zero waste" comes down to applying certain rules of the so-called 5R: refuse, reduce, reuse, recycle, and rot [5, 6].

The aim of the study is to present the problem of food waste on the food market and in households, as well as to show the consumer's approach to food waste and how to counteract this problem. On the other hand, the main practical aim of the work is to present the results of research on the scope of consumer knowledge and behavior towards the problem of food waste. In order to achieve the assumed aim of the work, own research was carried out with the use of the proprietary questionnaire. The respondents represented various age groups, place of residence, and they were also diversified by gender.

2 The Issue of Wasting Food

In the currently dynamically developing market and the constant change of civilization, satisfying physiological needs, such as eating food, remains unchanged. Over the past decades, eating habits have changed, but the need to consume food is still the same. For this reason, well-developed countries have begun to overproduce food on a large scale, due to excess food a significant part is spoiled. In poor countries, access to food is a problem, while in richer countries, wasting food is a problem [7]. After research in Europe, it turned out that about 179 kg of food are thrown away each year. According to Bilska et al. [8]: "in Poland about 235 kg per person is wasted per year. It is households that are responsible for the greatest food waste, producing 38 million tons of food waste per year, which is 42% of losses in the entire food chain in the European Union - production 39%, food services - 14%, retail and wholesale - 5%. Polish households waste 22.8% of food". Poland was ranked seventh among other European countries in terms of the amount of food wasted.

It is caused by inadequate management of food products and raw materials necessary for their production. Producers often do not take into account the costs incurred, the consumption of natural resources or human work [9]. The blame is also partly on the part of consumers - they buy too much food that they are not able to consume by the expiry date and, as a consequence, throw it into the trash. In recent years, more and more people acknowledge that they experience this practice. Despite the increase in the ecological approach to purchasing choices, the trend of showing your wealth to others by buying excess food is not diminishing. If nothing changes, we can expect a rapid increase in waste, which has a very harmful impact on the environment, and the loss of natural resources. In short, this can be summarized as waste. The reasons for this include increased prosperity, easier access to food and a wide range of products [10].

Food losses and food waste are used interchangeably, but they actually refer to different aspects of the same problem. The first concept includes losses at the level of storage, processing and production, while the second includes trade along with distribution and consumption. Production losses of food in terms of the agricultural sector refer to the accidental reduction in the quantity and quality of food before consumption. These losses are mainly related to weather anomalies, plant diseases, the presence of pests, as well as a change in cultivation / harvesting techniques [11].

The second term refers to the deliberate disposal of food that is fit for human consumption. Global food demand continues to increase; as previously mentioned – it is estimated that about 680 million people do not have access to food and suffer from malnutrition. In addition, the problem is projected to increase significantly due to the Covid-19 pandemic, which is still ongoing. Overall, the first wave of the pandemic disrupted the four pillars of food security (availability, access, use and stability) [12]. The society, fearing the impending threat, began to accumulate food in large amounts, regardless of whether there would be enough for the rest of the people. In March 2021, in many countries, including Poland, the shortages of durable food products, such as groats, pasta, canned food, could be seen on store shelves. Due to the sudden increase in demand, warehouses and wholesalers were unable to keep up with the import of stocks. One of the disruptions resulting from the outbreak of the pandemic on the food market was the impeded movement of people and limited stay in public places. There has been an increase in interest in cheaper food, of lower quality due to the increase in unemployed people as a result of the pandemic, as well as an increase in the prices of commonly selected products. It is known, however, that it is impossible to completely reduce the amount of wasted food and the losses arising in its production [13, 14].

3 Reasons for Food Waste

The most common food losses occur during production or cultivation, transport, processing, distribution and consumption. About 1.3 billion tons of edible food are estimated to be lost or wasted in the supply chain [15].

There are many reasons for this ever-increasing problem. The main factor that relates to each stage is mismanagement or lack of planning. In highly industrialized countries, the highest percentage of food is wasted at the stage of distribution and consumption, while in developing countries, at the stage of production [16]. In the case of the first group, it can be concluded that this is due to the excess production and the lack of a rational approach to purchases on the part of consumers. On the other hand, in the second group, it is mainly the result of the lack of skillful planning, modern production techniques and warehouses ensuring appropriate storage conditions [17].

It is also worth paying attention to the food quality and visual appearance; more and more often on store shelves food items that deviate from the commonly accepted standards by consumers can be found. Most people choose those products that look fresh and undamaged, with a long-term shelf life [18]. Mostly distributors and warehousemen are responsible for this, as they do not maintain adequate security measures during transport and storage of food [19]. This is due to the lack of competences and knowledge about a given topic of employees in these positions. Another reason is the failure of devices responsible for maintaining appropriate conditions of the stored food. The non-observance of the expiry dates of food stored in warehouses is also an oversight on the part of warehouse staff.

Markets and grocery stores lower the prices of these products in order to encourage the customer to buy, unfortunately a large part of them is wasted and thrown into the garbage [20]. Therefore, the causes of waste in the food industry should be analyzed. An inadequate marketing strategy, an incorrectly selected target group of recipients or inadequate inventory management affect the growth of unsold items. On the other hand, there are many organizations that accept food for recycling and give it to the poor and the poor. Some producers have no idea about the possibility of transferring products that are not suitable for sale to these organizations. This is one way in which it is possible to reduce the amount of food you throw away and also to reduce the number of people suffering from hunger. For example, in Poland, there is the Federation of Polish Food Banks that deals with helping the poor and homeless [21].

4 Material and Method

In order to assess the attitudes of consumers on the food market in terms of food waste, own research was carried out using the diagnostic survey method, which was carried out using a questionnaire distributed online. The main tool used for the study was a structured proprietary questionnaire. The questionnaire contained closed questions that were clearly developed so that they did not require additional comments, and the respondents were asked to indicate one correct answer from several available options, reflecting their attitudes and preferences.

The own-authored questionnaire contained both substantive questions in the analyzed research area as well as metric questions, enabling the obtaining of socio-demographic characteristics of the respondents due to various grouping variables (respondent's descriptive features, such as: gender, age, place of residence) – Table 1.

It was assumed that the research sample would be a minimum of 250 units - inhabitants of South-Eastern Poland - food consumers, which was successfully implemented in this study. A purposeful non-random selection of the research sample was used - the criterion of purposeful selection was the fulfillment by the participants of the study of the criteria defining the categories of grouping variables included in the record.

The Statistica, version 13.3 program was used to analyze the results of the research. The method of statistical analysis in the form of correspondence analysis was used,

Table 1 Socio-demographic profile of the population surveyed

Socio-demographic profile	Number of respondents	Percentage share [%]
Total	384	100.0
Gender		
Female	194	50.5
Male	190	49.5
Age		
Under 18 years of age	168	43.8
19–25 years old	70	18.2
26–40 years old	56	14.6
41–60 years old	58	15.1
60 years and more	32	8.3
Place of residence		
Rural area	112	29.2
City to 100.000 residents	94	24.5
100–300 residents	104	27.1
City with more than 300.000 residents	74	19.3

Source the authors' own studies

Table 2 Information resources' factors

Number of Dimensions	Eigenvalues and Inertia, Total Inertia = 0.29579, $\chi2 = 56.793$, df = 36, p = 0.01513				
	Singular value	Eigenvalues	Percentage of Inertia	Cumulative Percentage	χ^2
1	**0.361871**	**0.130950**	**44.27067**	**44.2707**	**25.14246**
2	**0.315850**	**0.099761**	**33.72643**	**77.9971**	**19.15411**
3	0.228797	0.052348	17.69743	95.6945	10.05083
4	0.112851	0.012735	4.30547	100.0000	2.44519

Source the authors' own studies

which is presented in Table 2. The use of this analysis will help to meet the research goals set by the authors.

5 Results and Discussion

Based on the information obtained from the respondents, in the initial phase of the research procedure, an analysis of correspondence between three groups of characteristics was carried out, i.e. the frequency of wasting food (5 groups of answers), the age of the respondents (5 groups of answers) and gender (2 groups). Respondents were asked if they ever wasted food, and if so, how often.

In order to present the configuration of points representing the input data, a two-dimensional factor space was chosen. The first factor allows to reproduce 44.27% of input data variation (i.e. total inertia), and the second factor 33.73% (Table 2).

The largest share in the creation of the two-dimensional factor space by the frequency of food wasted by the respondents had the answers that occasionally and never - I coordinate, and every day and 2–3 times a week- II coordinate. On the other hand, the greatest share in the creation of the two-dimensional factor space by age and gender was given by the answers given by women in the age group under 18, 19–25, 26–40 and over 65 - the 1st dimension, and the answers given by men over 65 years, 41–65, and those under 18 and women in the age range 41–65 - 2nd dimension (Fig. 1).

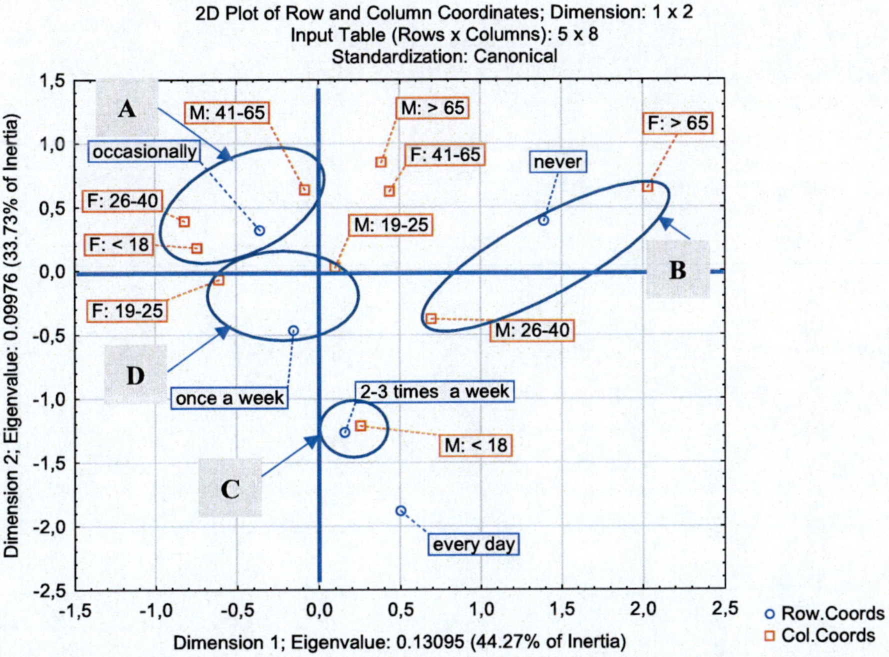

Fig. 1 The results of the correspondence analysis between the three groups of characteristics - the frequency of food waste, gender and age of the respondents. *Source* the authors' own studies

There are four distinct groups of respondents declaring specific answers in relation to food waste, these are the answers:

- declaration regarding occasional food wastage, which is made mainly by men aged 41–65, and women in the age group under 18, 19–25 and 26–40 (group A);
- declaration regarding the lack of food waste, which behavior is admitted mainly by women over 60 and men aged 26–40 (group B);
- declaration regarding food waste with a frequency of 2–3 times a week, which is made mainly by young men under 18 (group C);
- and the respondents' declaration of wasting food once a week, which is mainly done by young women and young men, i.e. those aged 19–25 (group D).

The greatest correlation occurs within the answers given in groups A and C, i.e. between the occasional wasting of food and 2–3 times a week (group C), here the relations in relation to the answers given by the respondents are quite strong. Due to the value of the indicator - i.e. the age and sex of the respondents, the two groups under consideration differ from the others (i.e. from groups B and D). The weakest link is in groups B and D, i.e. in groups where respondents admit that they never waste food (group B), and in the group where it happens once a week (group D).

Referring to Figs. 1 and 2, on the basis of the results carried out, it can be seen that the successive dimensions (which are orthogonal to the others) explain smaller

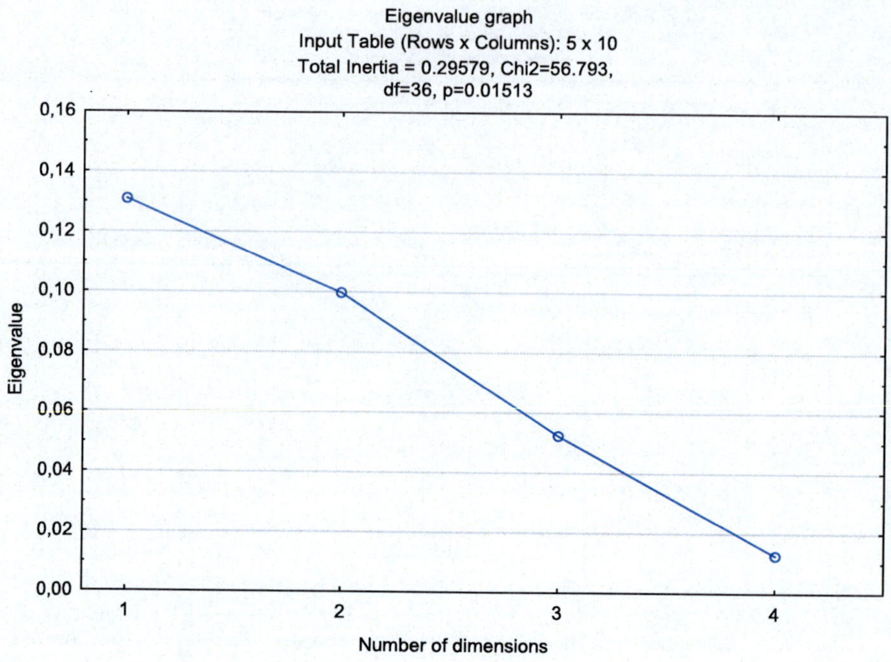

Fig. 2 Scree plot of eigenvalues. *Source* the authors' own studies

and smaller parts of the overall value of the chi-square statistic (i.e. inertia). Figure 2 shows four eigenvalues, due to the four dimensions distinguished in the statistical program, but for further - more in-depth analyzes – only two were selected. The reason was that the first dimension can reproduce as much as 44.27% of total inertia, and the second dimension included increases the percentage of explained inertia to 77.90% of total inertia. Therefore, based on this criterion, it is advisable to position the profiles in a two-dimensional space.

After determining the number of dimensions, in the next step the coordinates of the column profiles were calculated in a new orthonormal coordinate system defined by singular vectors. In order to interpret the coordinates of the points representing the columns, the row-column standardization method was used, where the coordinates are calculated from the matrix of column profiles. This standardization made it possible to obtain coordinates of points representing the respondents' attitude in terms of place of residence and gender in relation to the knowledge of the "zero waste" concept by the respondents. Figure 3 shows a categorized histogram of these relationship.

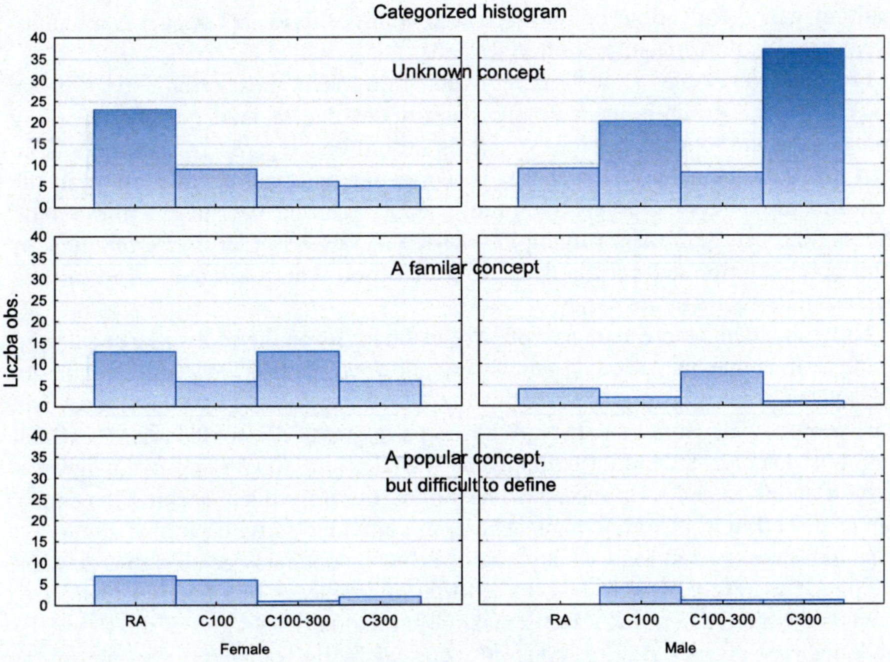

Fig. 3 Histogram categorized according to the knowledge of the concept of "zero waste", gender and place of residence of the respondents. Where: F – female, M – male; RA – rural areas, C < 100 – city with less than 100 th. residents, C100–300 – city with th. residents, C > 300 – city with more than 300 th. residents; *Source* the authors' own studies

6 Conclusion

Although the overwhelming number of women are still responsible for shopping in their households, the number of men performing this function is also increasing. Older people over 65 more often prepare a list of necessary products before going to the store /market, this is the result of problems with memory, more free time, thanks to which they do not buy unnecessary products. Young people aged 19–25 have divided opinions, some make a list and the rest do not have time to do so. Reasons for making a shopping list may be limited funds, in the case of students, or having your own family to support. The other part, on the other hand, is at risk of making larger purchases with unnecessary food products. Adults, self-supporters, and their families usually write down a list of items they need before they go to the market.

People living in the countryside occasionally throw their food into the trash, this is due to the fact that most of them have animals that eat food scraps. The number of people who practice throwing away food every day is satisfactorily low. The Covid-19 pandemic has left its mark and respondents are trying to reduce the amount of both bought and discarded food. However, Poles are famous for their hospitality and tradition, hence, for various holidays, large amounts of food are bought several times a year, which, unfortunately, is often wasted.

Fruit and vegetables as well as leftovers from meals most often end up in the basket, the first group is unstable and deteriorates due to inadequate or too long storage. The appropriate portions of meals should be prepared so that there is no need to throw them into the trash. It is worth paying attention to training in the matter of proper food storage, as not all products can be kept in the refrigerator. Excess food can be frozen. Buying in advance and using advertised promotions is not a good solution and results in the fact that food that is not consumed in time expires and ends up in the basket.

The respondents more and more often read the labels on the packaging of selected products, thus making informed purchasing and food choices. The most frequently checked information is product composition and expiry date. Nowadays, more and more people suffer from various types of food allergies and other diseases for which they must exclude individual components. In addition, advertising campaigns on harmful additives and substances also have their merit in this. Checking the expiry date of a food item is a good habit which, in combination with rational shopping, helps to reduce wasted food. A large part of the respondents declared that they are willing to buy products with a short shelf life, of course, at an attractive, low price, which is the main determinant when making purchases. As a result of the Covid-19 pandemic, society is trying to reduce the cost of living for the family, due to the constantly rising prices of food, energy, water, etc.

The respondents said that another important factor when choosing a product is its quality. When comparing the price and quality of two items, the one that pays off is usually chosen. The ecology of products is the least popular among the respondents. But as you know, products that come from, for example, Germany do not have much in common with the meaning of the word, due to the increased and frequent treatment with pesticides and plant protection products.

As experience grows, the number of people who say that they are doing nothing to prevent food waste is falling. This proves that at the stage of teaching they were made aware of this topic.

According to the respondents, the methods most frequently used by them to counteract food waste are to buy only the necessary products, segregate products with the use-by date and prepare a list of the products needed. If the respondents were honest in their answers, then in the future the problem of food waste should partially decrease. Another way mentioned by the respondents was to share the surplus of food with those in need, thanks to which the problem of hunger and wasted amount of food is eliminated.

The vast majority of respondents do not know the term or the concept of "zero waste", therefore the government should make the public aware of this issue and encourage the implementation of its principles. Young people encountered this concept through education and while using social media.

References

1. Parfitt J, Barthel M, Macnaughton S (2010) Food waste within food supply chains: quantification and potential for change to 2050. Phil Trans Royal Soc B: Biol Sci 365(1554):3065–3081
2. Ślusarczyk B, Machowska E (2019) Food Waste in the World and in Poland. Академічний огляд 1:91–100
3. Zaman AU (2015) A comprehensive review of the development of zero waste management: lessons learned and guidelines. J Clean Prod 91:12–25
4. Singh S, Ramakrishna S, Gupta MK (2017) Towards zero waste manufacturing: a multidisciplinary review. J Clean Prod 168:1230–1243
5. Murphy S, Pincetl S (2013) Zero waste in Los Angeles: is the emperor wearing any clothes? Resour Conserv Recycl 81:40–51
6. Wilde MD, Parry S (2022) Feminised concern or feminist care? reclaiming gender normativities in zero waste living. Sociol Rev 00380261221080110
7. Ober J, Karwot J (2022) Pro-ecological behavior: empirical analysis on the example of polish consumers. Energies 15(5):1690
8. Bilska B, Grzesińska W, Tomaszewska M, Rudziński M (2015) Marnotrawstwo żywności jako przykład nieefektywnego zarządzania w gospodarstwach domowych. Roczniki Naukowe Stowarzyszenia Ekonomistów Rolnictwa i Agrobiznesu 17(4)
9. Priefer C, Jörissen J, Bräutigam KR (2016) Food waste prevention in Europe–a cause-driven approach to identify the most relevant leverage points for action. Resour Conserv Recycl 109:155–165
10. Bednarczuk A, Śleszyński J (2019) Marnotrawstwo żywności w Polsce. Prob World Agric/Problemy Rolnictwa Światowego 19(1827–2020–143):19–30
11. Burlea-Schiopoiu A, Ogarca RF, Barbu CM, Craciun L, Baloi IC, Mihai LS (2021) The impact of COVID-19 pandemic on food waste behaviour of young people. J Clean Prod 294:126333
12. Laborde D, Martin W, Swinnen J, Vos R (2020) COVID-19 risks to global food security. Science 369(6503):500–502
13. Santeramo FG, Lamonaca E (2021) Food loss–food waste–food security: a new research agenda. Sustainability 13(9):4642
14. Santeramo FG, Lamonaca E (2021) Objective risk and subjective risk: the role of information in food supply chains. Food Res Int 139:109962
15. Nicastro R, Carillo P (2021) Food loss and waste prevention strategies from farm to fork. Sustainability 13(10):5443
16. Prusky D (2011) Reduction of the incidence of postharvest quality losses, and future prospects. Food Secur 3(4):463–474
17. Giza W, Wilk B (2021) Revolution 4.0 and its implications for consumer behaviour. Entrepre Bus Econ Rev 9(4):195–206
18. Zwanka RJ, Buff C (2021) COVID-19 generation: a conceptual framework of the consumer behavioral shifts to be caused by the COVID-19 pandemic. J Int Consum Mark 33(1):58–67
19. Chowdhury M, Sarkar A, Paul SK, Moktadir M (2020) A case study on strategies to deal with the impacts of COVID-19 pandemic in the food and beverage industry. Oper Manag Res 1–13
20. Del Giudice T, Stranieri S, Caracciolo F, Ricci EC, Cembalo L, Banterle A, Cicia G (2018) Corporate Social Responsibility certifications influence consumer preferences and seafood market price. J Clean Prod 178:526–533
21. Giuseppe A, Mario E, Cinzia M (2014) Economic benefits from food recovery at the retail stage: an application to italian food chains. Waste Manag 34(7):1306–1316

Technical and Social – Economic Impact and Remedy of a Large-Scale Mechanized Plantation Farming on the Host Community

Alex Folami Adisa, Peter Olugbenga Omotainse, Ayoola Abiola Babalola, Ayodele Gbenga Adisa, Omolola Mayowa Adeola, and Abidemi Toyosi Ajose

Abstract Agricultural technology contributes to increase in farm productivity, additional rural employment with antecedent effect of reduction in labour and costs which eventually lower food prices, requires careful planning because it addresses multiple and conflicting objectives. Socio-economic impact on community residents can be positive or negative. This study examined technical and the socio-economic impacts over a four year take off period of a mechanized large scale sugarcane crop production making use of data obtained from 2015 to 2018 cropping seasons, Sunti, Nigeria. Adequate steps were taken to solve 12 major occurred disturbances by the host communities due to the company's presence and operations by at the government and company's levels. The Presence of the Plantation has to some reasonable extent impacted positively on wellbeing and living standard of the communities in catchment areas by employing both skilled and unskilled labour, education, skill acquisition, infrastructure sectors and community crop farming, fishing activities and overall improves their economic life. All necessary measures were put in place for the future 17,000 ha irrigable farmland plantation activities not to put the communities in dangerous environmental condition by the sugarcane cultivation and 4,500 metric tons/day sugar processing factory operations with installation of equivalent wastewater treatment plant. Values of available fleet of tractors and equipment power ratings, draft, capacity and implements selection for field cultivation operations for 2017 and 2018 cropping seasons for 1,300 ha was determined and found adequate which took

A. F. Adisa (✉) · P. O. Omotainse · A. A. Babalola
Agricultural and Bio-Resources Engineering Department, Federal University of Agriculture, Abeokuta, Nigeria
e-mail: adisaaf@funaab.edu.ng

A. G. Adisa
Civil Engineering Department, Faculty of Engineering and the Built Engineering, University of Johannesburg, Johannesburg, South Africa

O. M. Adeola
Geography Department, Geoinformatics and Meteorology, University of Pretoria, Pretoria, South Africa

A. T. Ajose
The Flecther's School, Tuft University, Medford, MA, USA

S. Pascuzzi and F. Santoro (eds.), *Farm Machinery and Processes Management in Sustainable Agriculture*, Lecture Notes in Civil Engineering 289,
https://doi.org/10.1007/978-3-031-13090-8_31

into consideration precaution of negative impact of soil cultivation operation and equipment trafficability.

Keywords Agricultural · Socio · Economic effect · Plantation · Community

1 Introduction

Technical or agricultural technology can be defined as improved management practices relating to crops, livestock and fisheries, machinery/mechanization, infrastructural development, products of plant and animal breeding (biotechnology), introduction of new crops external inputs (chemicals, bio-control products and vet-nary products) and local inputs (soil amendments, mulches, to mention few). Agricultural technology is better presented as a key aspect of an effective system [3]. South and East Asia experienced rapid growth in agricultural productivity as a result of adoption of new agricultural technologies which increases incomes and wages, lower food price and generate new livelihood opportunities by providing basis for economic diversification under rainfed and irrigated agricultural practices [7].

Impact of technological change require identification of the most effective planning procedures to reduce poverty, decide the degree or appropriate level of agricultural investments and types of agricultural technology to be adopted with regard to available equipment, man power (both skilled and unskilled) and other relevant resources [3]. Technical or agricultural technology contributes to increase in farm productivity, additional rural employment with antecedent effect of reduction in labour and costs which eventually lower food prices. It requires careful planning because it addresses multiple and conflicting objectives [3].

There are potentials in proper agricultural investments on appropriate land to promote rural development and the expectation of such investments could be the engine of technical and economic development and food security for the host community and the country at large [5]. Socio-economic impact on a community is the impact of a change on residents of a community. The two words social and economic are seen as being inherently linked nature of social and economic impacts [4]. This is to assess the social impacts or influence on employment generation for local and imported people, social and cultural issues relating to the livelihood of the community and impacts on population, [4]. In a community where there are multiple types of land use, social and economic change may be occurring simultaneously, at times, it is usually difficult to identify what has really caused socio-economic change [4]. Due to various reported cases of disturbances at this project study location, there was need to examine the technical and the socio-economic impacts of mechanized large scale crop production and the remedy.

2 Methodology

This study was to examine the technical and the socio-economic impacts of the four-year period of mechanized large scale crop production with available data obtained for 2015 to 2017 while 2018 was ongoing cropping seasons. The study was supposed to be comprehensively done by analyzing trends in key agricultural industries; machinery and infrastructures put in place, employment and investment by the plantation sector; population living in rural areas and towns; socio-demographic characteristics of the population, including age structure, educational attainment, rate and type of employment and unemployment, number of new residents and household income; provision of key services including education, retail, government, health, community groups and rural land prices [4]. With data available, not all aforementioned factors and data analysis could be carried out hence this study was therefore carried out within the limited available information and data in the following sections.

2.1 *Project Study Location and the Host Community*

The sugarcane plantation, Sunti Golden Sugar Estates, a Flour Mill of Nigeria Plc subsidiary company is located in the tribal land of Nupe, northern bank of River Niger low land flood plain in Niger State, Nigeria. The host community social life was of a complete rural type, barely, no modern facilities/amenities before the takeoff of this plantation. The sugarcane plantation land acquired is about 15,000 ha out of which only 4,770 ha was found suitable for sugarcane cultivation due to the soil high water table located on the northern bank of a very big river, River Niger, Nigeria. The people are mainly peasant crop farmers and fisher men and women. There are plenty water ponds and channels covering a large portion of their land, hence fish farming and culturing is highly practiced by high percentage of the population, and they cultivate in small scale crops like rice, maize, melon, cassava and guinea corn.

2.2 *Company Organization of Sunti Golden Sugar Estates*

The Organogram of the Sunti Sugarcane Estate which is meant for sugarcane production for sugar refinery is shown in Fig. 1 as of 2018 cropping season. This study was majorly concerned with sugarcane related cultivation operations.

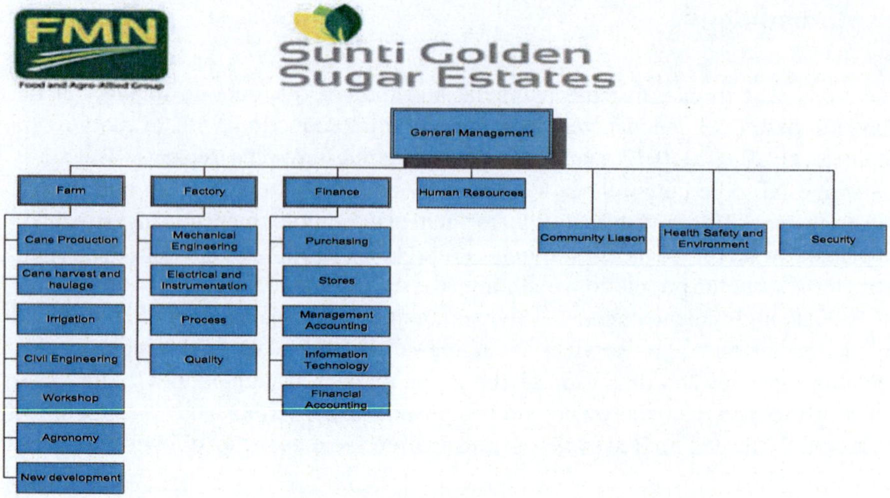

Fig. 1 Organogram of Sunti Golden Sugar Estate (*Source* Sunti Golden Sugar Estate personnel office, 20,180)

2.3 Findings

Disturbances of the Host Communities on the Company's Presence and Operations from 2015 to 2017 Cropping Seasons

Table 1 is summary of challenges and disturbances that occurred between the host communities and the company due to their presence and activities from 2015 to 2017.

Identification of challenges and problems faced by the host communities because of the sugarcane plantation's presence

This study was based on attempt to answer the following questions as much as possible and the challenges faced by the company from their host communities:

(i) What are the benefits of immediate community from the plantation?
(ii) Are they able to supply the needed labor, both skilled and unskilled?
(iii) Have the presence of the plantation improved the wellbeing, economic, social and infrastructural of the people?
(iv) Direct and indirect benefits being derived from the organization like crop farming, fish farming activities, good road, electricity, portable water, standard of living, modern education, skill acquisition, access to good health facility.
(v) What is the environmental impact on the irrigation and drainage farming activities, community road, air in the environment and drinking water like effect of application of chemical and fertilizer on stream water and atmosphere?

Table 1 Sunti Golden Sugar Estate host communities' disturbances on the company's presence and operations from 2015 to 2017

S/No	Date	Communities involved	Grievances/causes
1	11/2/2015	All host communities except Batagi village were involved	Issue of appointment of Human Resources/ Administrative Manager that is not an indigene of any of the host community made them to disturb the company's activities
2	24/6/2015	Kusogi	Appointment of one Kusogi indigene as a Senior Supervisor instead of their preferred position of Agronomist which did not go down well with them, hence resulted in disturbing the company's operation
3	5/8/2015		Company was challenged of siting borrow pits inside villagers' farms. The company was forced to stop work that day
4	9/12/2015	Kukpanti	Kukpanti community complained of being marginalized in company's recruitment of seasonal and permanent staff
5	19/12/2015	Kusogi	Company was forced to close the trench dug (for security purpose) by the company security management at golf one because the community claimed it was not needed by them
6	3/2/2016	Dzagun	Community kicked against the construction of dyke 3 (meant to control flood) which resulted in closure of some of the community fishing ponds
7	9/3/2016	Dzagun and Kusogi	The two communities Kicked against the suspension (disciplined for committing offence) of six field seasonal casual staff who are their indigenes
8	24/3/2016	Kusogi	Non-compensation for Kpata Dogo water point which was a source of water supply to the company sugarcane field and that it was being situated outside acquired land by the company

(continued)

Table 1 (continued)

S/No	Date	Communities involved	Grievances/causes
9	4–5/4/2016	Kanzhi	Community claimed that construction of dyke (flood control structure) was not needed and hence forced the company to stop work for two days
10	11–12/4/2016	Kusogi	Kicked against construction of haulage road (for sugarcane transportation) across Essan river which the community claimed was not needed because of its side effect on their fishing ponds around the area
11	16/1/2017	Host communities lead by Kusogi village	Vowed that Yoruba tribe staff should leave the company with emphases on Human Resources Manager and Sectional Manager. This resulted in destroying houses, offices and company staffs were beaten as well
12	22/3/2017	Kusogi community with support of men from Luasfi village	A man from Kusogi village was caught urinating at undesignated place at the factory premises (unhygienic behavior) and hence was apprehended and then reprimanded by the company Senior Security Officer. This triggered problem that resulted into death of three Kusogi people, eight company's vehicles were burnt, 151 ha of sugar cane field was set ablaze and four women were assaulted and rapped

Source Information supplied from Sunti Golden Sugar Company Liaison Department's record

Determination of available tractors/equipment power ratings, draft, capacity and size selection for field operations of 1,300 ha sugarcane farm for 2017 and 2018 cropping seasons

A lot of factors are involved or considered when selecting farm machinery like field efficiency, soil condition, topography, crop factors, moisture content, machinery capacity, size of job to be done, time availability just to mention few.

Effective field capacity is usually expressed on area basis calculated by Eqs. (1) and (2) while Eqs. (3) and (4) were used to calculate material capacity [6, 8]:

$$Ca = \frac{\mathrm{vw}\mathfrak{L}\mathrm{f}}{10}, ha/h \tag{1}$$

$$Ca = \frac{A}{\text{Za}T} \tag{2}$$

where: A – area under operation
Z_a – available working days
T – available working hours per day, h/day

Capacity expressed on material basis using Eqs. 3 and 4:

$$Cm = \frac{\text{vwY}\Omega\text{f}}{10}\ t/h \tag{3}$$

$$Cm = \text{CaY} \tag{4}$$

where: Ca – field capacity on area basis, ha/h
Cm– field capacity on material basis, t/h
v – machine travel speed, km/h
Y – crop yield, t/ha
w – machine working width,
Ωf – effective field efficiency, decimal from [6] (Table 15.1).

Required Machinery Width Calculation
Implement working width required was calculated from Eqs. (1), (2) and (3), [8].
Number of implements required calculation by using Eq. (5) [8]:

$$N = \frac{W}{w} \tag{5}$$

where: N – number of implements required
w – maximum available working width, m
W - total width required to do work within time available, m

Draft and Power Requirements
It is essential to determine adequate tractor or engine draft and power to meet the implement and machinery requirements for agricultural operations.
Draft requirement calculation by using Eq. (6) [6]:

$$D = F_i\left(A + Bv + Cv^2\right)wd \tag{6}$$

where: D – implement draft, kN, (Table 15.2 in [6]) are average.
F_i – dimensionless texture adjustment factor, (Table 15.2 in [6]).
i – 1 for fine, 2 for medium, or 3 for coarse textured soils.
A, B, and C – implement –specific constants (from Table 15.2 in [6]).
d – tillage depth, cm (use 1.0 for minor tillage tools and seeders).

v – travel speed, km/h (from Table 15.1 in [6])

Drawbar power calculation by using Eq. (7) [6]:

$$Pdb = \frac{\text{Dv}}{3.6} \tag{7}$$

where: P_{db} – drawbar power, kw

Power Take Off (PTO) Power Determination: This was estimated from [6] (Fig. 15.1). That is power relationships for agricultural tractors from [6] (Fig. 15.1) for 4WD, tractive condition for firm soil = 0.78 and for tilled soil = 0.75. And for self-propelled machines that has rotary power. Power – takeoff, PTO power required to power the implement from PTO shaft of the tractor or engine. PTO power requirements can typically be determined by using rotary power requirement parameters available from a number of sources. Implement power take-off power was calculated as [2] and [1] using Eq. (8):

$$Ppto = \text{a} + \text{bw} + \text{cCm} \tag{8}$$

where: Ppto – power-takeoff rotary power required by the implement, kw
w – implement working width, m
Cm – material feed rate on wet basis, t/h
a, b, c – machine specific constants (from Table 15.3 in [6])

For some machines, drawbar power requirement must be added to the rotary power requirement to obtain the total power requirement. Example is potato harvester – Table 15.3 in [6].

Hydraulic Power Required: Fluid power required by the implement from hydraulic system of tractor was calculated as follows using Eq. (9), [1]:

$$Phyd = \frac{\rho F}{1000} \tag{9}$$

where: Phyd – hydraulic power required by the implement, kw
F – fluid flow, L/s
ρ – fluid pressure, kpa

Electric Power Required: Electric power required to operate components of implement was calculated using Eq. (10) as follows, [1]:

$$Pel = \frac{IE}{1000} \tag{10}$$

where: Pel – electric power required by the implement, kw.

I – electric current, A
E- -electric potential, V

NOTE: The total engine power should be more than the total power to accelerate, overcome changes in topography and crop conditions apart from hydraulic control and electrical unit powering. About 20% may be required for power reserve.

3 Results and Discussion

3.1 Steps Taken to Solve Disturbances of the Host Communities from 2015 to 2017 on the Company's Presence and Operations

Sunti Golden Sugar plantation is covering three zones namely Kade, Sunflag and Kusopa zones. Niger State Government, Nigeria constituted two committees, one at state level comprising six members coordinating committee. Other is three sub committees for the three zones each comprising nine members for Kade, Sunflag and Kusopa zones. They have been committed with assignment of monitoring and mediating between the company and the communities for mutual and cordial coexistence for the plantation farming of 15,000 ha. Main while, armed security personnel are also stationed in strategic locations and infrastructural installations till the aggrieved host community members are pacified with the company's efforts to solve the causes of the grievances. Direct and indirect immediate community benefits from the plantation are discussed under the following sub-headings:

Efforts made to solve challenges and problems faced by the host communities because of the sugarcane plantation's presence
Some measures that company put in place to solve most of the raised questions and the challenges faced by the company from their host communities as highlighted in Table 1 are:

(i) Sinking bore holes for supplying drinking water for all the communities with a typical type shown in Fig. 2.
(ii) Schools' rehabilitation as shown in Figs. 3, 4 and 5 shows company continuous contribution for all the host communities to be benefited.
(iii) Training and employing men and women on operation of modern agricultural equipment. Figure 6 is showing one of the trained female operators on her tractor out of five females that were trained along with their several male counterparts. This also is a continuous exercise done periodically.

Figure 4 is showing a rehabilitated school building but the pulpies here does not have school uniforms because the parents could not afford it and the whole school was just having one teacher for all age group, awaiting the assistance of the company to help recruit more teachers and provide school uniforms.

Fig. 2 Typical bore hole water supply facilities for one of the communities

Fig. 3 A classroom building to be rehabilitated by the company in one of the communities

Availability of Unskilled Labour Within Plantation Catchment Area
Plantation coverage as of the time of this study was Kade axis community, comprising five villages namely Kusoji, Sunti, Dzagun, Kanzhi and Batagi with estimated population of 20,000 people. The community and other suburb communities were able to supply more than enough unskilled labour requirement for the plantation operations. During 2018 cropping season, 300 female labourers were employed to replace male counterparts for sugar cane manual planting and were found to be meticulously performing better. Five female were trained this cropping season as tractor operators, but only two of them decided to stay behind to do the work while 3 have gone to get married. Up to 500 male labourers were employed as fixed time workers or casual labourers who were involved in weeding, chemical application with knapsack

Fig. 4 A finished class room building rehabilitated by the company in one of the communities

Fig. 5 A rehabilitated classroom equipped with furniture, with a teacher and pupils here ware school uniform that was provided by the company and employed additional teachers

sprayers, manual fertilizer application, security guards, tractor operator mates, Estate cleaners to mention few.

Availability of Skilled Labour Within Plantation Catchment Area
Few skilled labour were available because larger number of the community members are not educated and have no required skilled in plantation operations. From the community, available few skilled labours were engaged as tractor operators, workers

Fig. 6 A trained lady tractor operator with her tractor brought to workshop for repair among five uneducated ladies that was trained as tractor operators from the host communities

in accounting unit, stores, data capturing, pump operators, factory workers, electricians, plumbers, boiler house operators, estate maintenance, medical personnel, some field sectional heads, field area managers, supervisors in workshop, factory maintenance unit, engineering department and few senior staff members.

Plantation Impact on Wellbeing and Living Standard of the Community

Community livelihoods were improved through wages earned to assist spinster to fulfill some marriage expenses, educational pursuit, modern facilities acquisition like television, radio, motorcycle to mention few. Improve housing infrastructure conditions and better living standard within the last four years of the plantation take off. As part of encouragement, some contracts within the indigenes financial and technical know- how capacities were awarded to them just to be part of infrastructural development within their community. Recently, 15 village heads benefited in contract award of construction of culverts and some classroom blocks under the supervision of competent company staff. Arrangement was on to construct health center for Kanzhi community.

Educational Sector

Education was an important aspect of community need. Hence schooling activities for the young ones was being attended to. The management especially the company Chairman was paying special attention to educational sector of the communities.

Sunti Golden Sugar Educational Trust Fund was already in place which was been chaired by the wife of the General Manager. Todate, three community schools have benefited with some blocks of classroom construction and rehabilitation with furniture under this program, viz: Batagi, Sunti and Dazgun villages. Also, at Sunti village, company provided children with uniforms, textbooks and employed additional paid class teachers. A Sunti village indigene was sponsored for his Higher National Diploma (HND) program in Agricultural Science in a College of Agriculture who after his National Youth Service Corps (N.Y.S.C.) service year was supposed to work in the company for five years as part of the condition for being sponsored.

Infrastructures

Five communities so far have benefited from bore holes with overhead tanks with piping and all powered by functional generators at Kusogi, Sunti, Batagi, Kukpanti and Kpatcheta. The communities took part in the project execution. Road network leading from trunk A road to Estate from Mokwa (33 km) and Kudu to Jaagi (20 km) are being maintained by the company. In the previous year, the host State, Niger State Government intervened on the two roads maintenance. The Estate Road network within the Estate is being maintained by the company. Electricity supply from National grid's work was already on the pipeline by the company's management to link the host communities for better and steady electricity power supply.

Impact on Community Farming and Fishing Activities

Larger percentage of the land being used before for communities' farmland and fishponds are now taken over by the sugarcane plantation. In place of these, the followings are the measures in the pipeline to alleviate their suffering:

(i) Out grower scheme was being put in place for the host communities and this will be implemented on currently earmarked 141 ha within the plantation as pilot scheme for the immediate five communities that are affected.
(ii) Aquaculture scheme for fishponds also were in pipeline to be established for the host five communities. A fishing consultant firm from Malawi has been commissioned to design a sustainable model for these communities. The arrangement was at 70% completion as of the time of this study in 2018. The two schemes will be on cooperative farmers and fishers' level based on the experience and 17 of the communities' members gained through their visit to similar schemes at Malawi previous year which was sponsored by the company.

3.2 Environmental Condition of the Location for Sugarcane Cultivation and Processing

A modern factory to process one million metric tons/year of sugarcane from 17,000 ha irrigable farmland into average 100,000 metric tons/year sugar was installed with equivalent wastewater treatment plant. There was need to determine how favorable and suitable to agricultural mechanization investment, the sugar cane production technology at this location of the followings:

(i) Soil condition- suitability of soil profile

- soil type generally is sandy loose loam with low water retention ability.
- soil depth generally ranged from 300 to 400 mm.
- water table is generally high, actual value not made available.
- land topography/terrain for mechanized/irrigated farming opera tion is generally flat.
- the size of available land area that is suitable for cane cultivation presently is 4,770 ha while 930 ha total lakes surface area out of 17,000 ha available land.
- the farm is located on the flood plain of a big river, River Niger, there is enough water for irrigation in the dry season but very great drainage problem in the rainy season which required construction of six drainage pumping station (DPS) along 35 km dike presently.

(ii) Weather condition –annual rainfall and period of the year at the farm location: -

- average rainfall intensity duration during land preparation & cane planting if done before April was found to be alright.
- soil water table and impact on cane crop grown is that 9,300 ha of the land has too high-water table which was not good for present cane varieties under cultivation with root depth between 500 to 1500 mm soil depth.
- agro-ecological condition of the environment is good for crops varieties that can tolerate high water table.

The presence of the company has made the road better maintained. Fertilizer and chemical applications were manually, and knapsack applied respectively, hence so far, it was not polluting the environment. Also from sugar processing factory, Fig. 7, wastewater from the factory was being treated, Fig. 8, hence not affecting stream water. In future when the company's operation will be fully mechanized, safety precautions/measures to control environmental and air pollution must be put in place.

3.3 Determined Tractor/Equipment Power Ratings, Draft and Capacity for Field Operations of 1,300 ha Sunti Farm for 2018 Cropping Season

Table 2 shows the values of the available fleet of tractors and equipment power ratings, draft, drawbar power, power take off (PTO) for both firm and tilled soils, hydraulic, electrical power, and unit numbers of equipment in the plantation were determined. Also, Table 3 shows the values of available fleet of implements power required range, implement size/working width and depth, required draft, drawbar and PTO and machinery capacity. Tables 2 and 3 adequately provides information

Fig. 7 A modern 4,500 metric tons/day sugarcane processing factory capacity at Sunti plantation, Sunti, Nigeria

Fig. 8 One million metric tons/year sugarcane processing industrial wastewater treatment plant, Sunti, Nigeria

for both power ratings, equipment size and capacity selection within the available fleet in the sugarcane plantation for field operation expansion from current 1,300 ha to the 4,770 ha available suitable land for sugarcane cultivation. This was done to take proper precautions against impact of soil cultivation/tillage operation and equipment trafficability resulting in soil texture and structure destruction. The available equipment was found to be adequately sufficient for the 1,300 has sugarcane cultivation.

Table 2 Determined values of available tractor/equipment power ratings for field operations of 1,300 ha Sunti farm for 2018 cropping season

S/No	Equipment/model	Power rating, Kw	PTO power, kw	Drawbar power, firm soil, Kw	Drawbar power, tiled soil, Kw	Hydraulic power, Kw	Electric power, Kw	Number of available tractor/equipment
1	Valtra tract.-T171	134.00	114. 71	89.47	86.03	39.36	2.21	4
2	Vatra tract.-T191	141.00	114.71	89.47	86.03	23.92	2.21	16
3	Vatra tract.-A950	71.00	49.87	38.90	37.40	11.83	0.78	4
4	Vatra tract.-BM125i	98.40	67.32	52.51	50.49	15.19	1.08	2
5	Vatra tractor-T194H	146.00	125.71	98.05	94.28	57.67	1.44	1
6	John Deere tract.-3050	67.60	62.87	49.04	47.15	13.57	1.32	3
7	John Deere tract.-2850S	64.00	49.50	38.01	37.13	13.87	1.32	1
8	John Deere tract.-4960	149.00	134.00	104.52	100.50	33.65	2.88	2
9	John Deere tract.-3650	85.30	77.00	60.06	57.75	26.03	1.32	2
10	John Deere tract.-2650	58.20	52.38	40.86	39.29	13.89	1.06	1
11	John Deere tract.-3350	74.60	67.60	52.73	50.70	13.57	1.32	1
12	John Deere tract.-8960	275.00	247.60	193.13	185.70	43.09	4.32	2
13	John Deere tract.-8970	298.40	264.30	206.15	198.23	49.37	4.32	4
14	John Deere tract.-7225 J	168.00	142.40	111.07	106.80	60.91	2.16	3
15	JD Cane harvester-3520	251.0	208.33	162.50		125.36	4.80	1
16	Cane loader-BM100	74.60	67.14	53.37	50.36	15.19	2.88	3
17	Cane loader-Bell	49.00	44.10	34.40	33.18	30.45	0.49	1
18	Cane loader-1850	78.00	70.20	54.76	52.15	33.08	2.88	2
19	John D. cane loader-6068 T	138.00	114.54	89.34	85.91	40.36	1.92	1
20	Cat. Bulldozer-D7G	164.00	147.00	114.66	110.25	63.85	0.84	5

(continued)

Table 2 (continued)

S/No	Equipment/model	Power rating, Kw	PTO power, kw	Drawbar power, firm soil, Kw	Drawbar power, tiled soil, Kw	Hydraulic power, Kw	Electric power, Kw	Number of available tractor/equipment
21	Cat. Bulldozer-D8R	245.00	221.13	172.48	165.85	120.66	1.20	5
22	Excavator-325C	141.00	115.20	89.86	86.40	34.46	1.56	4
23	Excavator-320L	97.00	87.30	68.09	65.48	117.31	1.20	2
24	Excavator-JS2052C	106	85.86	66.97	64.40	125.85	1.80	1
25	Motor grader-140H	138.00	124.2	96.88	93.15	82.83	1.80	4
26	Compactor-VM116drum-CP563E	85.00	76.50	59.67	57.38	129.61	2.16	5
27	Paywheel loader-432ZX	112.00	100.00	78.00	75.00	74.22	1.08	1
28	Backhoe loader JCB-3DX	68.60	61.74	48.16	46.31	46.66	1.08	2
29	Backhoe loader-JCB-3CX	55.00	49.50	38.61	37.13	69.01	1.14	1
30	Backhoe loader-JCB-4CX	74.60	57.88	45.15	43.41	69.01	1.14	1
31	Wheel loader-938G	134.00	111.22	86.75		22.54	1.92	1
32	Wheel loader-928G	116.00	96.28	75.10		17.44	1.92	2
33	Wheel loader-950H	149.00	123.67	96.46		30.89	1.92	2
34	Ditch witch-RT55	45.00	37.33	29.13		41.78	0.12	4
35	TelehandlerTH-TH460B	55.00	45.65	35.61		47.07	1.08	1
36	Fork lift-7FDU30, 3 tons	40.00	33.20	25.90		38.63	0.58	1
37	Fork lift-DP35N, 3.5 tons	40.00	33.20	25.90		38.63	0.84	1
38	Fork lift-DP30N, 3 tons	40.00	33.20	25.90		38.63	0.58	2
39	Fork lift-32 T-SVE TRUCK	40.00	33.20	25.90		38.63	0.58	1

(continued)

Table 2 (continued)

S/No	Equipment/model	Power rating, Kw	PTO power, kw	Drawbar power, firm soil, Kw	Drawbar power, tiled soil, Kw	Hydraulic power, Kw	Electric power, Kw	Number of available tractor/equipment
40	Crane truck-JIANG HUAN-12 T	132.00	109.56	99.00				1
41	Crane truck-BEDFORD	180.00	149.40	116.53				1
42	Crane truck-IVECO	307.00	254.81	196.20				1
43	Crane truck-HOWO	251.00	208.33	160.41				1
44	Crane liebher-60 T-LIEBHER	270.00	224.10	172.56				1
45	Crane liebher-70 T	330.00	273.90	210.90				1
46	Tipper lorry-MP380E38H EURTRAKKER	283.0	234.89	180.87				7
47	Hydra Crane-15XW	80.90	67.15	52.37				2
48	Fuel bowser - FKWT-3000L	52.20	43.33	33.36				2

Table 3 Determined values of available field implement capacity and power rating used for 1,300 ha Sunti farm plantation for 2018 cropping season

S/No	Equipment/model	Tractor power rating required, kw	Disc diam., cm/number of discs	Working width, m	Tillage Depth, cm	Implement draft, KN	Drawbar power (kW)	PTO required for firm soil, Kw	PTO required for tilled soil, KW	Area capacity, ha/h	Material capacity, t/h
1	Mounted disc plough, FKMDP- 2	37.3 – 44.8	66.04/2	0.55	33	4.02	7.81	10.01	10.41	0.33	5.51
2	Mounted disc plough, FKMDP- 3	48.5 – 59.7	68.58/3	0.925	33	6.76	13.14	16.84	17.52	0.55	9.26
3	Rome harrow- TCW - 40	66.4 – 85.0	60.96/40	4.7	20.3	25.78	50.13	64.26	66.83	2.63	44.30
4	Rome harrow- TBW - 40	117.1 – 136.5	66.04/5 m	5.5	22.9	43.28	84.17	107.90	112.22	3.08	51.84
5	Tractor mounted Boom sprayer- ALFA 1000/FKMCP-6	52.258.9 – 67.1	38 (nozzles)	18	1	12.16	33.79	43.33	43.33	11.7	
6	Disc Ridger- FKDR - 4	52.2 – 67.1	66.04/4	0.8	33	4.79	6.65	8.53	8.87	0.34	5.72
7	Single row ridger- FKDR - 2	37.3 – 48.5	66.04/2	0.6	33	3.59	4.99	6.39	6.65	0.26	4.29
8	Valtra Mould former (mouldboard plough)- BT MD2	41.0–52.2	3 row sweep	1.05	35	28.17	54.78	70.23	73.04	0.62	10.51
9	Row marker- Cultivator sweep	67.14	6 row sweep	12	10	13.04	43.47	55.73	55.73	11.52	

(continued)

Table 3 (continued)

S/No	Equipment/model	Tractor power rating required, kw	Disc diam., cm/number of discs	Working width, m	Tillage Depth, cm	Implement draft, KN	Drawbar power (kW)	PTO required for firm soil, Kw	PTO required for tilled soil, KW	Area capacity, ha/h	Material capacity, t/h
10	Mechanical weeder- John Deere rotary hoe weeder	80.89	6 row weeder	12	10	13.47	52.37	67.14		12.60	
11	Ditch witch implement- Ditch witch – RT55	45		0.3	1.58		29.13		37.35		
13	Fertilizer spreader/broadcaster- FKFS - 400	14.92		14.02	1						
14	Bowser- FKWT-3000L	37.3 – 52.2	500/3000L	1.88	1	0.00					
18	Valtra Light duty tractor mouldboard plow- BT MD3	52.2–63.4	NA/2 discs	0.70	18	9.66	18.78	24.08	25.04	0.42	7.01
19	Medium duty tractor plow- FKRPDH-8	74.6–93.3	26.4/8	1.70	25	32.66	63.50	81.41	84.67	1.01	17.06
20	Heavy duty tractor plow (disc)- Shengxuan 1LTF-630Q (12 plow)	89.52 -119.4	26.8/8	1.8	30	41.40	80.49	103.20	107.33	1.07	18.02

(continued)

Table 3 (continued)

S/No	Equipment/model	Tractor power rating required, kw	Disc diam., cm/number of discs	Working width, m	Tillage Depth, cm	Implement draft, KN	Drawbar power (kW)	PTO required for firm soil, Kw	PTO required for tilled soil, KW	Area capacity, ha/h	Material capacity, t/h
21	Rome Light duty tractor harrow-TBW-16 wheel offset disk harrow	36.6–55.2	27.9/16	2.1	22.9	16.34	45.38	58.17	60.50	1.79	30.04
22	Rome Medium duty tractor harrow-TRCH-10 wheel offset disk harrow	123.1–141.7	50.8/10	2.3	33	25.78	71.62	91.82	95.49	1.96	32.90
23	Rome Heavy duty tractor harrow- TRC-20 wheel offset disc harrow	279.8–335.7	50.8/10	4.9	33	54.93	152.57	195.61	203.43	4.17	70.10
24	Tandem disc harrow (primary)	68.86–75.00	45/20	2.55	25	26.75	44.58	57.15	59.44	1.22	20.60
25	Tandem disc harrow (secondary)	38.49- 45.00	35/20	2.55	20	14.95	24.92	31.95	33.23	1.22	20.60

4 Conclusion

The study carried out was to examine the technical and the socio-economic impacts of the change over a four-year period this mechanized large scale crop production took off making use of available data obtained for 2015 to 2017 cropping seasons. This study was able to discover negative and positive technical and socio-economic impacts, which was reasonably addressed. Adequate steps were taken to solve disturbances of the host communities because of negative impacts of the company's presence and operations by putting in place peace committees at both government and company's levels. In term of employment, efforts were made to solve challenges and problems faced by the host communities by employing as much as possible both skilled and unskilled labour within plantation catchment area. The presence of the plantation has to some reasonable extent impacted positively on wellbeing and living standard of the communities in areas of education, skill acquisition, infrastructure sectors and Community farming and fishing activities and overall improves their economic life. All necessary measures are being put in place for the plantation activities not to put the communities in dangerous environmental condition by the sugarcane cultivation operation and sugar refinery. Further efforts were taken to determine values of available tractor and equipment power ratings, draft, and capacity for field operations of 1,300 ha Sunti farm for 2018 cropping season. Also implements power ratings, equipment size and capacity selection of the available fleet in the sugarcane plantation for field operation of 1,300 ha cultivation was determined and found adequately sufficient for the sugarcane cultivation which took into consideration precaution of negative impact of soil cultivation operation and equipment trafficability.

Acknowledgements We hereby acknowledge The World Bank Sponsored Project of African Centre of Excellence in Agricultural Development and Sustainable Environment (CAEDESE), Federal University of Agriculture, Abeokuta, Nigeria for sponsoring my participation of one month Faculty Outreach (2018) at Flour Mills Nigeria Plc, Sunti Golden Sugar Company, Niger State, Nigeria. Also acknowledged are the Sunti Golden Sugar Company Staff team who contributed immensely to the success of this study comprising Mr. Andries L. Du Pisani (General Manager), Engr. Samuel F. Okunola (Community Liaison Manager), Mr. Rasheed A. Kuranga (GIS Analyst) and Alh. Ahmadu Adamu Lwafu (Community Liaison Officer).

References

1. ASAE EP496.3 (2006) Agricultural Machinery Management. American Society of Agricultural and Biological Engineers, ASABE Standard, pp 383–390
2. Coates WE (2002) Agricultural Machinery Management. Published by College of Agriculture and Life Sciences, University of Arizona, USA. ACIS, pp 1–9
3. Odi L, Tripp R (2005) Technology and its contribution to pro-poor agricultural development. UK Department for International Development (DFID), ODI, London, pp 1–23

4. Schirmer J, Parsons M, Charalambon C, Gavran M (2005) Socio-economic impacts of plantation forestry in the great southern region of WA, 1991 to 2004. Bureau of Rural Sciences, Forest and Wood Products Research and Development Corporation, Australia. Report Produced for FWPRDC project PNO4.4007, pp 1–3, 1–128
5. Shete M, Rutten M (2005) Impacts of large-scale farming on local communities' food security and income levels – empirical evidence from oromia region, Ethiopia. Land use policy 47:282–292
6. Srivastava AK, Goering CE, Rohrbach RP, Buckmaster DR (2006) Engineering Principles of Agricultural Machines. America Society of Agricultural and Biological Engineers, USA. ASAE publication 801M0206, 2 edn, pp 525–549
7. Timmer CP (1988) The agricultural transformation. In: Chenery H, Srinivasan T (eds) Handbook of Development Economics,Amsterdam, North Holland, vol 1
8. Yousif LA, Dahab MH, El-Ramlawi HR (2013) Crop-machinery management system for field operations and farm machinery selection. J Agric Biotechnol Sustain Dev 5(5):84–90

Assessment of Potential Territorial Suitability for the Sustainable Production of the "Acquaviva Red Onion" (*Allium Cepa* L.): A Study Using Overlapping Approach

Giuseppe Ruggiero, Marco Parlavecchia, Francesco Santoro, and Simone Pascuzzi

Abstract The onion (Allium cepa L.) is one of the most important horticultural crops worldwide; it is widely appreciated by consumers for its nutraceutical effects. In the extreme South of the Murgia area (Apulia, Italy), the Acquaviva red onion has been cultivated for a very long time, as an excellence of this territory. The production of this product typically extends across a rather limited and substantially confined area within the municipality of Acquaviva delle Fonti. Therefore, this circumscribed oasis of production demands territorial planning policies aimed at valorizing its sustainable production by safeguarding the soils most suited to this typicality of production, whose erosion would result in the disappearance of a product with high territorial value. In order to assess the potential territorial suitability for the Acquaviva red onion, data on climate (average temperature and precipitation), soil (texture, pH, electrical conductivity and organic matter content), topography (altitude and slope) and land use were found for each district, through preliminary surveys, and using the Informative Territorial System (SIT Puglia) database. Territorial suitability ranges were established, based on the onion cultivation needs and the indications of the Designation of Municipal Origin (DE.C.O) production specification and the Slow Food production specification for Acquaviva red onion. Using the GIS software Arcmap Arcgis 10.5, all information layers were overlapped, obtaining the maps of suitability for each feature. Finally, from the overlap of these, it was possible to obtain an overall map of the areas with a greater territorial suitability. The results obtained suggest that accurate territorial planning can be integrated with the contents of a regional development program that is attentive to the valorisation of this typical local product. Beyond the purposes described above, this method could be extended to conterminous territorial areas to verify the possible expansion of the optimal production area.

G. Ruggiero (✉) · F. Santoro · S. Pascuzzi
Dipartimento di Scienze Agro Ambientali e Territoriali, Università degli Studi di Bari, Bari, Italy
e-mail: giuseppe.ruggiero@uniba.it

M. Parlavecchia
Consiglio per la Ricerca in Agricoltura e l'Analisi dell'Economia Agraria, CREA-AA Bari, Bari, Italy

S. Pascuzzi and F. Santoro (eds.), *Farm Machinery and Processes Management in Sustainable Agriculture*, Lecture Notes in Civil Engineering 289,
https://doi.org/10.1007/978-3-031-13090-8_32

Keywords Land suitability · Territory · Onion · GIS · Apulia

1 Introduction

Worldwide, the onion (Allium Cepa L.) is the second horticultural crop in terms of production and it is widely appreciated for its nutraceutical benefits [1]. Additionally, in Italy, the onion is one of the most important vegetables and it is widespread throughout the national territory with a great number of varieties strongly connected to specific production areas. In fact, every Italian region has numerous typical productions of onion not always recognized by quality certifications. In Apulia there are different varieties distributed in different production areas, such as Barletta's Precocissima White Onion, Margherita di Savoia White, August's Bianca grossa or Agostegna, Rossa and Brown Flat Barletta, and Zapponeta Onion. In particular, in the extreme south of the Murgia area, the Acquaviva Red Onion, cultivated since time immemorial, excels. This local product is known for its organoleptic and healthy qualities and is typically spread in a production area rather limited and exclusively confined to the municipality of Acquaviva delle Fonti, in the province of Bari [2]. It is therefore an oasis of production, which imposes territorial planning policies aimed at safeguarding the soils most suited to the typicality of production, the erosion of which would result in the disappearance of a product with a high territorial value. This risk may be averted by protective and rational choice initiatives concerning alternative land uses. Proper spatial planning can, inter alia, be integrated into the content of a regional development programme that focuses on the valorisation of typical local products. Ultimately, in light of the established symbiotic relation between the produce and its surrounding territory, such endeavors would allow for the implementation of more sustainable production methods. Sometimes, territorial vocation has been studied only with respect to some factors, such as climate characteristics [3]. Recently, the cartographic overlay using GIS based software, allowed to carry out several territorial studies [4–8]. In addition to this, different GIS approaches have been useful to define the territorial vocation for crops, regarding various territorial features, such as soil quality features, climatic characteristics, land topography, and land use [9–11]. This work, therefore, aims at exploring and establishing the productive vocation of the municipal territory of Acquaviva delle Fonti for the red onion produce, identifying areas with different production potential both in terms of quality and in relation to quantitative aspects. Territorial productive vocation consists in the predisposition of a specific territory to host successfully a crop. Vocation, therefore, indicates the efficiency of the symbiotic relationship between a specific crop and the territory that hosts it. The creation of a vocational map aims to differentiate the productive territorial potential by identifying both the areas that allow the optimal development of the crop and those in which such vocations are scarce or even null.

2 Materials and Methods

In order to acquire the essential data to characterize the productive peculiarity of the Acquaviva Red Onion, the state of the art was studied through the study of technical-scientific sources, merchandise sources and cartographic sources. In particular, the technical-scientific bibliography concerning the agronomic characteristics of the onion was consulted. Among the different scientific texts and works examined are: "A come Agronomia" [12], "Manuale di Corretta Prassi per la produzione integrata della cipolla" [13] and "Metodi di Valutazione dei suoli e delle terre" [14]. With reference to product sources, the Production Disciplinary of the Acquaviva Red Onion elaborated by the "Consortium for the protection and valorization of the Acquaviva Red Onion" [15], and the Acquaviva Red Onion Integrated Production Disciplinary drawn up by the "Consortium for the promotion, protection and enhancement of the typical qualities of Apulia" [16] have been analysed. These references have been compared with similar elaborations related to other local productions such as the onion of Tropea (Production Specification Onion of Tropea), and that of Margherita di Savoia (Production Specification for Margherita di Savoia onion PGI).

The cartographic elaborations have required the acquisition of a chart able to represent the toponymy and agronomic inter-communal subdivision in order to more effectively correlate the analyses and the results to the same territory. The chart was built starting from the framework of cadastral union to which the subdivision of territorial districts (contrade) was superimposed (Fig. 1). This choice is justified by the possibility of correlating analyses and subsequent vocational attributions to territorial units (districts) easily identifiable by place names and agronomic characteristics. In fact, the districts allow easy-to-interpret correlations between the geography of the territory and its production variability. For the analysis of the territorial characteristics deemed essential for the study, official cartographic sources have also been used, such as the site of the Territorial Information System (SIT Puglia). The cartographic elaborations have been conducted in GIS environment through the use of the software Arcgis Arcmap 10.5.

The study has been articulated in several phases, synthetically represented in Fig. 2. Analysis of the sources above-mentioned allows to identify the indices qualifying the production performance of the Cipolla Rossa di Acquaviva in the area under analysis and the corresponding quality classes. Figure 3 indicates three classes of production capacity (i.e., high, medium, low) defined through value ranges attributed to various production indices, further grouped into the commercial, organoleptic, and field performance category.

Additionally, territorial factors that influence the production capacity of the onion in the same territory were identified and subsequently reported on the cartographic base (see Fig. 1). The spatial reference framework featuring such factors was developed by, firstly, adopting the altimetry map and the land use map from SIT Puglia.

Secondly, additional qualifying factors consisting in soil maps (texture, pH, electrical conductivity and soil organic matter) and maps of average temperatures and

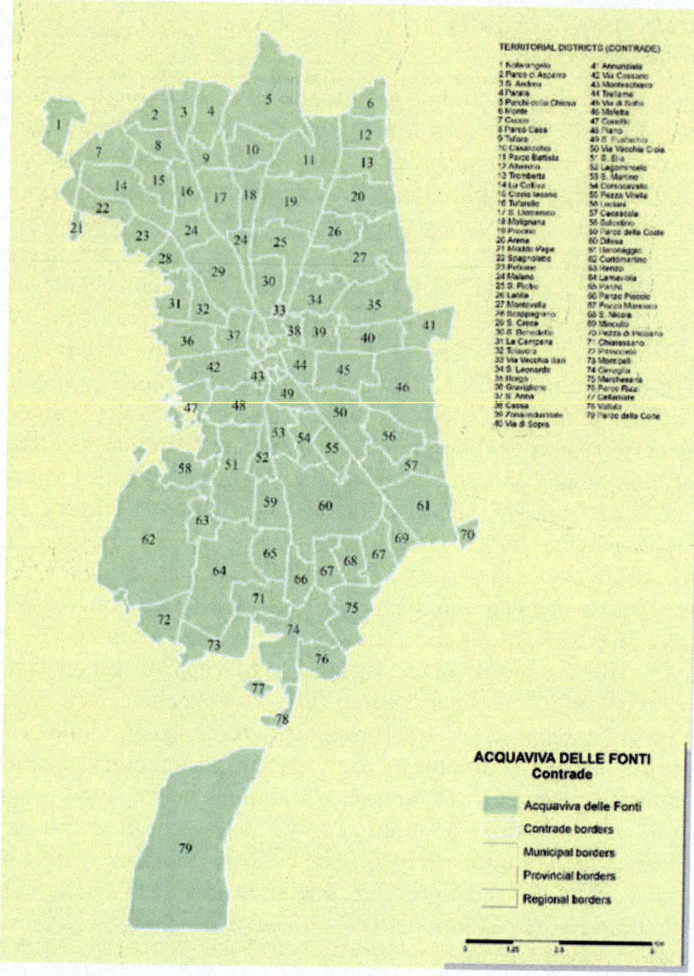

Fig. 1 Chart of Acquaviva delle Fonti with the territorial districts subdivision (contrade)

precipitation were developed through analyzing data from soil analysis and the Civil Protection Service of the Apulia Region respectively. Such thematic maps were then subsumed under the three classes of productive vocation: high, medium and low (Table 1), ultimately producing the maps of territorial vocation. In other words, from the maps it is possible to read, for each territorial factor considered qualifying, the level of territorial productive vocation of each district for the cultivation of the red onion of Acquaviva delle Fonti. Furthermore, the areas with different vocations are represented by different colors: in green are reported the districts particularly suited for the cultivation of onion, in yellow the medium-suited areas and in red the areas not suitable for the cultivation of the product. Finally, the cartographic overlay of

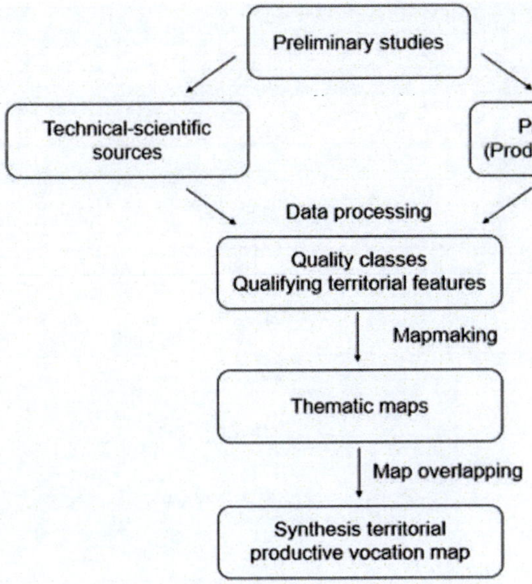

Fig. 2 Study phases scheme

the thematic maps so produced has allowed to obtain a synthesis of the productive vocation of the onion of Acquaviva across the territory (Fig. 4).

The identification of production indices and territorial qualifying factors carried out according to the cultivation needs found in the literature and the indications of the production specification of the onion of Acquaviva and the specification Slow Food, has made it possible to differentiate the territorial quantitative–qualitative profile of the onion in the Acquaviva delle Fonti countryside by identifying three ranges of territorial vocationality as per Table 1. With regard to the selection of the territorial qualifying factors, those considered most significant for the determination of the quantitative–qualitative aspects of the product were identified, and listed below: average temperature and precipitation (climatic factors); texture, pH, electrical conductivity, soil organic matter (pedological factors); altimetry (topographical factors); land use.

In relation to climatic factors, the annual average values of temperatures and rainfall have been taken into account. The lack of official thermo-pluviometric stations within the area under study, led to the use of other databases, namely those provided by the Civil Protection Service of the Apulia Region and the website www.climate-data.org. Both provide data for the entire municipal area without any internal differentiation. The climate data were thus processed through a comparison between the two sources used, referring to the period corresponding to the crop cycle of the onion (December to June). In the knowledge that pedological factors are decisive for the productive results of quality, it was considered necessary to evaluate the variability of the nature of the soil within the municipality. In the absence of official sources, a survey has been conducted to take the results of physical–chemical analysis of the

QUANTITATIVE INDICES			
Bulb weight (g)	300	300 – 600	> 600
Bulb diameter (cm)	< 10	10-12	> 12
Average yield (q/ha)	80-100	100-150	150-200
Harvest period	EARLY	MEDIUM	LATE
QUALITATIVE INDICES			
Bulb form	SFERIC BULB		FLAT BULB
Bulb color	SCARLET RED	PURPLE RED	COPPERY RED
Flesh color	PINKYSH RED	WHITISH	PURE WHITE
Sweetness	SWEET	SWEETISH (SPICY AFTERTASTE)	
Smell (back-nasal way perception)	AROMATIC		SOFT SMELL
Flesh consistency	SOFT	COMPACT	CRUNCHY
Size uniformity	NON UNIFORM		UNIFORM

- high production capacity ➡

- bulb weight: 300 g
- bulb diameter: 10-12 cm
- average yield q/ha: 150-200
- early harvest
- flat bulb
- purple red bulb, sweet and uniform
- soft and pinkysh red flesh
- aromatic smell

- medium production capacity ➡

- bulb weight: 300-600 g
- bulb diameter: < 10 cm
- average yield q/ha: 100-150
- medium harvest
- sferic bulb
- scarlet red bulb, sweetish and non uniform
- crunchy and whitish flesh
- soft smell

- low production capacity ➡

- bulb weight: > 600 g
- bulb diameter: > 12 cm
- average yield q/ha: 80-100
- late harvest
- sferic bulb
- coppery red bulb, sweetish and non uniform
- crunchy and pure white flesh
- soft smell

Fig. 3 Classes of production capacity defined through value ranges attributed to various production indices

land carried out during the two-year period 2019–2020 by local laboratories commissioned by farms. The collection of analysis certifications was carried out in order to cover the entire territory as well as the individual districts, with a significant number of analytical results. The characteristics taken into consideration concern the soil texture and structure, the total organic matter content and the depth of the soil depth. Among the various factors that determine the topography of a territory, altimetry is undoubtedly the characteristic that influences the productive performance of crops because it helps to determine climatic conditions with particular reference to temperature. In order to design the map of the productive vocation of the onion connected to the altimetries, the relative excerpts of information from the SIT Puglia have been

Table 1 Territorial features ranges of vocationality

Territorial features	High vocationality	Medium vocationality	Low vocationality
Climatic features			
- average temperature (°C)	15–25	10–14	<10/>25
- precipitations (mm/year)	400–700	700–800	<400/>800
Pedological features			
- texture	Loamy Sandy loam	Clay loam Clay sandy loam Clay silty loam	Others
- pH	6,5–7,4	7,5–8,5	<6,5/>8,5
- ECe (mS/cm)	<1,2	1,2–1,8	>1,8
- SO (%)	>1,0	0,5–1,0	<0,5
Topographical features			
- altitude (mamsl)	<300	300–400	>400
Land use features			
- consociation	>3	2–3	<2

acquired, subdividing, with appropriate elaborations, the territory in three altimetric bands. The use of soil was considered a useful indicator in assessing the productive potential of the red onion of Acquaviva through the agronomic productive correlations between the prevailing crops and the cultivation of onion in a given area. Also, for this character the cartographic source of the informative system of the SIT Puglia has been used.

3 Results and Discussion

The production of the map of the districts, used as a base map for the analyses and territorial elaborations carried out, is the first result of this study. It is a map that includes about 13,000 ha of municipal territory, divided into districts on a cadastral basis (framework of union). It displays location of districts with the corresponding sheets of map and extension (ha). From the cartographic elaboration (overlapping of the contradas to the sheets of map) they have been perimeter 79 contrade (Fig. 1). The territorial significance of the districts, already expressed above (method) is confirmed by the mean surface area of 163 ha per unit.

As reported by Grassano et al. (2011) the territory evaluation for a special production is determined by investigating the agents related to soil properties, climate, topography factors and by local expert opinions [17]. The map of territorial vocationality related to the average temperature in reference to the cultivation cycle of the onion highlights only two bands of the three possible ranges, and these are the

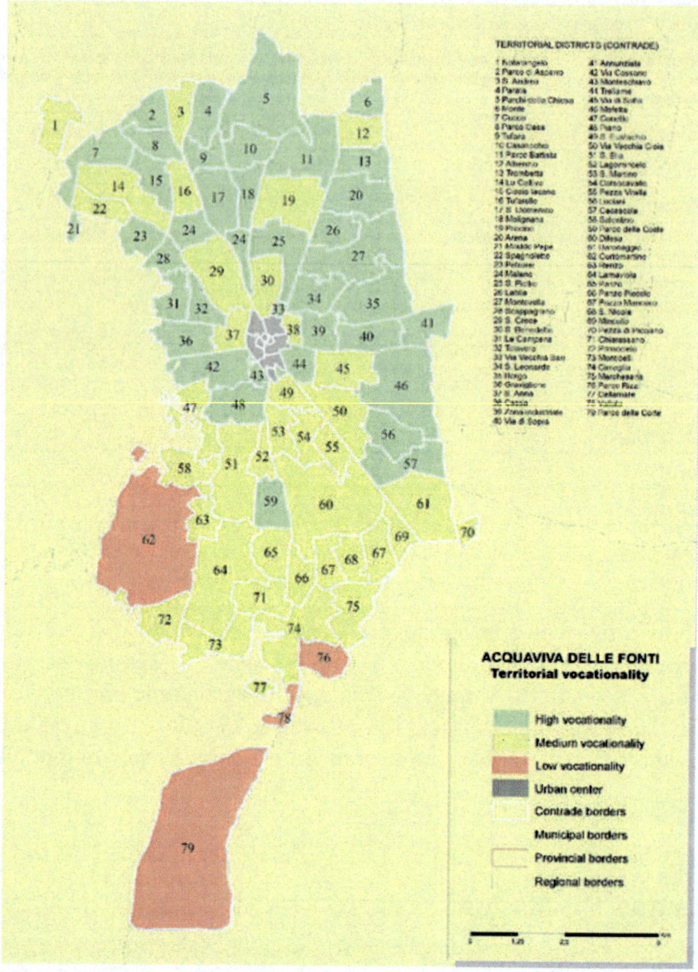

Fig. 4 Territorial vocationality synthesis map

high and medium level. With reference to the thematic map related to average rainfall, defined in relation to the production period of the onion, homogeneity was found, as the whole territory is characterized by the only class with high vocationality (400–700 mm/year). It is, therefore, a map that, in the inter-map overlapping process (Fig. 4), will not have an impact on the territorial differentiation of productive vocationality.

The thematic maps relating to pedological aspects (texture, pH, electrical conductivity and soil organic matter) show greater territorial differentiation [18]. As for the soil texture, 25 districts (32% of the municipal territory) fall into the high-vocationality range, 52 districts (66% of the municipal territory) in medium vocationality and only 2 districts (Cimaglia and Marchesana) have a low vocationality. The latter are located at the extreme south of the countryside of Acquaviva delle Fonti, while those falling in the middle belt occupy the central north–south axis of the municipal territory. In the map of territorial vocationality linked to pH it is possible instead to identify the East–West transversal bands with different vocations. Altogether 43 districts (54% of the territorial surface) show a high vocationality, while the remaining districts have an medium vocationality (35 districts); the exception is the district Cessa where there was a low territorial vocationality. The thematic chart of electrical conductivity, shows a lower variability in vocations, as 75% of the territory presents itself to medium vocationality. The other two classes are present with different percentages (16% of the territory to high vocationality and 9% to low), specifying that the areas to high vocationality are concentrated in the south-east portion of the communal territory. With reference to the content of soil organic matter, the territory of Acquaviva delle Fonti shows good levels that define only two categories of vocationality: high vocationality (content of organic matter > 1.0%) affecting 45 districts (57% of the communal territory) and medium that defines 34 districts (43% of the territorial surface). The analysis of altimetry has made it possible to draw up a thematic chart of vocationality that divides the territory into three transversal bands. The territory to the north is defined by a high vocationality (55% of the communal territory); the central portion is defined by an medium value of vocationality and interests approximately 33% of the territory; finally, the third band that constitutes the less extended portion is relegated to the extreme south. Finally, considering the territorial thematic map related to land use, it has been shown that a large part of the territory has medium values of vocationality (68% of agro-municipal) and is associated mainly with tree crops. While high vocationality has been recorded in the districts where differentiated agronomic uses prevail, low vocationality was found in areas presenting a narrowed range of cultivations, such as olive groves and arable land. The synthesis map (Fig. 4) demonstrates that a large part of the territory (95%) shows a vocation for the production of red onion between medium and high values. More specifically, the northern portion of the agro (50% of the land area) has a high productive vocation. All the results obtained are not comparable to previous studies, because in literature there are not similar analysis about the study area and the red onion crop.

4 Conclusions

This study confirmed the vocationality of the territory of Acquaviva delle Fonti for the production of red onion, verifying a diversification of the same within the municipal territory. The zoning obtained showed that the productive potential of the red onion is still poorly exploited because only a small part of the territory manifesting high productive vocation, is in fact used. Notably, the degree of under-utilization increases particularly in areas with medium vocations. These considerations call for territorial development policies aimed at enhancing the product by increasing the areas involved. It also seems essential to launch agronomic programming of farm production in order to make the best use of the ability to adapt the onion and its versatility to be used as an intermediate crop while overcoming the limitations linked to the necessary crop rotation. Finally, the authors recognize that the method developed can be used, through appropriate corrections, to evaluate the productive vocationality of other territories in relation to agricultural products similar to the Red Onion of Acquaviva delle Fonti.

References

1. Ricciardi L, Mazzeo R, Marcotrigiano AR, Rainaldi G, Iovieno P, Zonno V, Pavan S, Lotti C (2020) Assessment of genetic diversity of the "Acquaviva Red Onion" (*Allium cepa* L.) Apulian landrace. Plants 9:260
2. Elia A, Santamaria P (2013) Biodiversity in vegetable crops, a heritage to save: the case of Puglia region. Ital J Agron 8(4):21–34
3. Gentilucci M, Barbieri M, Burt P (2018) Climate and territorial suitability for the vineyards developed using GIS techniques. In: Conference of the Arabian journal of geosciences, pp 11–13
4. Ruggiero G, Parlavecchia M, Dal Sasso P (2019) Typological characterisation and territorial distribution of traditional rural buildings in the Apulian territory (Italy). J Cult Herit 39:278–287
5. Parlavecchia M, Pascuzzi S, Anifantis AS, Santoro F, Ruggiero G (2019) Use of GIS to evaluate minor rural buildings distribution compared to the communication routes in a part of the Apulian territory (Southern Italy). Sustainability 11:4700
6. Parlavecchia M, Pascuzzi S, Santoro F, Ruggiero G (2021) Minor rural building heritage and territorial features in Local Action Group Sud Est Barese area (Southern Italy). In: Engineering for rural development, Jelgava, 25–28 May 2021
7. Statuto D, Cillis G, Picuno P (2017) Using historical maps within a GIS to analyze two centuries of rural landscape changes in Southern Italy. Land 6(3):65
8. Ety NJ, Rashid MS (2020) Spatiotemporal variability of erosion and accretion in Ganges River using GIS and RS: a comparative study overlapping Rennel's map of 1760s. Environ Dev Sustain 22:3757–3775
9. Jayasinghe SL, Kumar L, Sandamali J (2019) Assessment of potential land suitability for tea (Camellia sinensis (L.) O. Kuntze) in Sri Lanka using a GIS-based multi-criteria approach. Agriculture 9:248
10. Ostovari Y, Honarbakhsh A, Sangoony H, Zolfaghari F, Maleki K, Ingram B (2019) Ecol Ind 103:479–487
11. Pilevar AR, Matinfar HR, Sohrabi A, Sarmadian F (2020) Integrated fuzzy, AHP and GIS techniques for land suitability assessment in semi-arid regions for wheat and maize farming. Ecol Ind 110:105887
12. Giardini L (2003) A come Agronomia. Pàtron Editore, Bologna
13. Tei F (2004) Manuale di Corretta Prassi per la produzione integrata della cipolla. Regione Umbria, Italy
14. Aramini G, Colloca C, Corea AM, Paone R (2006) Metodi di valutazione dei suoli e delle terre. Edizioni Cantagalli, Siena
15. Consortium for the protection and valorization of the Acquaviva Red Onion (2014) Production Disciplinary of the Acquaviva Red Onion
16. Consortium for the promotion, protection and enhancement of the typical qualities of Apulia (2013) Acquaviva Red Onion Integrated Production Disciplinary

17. Grassano N, Tedone L, Verdini L, De Mastro G (2011) Evaluation of rapeseed cultivation suitability in Apulia with GIS-multicriteria analysis. Ital J Agron 6(2):e16
18. Anifantis AS, Camposeo S, Vivaldi GV, Santoro F, Pascuzzi S (2019) Comparison of UAV photogrammetry and 3D modeling techniques with other currently used methods for estimation of the tree row volume of a super-high-density olive orchard. Agriculture 9:233

Analysis of Market Behaviour on the Organic Food Market in Terms of Environmental Protection and Consumer Environmentalism

Monika Stoma and Agnieszka Dudziak

Abstract The growing awareness of the population in various areas and aspects of life means that we pay more and more attention to what we put on our plates. Therefore, food without preservatives is gaining popularity, because products that contain harmful substances adversely affect our health. Products with labels confirming that they are ecological are appearing more and more often in stationary and online stores. There are also special zones with such products. It should be added that the access to ecological products is increasing every year, which consequently increases the possibility of choosing the best products.

The aim of the research was to analyze consumer practices and behaviors in the field of buying food in relation to the issue of sustainable development, especially preferences regarding organic products, as well as to analyze changes in consumer behavior over the last 5 years. The main reason for taking up this topic was the observed increase in interest among modern buyers in buying organic products, as well as checking whether an upward trend can also be observed when it comes to consumer awareness in this topic.

This work is theoretical and empirical in nature. For the purposes of this study, a questionnaire survey was conducted. The respondents answered questions about changes in the food market in terms of environmental protection.

Keywords Organic products · Organic food market · Market behavior · Organic farming · Sustainable production · Consumer environmentalism · Environmental protection

---

M. Stoma · A. Dudziak (✉)
Department of Power Engineering and Transportation, Subdepartment of Logistic and Business Management, Faculty of Production Engineering, University of Life Sciences in Lublin, 20-612 Lublin, Poland
e-mail: agnieszka.dudziak@up.lublin.pl

S. Pascuzzi and F. Santoro (eds.), *Farm Machinery and Processes Management in Sustainable Agriculture*, Lecture Notes in Civil Engineering 289,
https://doi.org/10.1007/978-3-031-13090-8_33

1 Introduction

Food is one of the most important areas of consumption responsible for the largest share of the environmental impact. According to Reisch et al. [1] food production and consumption are associated with several important environmental impacts, such as greenhouse gas emissions, water pollution and biodiversity loss, which will be exacerbated in the future by a growing world population.

All over the world a growing demand for organic food products can now be noticed. This may result, on the one hand, from the perception of organic food by consumers as being safer, nutritious and trustworthy than traditional food [2–4], and on the other, from the fact, that eating organic food can be a way to maintain health and fight the progressive harmful effects of pesticides and chemical components, characteristic for the modern economy, mainly based on industry [5, 6].

The organic food market is constantly changing due to the growing popularity of organic products. The European organic food market is diverse and has a high sales rate. The most dynamic development of ecological products can be observed on the German and Swiss markets as well as in the Scandinavian countries. Whereas, the Polish market of organic products is still at a lower level than the above-mentioned countries, although an increase in the popularity of organic products can already be observed here, which also implies an increasing selection of this type of food product range.

Organic food is available in various places and points of sale, as it can be purchased from local farmers, at marketplaces, in specially specialized stores and in large-format stores. However, the largest selection of organic products can be found in specialized stores, as they are focused only on this type of product. In addition, online platforms are being created, where the choice of organic food is also often very large.

The increase in the competitive advantage of organic food over conventional food is thus influenced not only by the increase in environmental awareness of consumers and attention to the quality and composition of products, but also by the increased availability, attractiveness and diversity of this type of product.

Hence, the aim of this article and the research conducted was to analyze consumer practices and behaviors in the field of buying food in relation to the issue of sustainable development, especially preferences regarding organic products, as well as to analyze changes in consumer behavior over the last 5 years.

2 Organic Farming and Organic Food

Organic food is also often referred to as "eco" and "bio". According to Tanase et al. [7] there is no one universal definition of this concept, moreover, it happens that consumers interpret it completely differently than researchers; because consumers do not need regulations, so they often consider ecological products as those products that are environmentally and health-friendly, and not only those that are certified and labeled in accordance with applicable regulations. It should be added that the concept of an organic product is also multifaceted for the scientific community itself. Nevertheless, it is a term for food that comes from organic farming, where the primary goal is to care for the environment and the welfare of farm animals.

Organic farming, like organic food, is of interest to many researchers - both theoreticians and practitioners, hence different authors or organizations, when defining this concept, emphasize its various aspects. According to International Federation of Organic Agriculture Movements (IFOAM) organic farming is „a production system that sustains the health of soils, ecosystems and people. It relies on ecological processes, biodiversity and cycles adapted to local conditions, rather than the use of inputs with adverse effects" [8]. Organic farming combines tradition with innovation and science to benefit the environment and promote a good quality of life for all parties involved [9, 10].

In turn, Domagalska and Buczkowska [11] point out that organic farming is a system of food management and production in the most beneficial way for the environment and climate. This system is designed to protect natural resources and ensure the welfare of livestock. Organic food production has a social function, provides consumers with organic products, and also contributes to the environment, because it protects plants and animals, and also affects the development of rural areas. It should be added that the compliance with the requirements of organic farming is supervised by regulations in the form of ordinances and directives.

In general, various scientists [3, 12] consider organic farming to be an organic production system that is produced without the use of fertilizers (pesticides and herbicides) and other chemicals to support the environment and protect human health; hence, this type of agricultural practice is characterized by a high degree of environmental sustainability.

It should be added that the overall strategies in the field of organic farming include the application of practices that (a) grow healthy plants with good defense capabilities, (b) stressing pests, and (c) enhancing populations of beneficial organisms [13]. Whereas, the main goals of organic farming are: maintaining the biological balance of the environment, maintaining good soil fertility by caring for the biological quality of humus, as well as balancing plant and animal production by closing the matter cycle [14].

Summarizing the above, it can be stated that in "eco" food there are no additives of chemical fertilizers or plant protection products. In the case of animals, antibiotics are not administered, no organisms are artificially modified, and the products do not contain artificial dyes, flavors or fragrances. Sustainably produced products are produced in compliance with all ecological and social requirements, to a much greater extent than traditional products [15].

Regardless of the definitions, this topic is an important issue not only for producers and consumers, but also for the market and the economy, society as a whole, authorities and the environment.

3 Environmental Protection and Consumer Environmentalism

As already mentioned, both the production and consumption of food have a significant impact on the natural environment. Hence, the concept of "consumer environmentalism" has appeared in scientific studies. This means that recently consumers are more likely to buy organic products, on the one hand because they are fresh, healthy, of good perceived quality and contain more nutrients, and on the other because they respect more both human health and the environment [3, 16, 17]. This may be due to the fact that modern consumers pay more attention to the increase in environmental protection efforts and the impact of the destruction of fauna and flora.

The very concept of consumer environmentalism is defined differently by different authors, paying attention to its other aspects or emphasizing its various features. Mutiara andArai [8] defined consumer environmentalism as an individual's readiness to maintain ecological purchasing behavior, which in particular reflects the consideration of less pollution. However, King et al. [6] found that the more a person is sensitive to healthy and organic consumption, the greater their intention to buy this type of organic product. Tarkiainen and Sundqvist referred to this phenomenon in a similar way [18] writing that the more a consumer is interested in supporting his environment and protecting his health, the more he becomes involved in purchasing organic food. This commitment should be long-term to lead to a positive attitude and purchase intention [4]. Nasif adds that the term covers various and separate elements of consumer behavior and their environmental attitudes [19].

Consumers can therefore be important, active contributors to a sustainable society by making food choices that are both healthy and that are produced in compliance with environmental and socially ethical standards. They can be considered as one of the key factors for sustainable development and can play a key role in the transition to sustainable food systems [20, 21]. Sustainable consumer consumption, through the purchase of sustainable and ecological products, can lead to the reduction of the amount of waste, consequently minimizing its impact on the environment [20, 22].

4 Material and Research Methods

In order to assess the attitudes of consumers on the food market in terms of environmental protection, own research was carried out using the diagnostic survey method. The main tool used for the research was a structured proprietary questionnaire. The questionnaire contained closed questions that were clearly developed so that they did not require additional comments, and the respondents were asked to indicate one correct answer from several available options that best or most precisely reflects their attitudes, preferences, feelings and behavior. In addition, several questions were multiple-choice questions, also with the possibility of giving your own, different answer, in a situation where none of the proposed variants fully reflect the attitudes or feelings of the respondent.

The own-authored questionnaire contained both substantive questions in the analyzed research area, as well as metric questions, enabling to obtain the socio-demographic characteristics of the respondents due to various grouping variables (respondent's descriptive features, such as: gender, age, place of residence, education, professional status and financial situation of the household) – Table 1.

It was assumed that the research sample would be at least 300 units - inhabitants of South-Eastern Poland - modern consumers. A purposeful non-random selection of the research sample was used - the criterion of purposeful selection was the fulfillment by the participants of the study of the criteria defining the categories of grouping variables included in the metric part of the questionnaire.

The research was carried out in two stages: pilot studies and main studies. The pilot study was conducted in a group of 10 people. Its purpose was to check the communicativeness, clarity, unambiguity and completeness of the questions and variants of answers contained in the research questionnaire. Following the results of the pilot survey, some questions were modified to make them more understandable to the respondents. New answers have also been added to some questions; on the other hand, several answers - which, in the opinion of the respondents, were unnecessary or did not fit the given question - were removed. The pilot study also raised further questions that were added to the survey questionnaire.

346 people participated in the main study. People participating in the study were informed about the purpose of the study and took part in it voluntarily. Before starting the study, they were informed about the purpose of the research and were assured of the anonymity of the survey.

MS Excel was used to analyze the research. Moreover, the statistical analysis of the chi-square distribution function was used, which is presented in Table 2. The use of this analysis will help to answer the research problems posed by the authors.

5 Results and Discussion

As already mentioned, 346 people took part in the survey. The exact socio-demographic characteristics of the respondents, due to the assumed grouping variables, are presented in Table 1.

As can be seen from the data presented in Table 1, the research involved respondents differentiated according to the assumed grouping variables - both women and men, of different ages (although mainly 18–60), with different education (mainly secondary and higher), living in areas with varying degrees of urbanization, mainly working people (working for someone, running their own business, or working and studying at the same time), declaring an average or good financial situation.

After the respondents were characterized in terms of selected grouping variables, the content of the questionnaire was analyzed. The obtained results are presented in a descriptive form, as well as in graphical form in Figs. 1 and 2 and in Table 2.

Table 1 Socio-demographic characteristics of the respondents

Grouping variable	Number of respondents	Percentage share [%]
Gender		
Female	145	42
Male	201	58
Age		
<18	7	2
18–25	90	26
26–40	111	32
41–60	118	34
>60	20	6
Place of residence		
Rural areas	145	42
City to 50.000 residents	69	20
City 50–300 th of residents	59	17
City > 300 th of residents	73	21
Level of education		
Primary education	17	5
Vocational education	35	10
Secondary education	128	37
Higher education	166	48
Professional status		
Professionally inactive student	42	12
Professionally active student	45	13
Professionally active ("hired" employee)	173	50
Self-employed	38	11
Unemployed	17	5
Pensioners	31	9
Financial situation		
Very good	31	9
Good	180	52
Average	118	34
Weak	14	4
Very weak	3	1

Table 2 Relationship between age and buying organic food

Age	Definitely yes	Rather yes	Rather no	Definitely no	I have no opinion	Total
	n	n	n	n	n	n
	%	%	%	%	%	%
<18	0	1	1	1	1	4
	0,00%	25,00%	25,00%	25,00%	25,00%	100,00%
18–25	1	20	12	11	1	45
	2,22%	44,44%	26,67%	24,44%	2,22%	100,00%
26–40	5	21	23	5	1	55
	9,09%	38,18%	41,82%	9,09%	1,82%	100,00%
41–60	3	27	22	5	1	58
	5,17%	46,55%	37,93%	8,62%	1,72%	100,00%
>60	2	7	0	2	0	11
	18,18%	63,64%	0,00%	18,18%	0,00%	100,00%
Total	11	76	58	24	4	173
	6,36%	43,93%	33,53%	13,87%	2,31%	100,00%

Statistical analysis: $Chi^2 = 27{,}8283$; $p = 0{,}0331$

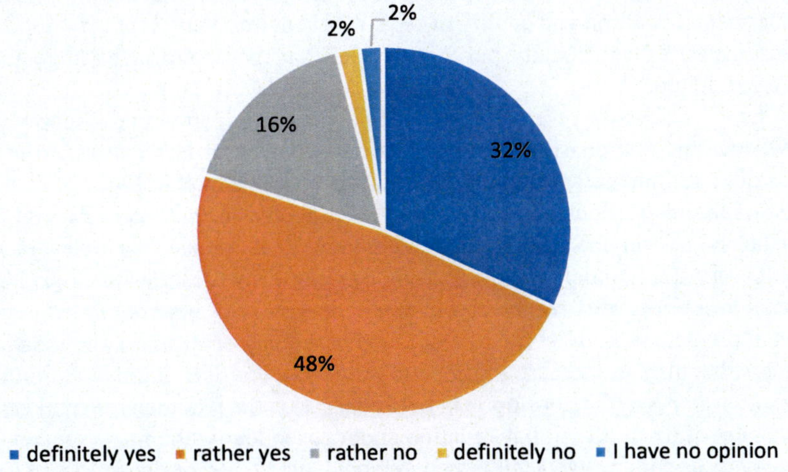

Fig. 1 Changes in consumer behavior over the last 5 years in the opinion of the respondents

At the beginning, the respondents were asked a question about their interest in the state of the environment in their place of residence, and as it turned out from the results obtained - 81% of respondents declared such interest to a greater or lesser extent. It would seem, therefore, that the level of interest and environmental awareness of the respondents is high, which is very optimistic, however, it should be noted that 17% are not interested in this problem, and 2% have no opinion on this subject.

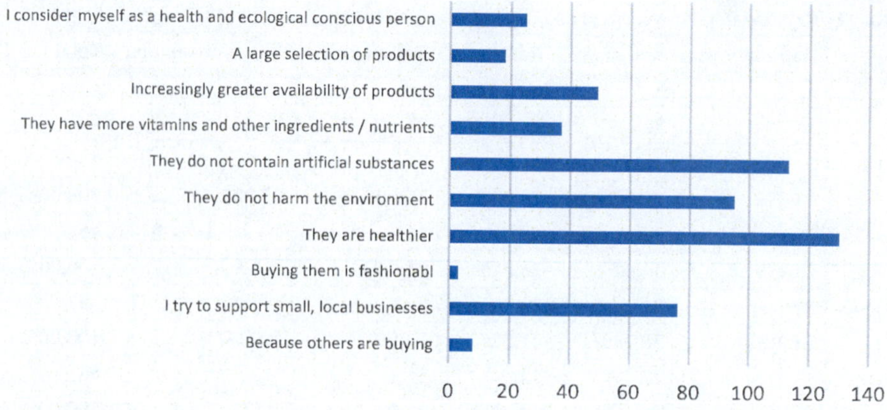

Fig. 2 The main reasons for buying organic products

Then, the respondents were asked for their opinion on the impact of consumer market behavior on the environment. The vast majority of respondents (90%) believe that market behavior affects the state of the environment. Only 7% believe that they rather do not influence, and 3% do not have an opinion on this subject. None of the respondents expressed a firm opinion about the lack of influence of consumer behavior on the condition and quality of the natural environment. It is very important that consumers are aware that the purchasing decisions they make have a huge impact on the environment.

In order to achieve the research goals, the people participating in the survey were asked whether their purchasing behavior has changed over the last 5 years, especially in terms of environmental protection. The results are presented in Fig. 1.

Most of the respondents, as much as 80%, replied that they had changed their market behavior over the last 5 years. This may be influenced by many factors, including: a greater number of actions promoting a healthy lifestyle and raising environmental awareness of consumers, a different approach to matters related to environmental protection, both on the part of the government, enterprises and individual consumers, but also an increase in the number of retail outlets, a greater amount of assortment in stores or the development of purchases on the e-commerce market. The improvement of the professional situation and, consequently, the financial situation of the society, as well as the development of social media and various modern forms of marketing activities seem to be no less important.

Then, reference was made to organic products and the respondents were asked whether they were interested in this type of product. Almost 60% of respondents (48% - rather yes, 11% - definitely yes) declared that they are more or less interested in the subject of organic food. It may be influenced by the general increase in environmental awareness of the society and the increasing popularity of these products. On the other hand, almost 40% are not interested in a given topic.

The next question concerned the respondents' declarations regarding the purchase of organic food by them. The results were distributed quite evenly because nearly half of the respondents (49%) declared that they buy organic food - this may be due to the greater awareness of the respondents, greater availability and popularity of these products as well as the increasing selection of the assortment. However, on the other hand, 48% of respondents replied that they do not buy eco products. 3% of the respondents did not have or did not want to express their opinion on the analyzed issue.

In order to deepen the considerations, the relationship between age and the declaration of buying organic food was examined (Table 2).

The further analysis carried out (Table 2) shows that organic products are most often bought by people over 60 years of age. The smallest group that does not buy organic food are people between 18 and 25 years of age. There is also a tendency to notice that the older people are, the more likely they are to buy organic products. The demonstrated differences were statistically significant ($p = 0{,}0331$).

Figure 2 shows the main reasons - indicated by respondents - influencing the purchase of organic products.

As many as 130 people replied that they buy organic products because they are healthier than conventional products and do not contain artificial substances (113 Nowadays, consumers are more and more aware of and care about their health. 95 people buy these products because they do not harm the environment, and 76 people buy them because they support local farmers. The fewest people buy organic products because they are fashionable. In addition, 49 people declared that they buy these products as they are increasingly available in various stores and other points of sale.

According to Wojciechowska-Solis and Soroka [23] when choosing ecological products, consumers were most often guided by health considerations and the lack of harmful substances in these products. Other authors have also studied the motives of organic food consumption [3, 5, 24] – particular attention was paid to the increased awareness of the relationship between diet and health [25], the interest in food safety [26] and environmental benefits [16, 17]. Therefore, consumers consider organic products from organic farming to be higher-quality, healthier and more environmentally friendly products [24, 27].

Comparing to the research conducted by the authors, the study group is also guided by the same factors. However, we are glad that consumers associate the purchase of ecological products with a reduced negative impact on the natural environment.

The respondents were also asked about the frequency of purchasing ecological products by them. According to the data obtained, 31% of respondents buy organic products 3–4 times a month, 20% 1–2 times a week, 16% as much as 3–4 times a week. The smallest group are people who buy organic products every day (3%). People who do not buy organic products are a large group - 23%. Research conducted by Nestorowicz and his team [28] showed that 35% of respondents do not buy organic food, 23% do it once a year or less frequently, 18% every three months, and the remaining respondents do so at least once a month (15%) or at least once a week - 7%. Thus, it can be observed that in our own research, consumers declared a much higher frequency of purchasing organic food. This may result from the passage of time and, as a consequence, changes in consumer behavior in the area of purchasing organic products, which can be observed over the last 5 years.

6 Conclusions

Responding to the research problems posed, it can be stated that the people taking part in the study are quite conscious consumers in terms of purchasing organic products and environmental protection. Consumers have a strong interest in the state of the environment in their place of residence; they are also aware that the purchasing decisions they make affect the natural environment. Consumers also declared that their purchasing behavior in terms of environmental protection has changed significantly over the last 5 years.

On the basis of the conducted research, it was also found that:

- inhabitants of South-Eastern Poland are interested in the subject of organic food to a greater or lesser extent;
- the level of organic food purchases is at an average level, as only half of them declared that they buy organic products;
- the older people are, the more likely they are to buy organic products;
- it would seem that the increasing popularity and fashion affects the increase in the purchase of organic products, but for consumers the most important are the health benefits of these products, as well as their lower negative impact on the environment.

What seems disturbing, however, is the fact that in many of the questions it was possible to observe a certain, sometimes quite significant, percentage of answers such as: "I'm not interested in it" and "I have no opinion" - it seems that such a structure of results should imply the introduction of more educational actions or promoting pro-ecological behavior and consumer environmentalism in order to raise the level of consumer awareness of sustainable consumption. As emphasized by various authors in their works,[29, 30] - although people's awareness of food sustainability is increased, the inclusion of sustainable development in everyday food choices remains marginal and food may be a consequence of deliberate or unconscious actions. It is therefore a kind of challenge for the authorities, both at the central and local level, but also for enterprises.

References

1. Reisch L, Eberle U, Lorek S (2013) Sustainable food consumption: an overview of contemporary issues and policies. Sustain Sci Pract Policy 9:7–25

2. Popa ME, Mitelut AC, Popa EE, Stan A, Popa VI (2019) Organic foods contribution to nutritional quality and value. Trends Food Sci Technol 84:15–18
3. Hsu C-L, Chen M-C (2014) Explaining consumer attitudes and purchase intentions toward organic food: contributions from regulatory fit and consumer characteristics. Food Qual Prefer 35:6–13
4. Rahman I (2018) The interplay of product involvement and sustainable consumption: an empirical analysis of behavioral intentions related to green hotels, organic wines and green cars. Sustain Dev 26:399–414
5. Ghali-Zinoubi Z, Toukabri M (2019) The antecedents of the consumer purchase intention: sensitivity to price and involvement in organic product: moderating role of product regional identity. Trends Food Sci Technol 90:175–179
6. King T, Cole M, Farber JM, Eisenbrand G, Zabaras D, Fox EM, Hill JP (2017) Food safety for food security: relationship between global megatrends and developments in food safety. Trends Food Sci Technol 68:160–175
7. Tanase LD, Roşca MI, Jurcoane A (2012) The perception on ecological products-a research on the urban consumer. Ann Univ Oradea Econ Sci Ser 21
8. Mutiara VI, Arai S (2017) The challenges in organic agricultural products market in Southeast Asia. Rev Agric Sci 5:36–44
9. Gopalakrishnan O (2006) In Proceedings of the IFOAM International conference on animals in organic production 2006, Saint Paul, Minn, IFOAM
10. Halberg N (2012) Assessment of the environmental sustainability of organic farming: definitions, indicators and the major challenges. Can J Plant Sci 92:981–996
11. Domagalska J, Buczkowska M (2015) Ecological agriculture - opportunities and development prospects. Probl. Hig. Epidemiol. 96:370–376 [in Polish]
12. Chekima Brahim, Igau Aisat, Wafa SAWSK, Chekima K (2017) Narrowing the gap: factors driving organic food consumption. J Clean Production 166:1438–1447. https://doi.org/10.1016/j.jclepro.2017.08.086
13. Magdoff F (2007) Ecological agriculture: principles, practices, and constraints1. Renew Agric Food Syst 22:109–117
14. Sołtysiak U (1995) Organic farming: from producer to consumer. Stowarzyszenie Ekoland. [in Polish]
15. Nestorowicz R (2015) Food product labeling and sustainable consumption. J Agribus Rural Dev. [in Polish]
16. Ghvanidze S, Velikova N, Dodd TH, Oldewage-Theron W (2016) Consumers' environmental and ethical consciousness and the use of the related food products information: the role of perceived consumer effectiveness. Appetite 107:311–322
17. Laureti T, Benedetti I (2018) Exploring pro-environmental food purchasing behaviour: an empirical analysis of Italian consumers. J Clean Prod 172:3367–3378
18. Tarkiainen A, Sundqvist S (2009) Product involvement in organic food consumption: does ideology meet practice? Psychol Mark 26:844–863
19. Nasif N (2021) A reconsideration of the environmentalism construct for consumer research. Interdiscip Environ Rev 21:220–237
20. Sargant E (2014) Sustainable food consumption: a practice based approach. Wageningen University and Research
21. Vittersø G, Tangeland T (2015) The role of consumers in transitions towards sustainable food consumption: the case of organic food in Norway. J Clean Prod 92:91–99
22. Azzurra A, Massimiliano A, Angela M (2019) Measuring sustainable food consumption: a case study on organic food. Sustain Prod Consum 17:95–107
23. Wojciechowska-Solis J, Soroka A (2016) Criteria for purchase of organic food by polish consumers in relation to world consumption. Probl Rol Świat 16. [in Polish]
24. Lee WJ, Shimizu M, Kniffin KM, Wansink B (2013) You taste what you see: do organic labels bias taste perceptions? Food Qual Prefer 29:33–39
25. Asioli D, Aschemann-Witzel J, Caputo V, Vecchio R, Annunziata A, Næs T, Varela P (2017) Making sense of the "clean label" trends: a review of consumer food choice behavior and discussion of industry implications. Food Res Int 99:58–71

26. Maeda-Yamamoto M (2017) Development of functional agricultural products and use of a new health claim system in Japan. Trends Food Sci Technol 69:324–332
27. Marian L, Chrysochou P, Krystallis A, Thøgersen J (2014) The role of price as a product attribute in the organic food context: an exploration based on actual purchase data. Food Qual Prefer 37:52–60
28. Nestorowicz R, Rogala A, Jerzyk E, Pilarczyk B (2016) Consumer ethnocentrism in a local approach as a factor in the development of the organic food market in Poland. Rocz Nauk Stowarzyszenia Ekon Rol Agrobiznesu 18. [in Polish]
29. Verbeke W, Roosen J (2009) Market differentiation potential of country-of-origin, quality and traceability labeling. Estey Cent J Int Law Trade Policy 10:20–35
30. Verain MC, Bartels J, Dagevos H, Sijtsema SJ, Onwezen MC, Antonides G (2012) Segments of sustainable food consumers: a literature review. Int J Consum Stud 36:123–132

Development of a New Unbiased Impact Index on Water for Pesticides Use – A Case Study in Belgium for Herbicides Application on Sugar Beet Crop

Guillaume Bergiers, Bastien Durenne, Bernard Weickmans, and Bruno Huyghebaert

Abstract Reduction of pesticides use is one of the main objectives of the Farm to Fork Strategy developed in the framework of the European Green Deal, aiming to make food systems fair, healthy and environmentally-friendly [4]. The operational target is a 50% reduction in use and risk of chemical pesticides by 2030. The Walloon Region in Belgium is committed to a strong prevention policy to reduce the impact of pesticide residues on human health and the environment, with a particular focus on water resources. This target will be measured based on i) the quantities of active substances by using an unbiased index and ii) the hazard properties of plant protection products (PPP) gathered in a negative list for the water resource. In order to be able to monitor PPP applications at farm level, several indicators have been used, such as the French widespread indicator "TFI" (Treatment Frequency Index). Nevertheless, it was demonstrated that this index implies a systematic bias in the evaluation of active substances used. Therefore, to quantify the pesticides used by farmers, a derived unbiased index called "ISAC" based on Active Substances per Crop has been developed and is now applied to compare several weeds control strategies. A focus on water resources preservation can be realized by selecting problematic active substances (such as chloridazon, metamitron, dimethenamid-P, ethofumesate, lenacile, …) into the ISAC, therefore obtaining an index of pressure on water resources (ISAC'eau). Theses indexes have been applied on Walloon farming practices related to weeds control through herbicides use on sugar beet crop. In conclusion, the use of the

G. Bergiers · B. Durenne · B. Weickmans · B. Huyghebaert (✉)
Soil, Water and Integrated Production Unit; Sustainability, Systems and Prospectives Department, Walloon Agricultural Research Centre (CRA-W), 5030 Gembloux, Belgium
e-mail: b.huyghebaert@cra.wallonie.be

G. Bergiers
e-mail: g.bergiers@cra.wallonie.be

B. Durenne
e-mail: b.durenne@cra.wallonie.be

B. Weickmans
e-mail: b.weickmans@cra.wallonie.be

S. Pascuzzi and F. Santoro (eds.), *Farm Machinery and Processes Management in Sustainable Agriculture*, Lecture Notes in Civil Engineering 289,
https://doi.org/10.1007/978-3-031-13090-8_34

Unbiased Impact Index allows relevant comparisons between different strategies that can in fine help to propose alternatives for reducing herbicides use in sugar beet and their impact on the water resources. Obviously, it is necessary to investigate how to manage weeds with less or no herbicides in arable farming while maintaining the productivity, which implies considering agronomic performance such as yield, and crop net profit.

Keywords Herbicides · Sugar beet · ISAC · Water resources

1 Introduction

From a regional point of view, Wallonia is committed to a strong prevention policy to the reduction of the risks linked to plant protection products (PPP) use, especially for human health and the environment. Pesticides have strongly contributed to the intensification of agriculture and the increase in food production, but it has also caused contamination of the environment by the diffusion of active substances residues and/or of their associated metabolites [11]. At the Walloon level, the water resource is carefully monitored and especially the catchments for the production of drinking water. A significant increase in pesticide contamination of water catchments has been reported by these monitoring in recent years across Europe [6]. Herbicides are largely concerned by groundwater resources contaminations [9], which is also the case in Belgium, in particular with radicular herbicides used in sugar beet crop.

In order to quantify and compare the PPP use from data at farm level, the TFI (Treatment Frequency Index) has been widely promoted for a few years now in France [5]. This index turned out to be biased, because it is based on commercial products available at national scale, which can contain more than one active substance (a.s.). per product, consequently driving farmers to misguided decisions. In fact, artificial reductions of herbicides use (based on the TFI) have been observed, while sales of herbicides a.s. have increased. Therefore, to avoid the pitfall of TFI, a new index has to be set up to compare PPP uses and allowing putative reductions of use. The so-called ISAC (Index of Active Substances per Crop) is based on the a.s. included in the commercial products used by farmers. The ISAC allows an unbiased characterization of the use of PPP, and is normalized by a maximal dose defined for each a.s. [3]. This standard dose per active substance is legally registered at the national level. This way, an emphasis on specific a.s. is made possible. In Wallonia and within research activities dedicated to water resources preservation, a focus was done on drinkable water resources by selecting the problematic a.s. The ISAC applied to active substances detected in water and identified in Wallonia becomes an index of impact on water resources (ISAC'eau). The PPP use for 19 different farmer's parcels were collected and analyzed. The dataset was dedicated to herbicides application on sugar beet. In Wallonia, sugar beet is one of the main crop, providing high yields exceeding regularly 90 tons/ha/years, but in return requiring intensive use of PPP and specifically herbicides. This crop is therefore of interest to promote the PPP use's reduction and alternatives such as mechanical weed control that is especially effective for young annual weeds [2].

Collecting data from the field is an essential step to characterize actual PPP use and examine its diversity. Based on these data, ISAC and ISAC'eau were obtained and compared. This comparison using ISAC allows us to see the variety of PPP uses, as well a benchmarking of the agronomical practices, and identify the most virtuous. In sugar beet, the reduction of herbicides doses or the application of non-chemical weed measures is a major agronomic challenge, because the yield is intensively impacted by competitions with weeds [2]. Finally, for ISAC'eau in this case study, focusing on herbicides use on sugar beet only five a.s. were applied (chloridazon, dimethenamid-P, ethofumesate, lenacil and metamitron). These five a.s. have a high solubility, which is an important factor to explain their presence in groundwater and also their use as root herbicide.

2 Material and Methods

2.1 *Collection of PPP Use Quantities and Dataset Analysis*

Data from the field were collected in farms situated in Wallonia (in the southern part of the Sambre-Meuse axis, which is a very productive area). Walloon voluntary farmers were asked to provide for selected field the list of the herbicides applied as well as their doses, date of application, surface treated, and if possible the reason of the application. Once all data were collected, a keen analysis is realised in order to verify the reliability and the coherence of the data.

2.2 *QAC and ISAC*

The doses per hectare of herbicides is needed to calculate the Quantities of active substances Applied per Crop (QAC) which is defined per hectare ($g.ha^{-1}$).This decomposition of each commercial product into its formulation is made in order to avoid the pitfall of the Treatment Frequency Indicator (TFI). TFI is clearly based on commercial products contrary to the ISAC index which corresponds to an Index of Active Substances by Culture. ISAC is equal to the QAC divided by the corresponding DMA, i.e. the maximum dose authorized by a.s. per hectare ($g.ha^{-1}$) for a specific crop. This step normalizes all PPP uses, which makes it an a-dimensional measure. All DMA must be validated for each a.s. on sugar beet and is registered at national level. The value of the ISAC represents the percentage of the DMA applied. As DMA is a legal value, an ISAC above 1, is an illegal overdosage of the a.s.

2.3 ISAC'eau

In order to focus on a specific issue, corresponding to water protection, a category of PPP containing a.s. can be isolated from the ISAC (which consider all the a.s. applied). Therefore, ISAC'eau was specifically developed for the protection of water resources (groundwater), with a focus on the 22 problematical a.s. predefined for the Walloon region (2,4-D, aclonifen, bentazone, bifenox, chloridazon, chlortoluron, cypermethrine, diflufenican, dimethenamid-P, ethofumesate, flufenacet, lenacile, MCPA, mecoprop-P, metamitron, metazachlor, metobromuron, metribuzin, pendimethalin, prosulfocarb, S-metolachlor and terbuthylazine) This short list of 22 a.s. is the result of the analysis at the regional level of all identified worrying water's contaminations (according to the drinking-water norms and standards for contamination and its frequency in water resources).

3 Results and Discussion

The chemical weed treatment of sugar beet is driven by the low doses system developed in Europe during the 90's. It consists in repeating small doses of PPP to control weed in sugar beet using foliar or radicular herbicides active substances. The number of treatments can vary from 3 to 6 according to the weed pressure and the meteorological conditions before and after PPP application. The ISAC and ISAC'eau bring multiple and clear information about PPP's uses allowing a comparison between different farmer's practices. The weed pressure (specific weed or family) and its time of emergence drives the choices of active substances. For instance, many annual weed species have well-defined periods of emergence [2]. Therefore, the date of the application of a.s. depends on such timing and recommendation. For instance, chloridazon, metamitron and dimethenamid-P are radicular herbicides and lenacil can be sprayed in pre-emergence and has a clear action of photosynthesis inhibition on annual dicotyledons.

3.1 Presentation of ISAC on Sugar Beet

The main objective of the ISAC is to compare different practices amongst famers. The first graph (Fig. 1) represents the ISAC for 19 different agricultural parcels of sugar beet in Belgium for the year 2020. A variability in the herbicide application schemes can be observed. The maximal values are around an ISAC of 4, while the minimum is around an ISAC of 2. Between those two extremes, a 50% reduction of herbicides use is observed. Many agronomical reasons can explain this significant difference. Therefore, in the analysis it's of prime importance to consider different factor, such as the history of the parcel, the weed pressure, the meteorological conditions, the frequency of treatment ... All these factors do influence the choice of the commercial

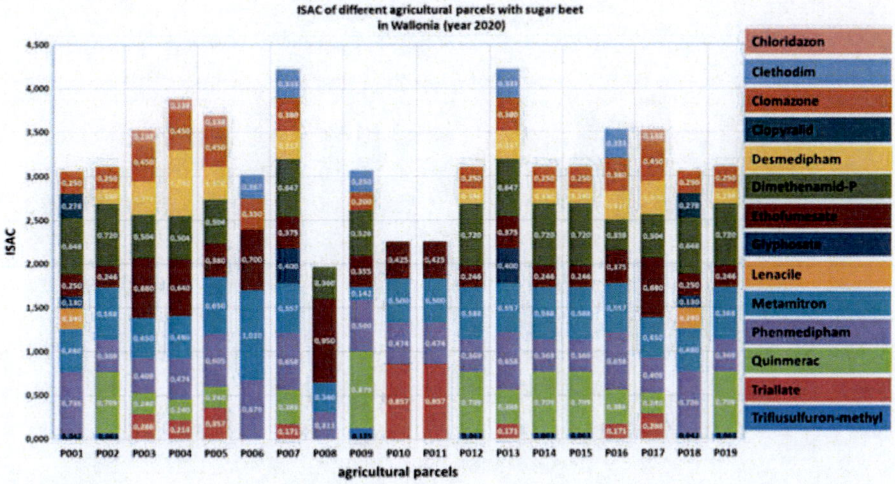

Fig. 1 Histogram showing the ISAC for 19 different agricultural parcels of sugar beet in Belgium for the year 2020

product (with a specific formulation), consequently the ISAC. With the ISAC, it is possible to assess the full treatment. In some cases a treatment can be replaced by a mechanical weed control, consequently decreasing the PPP consumption, and therefore the ISAC.

3.2 Use of ISAC'eau to Focus on Water Preservation

Focusing the analysis only on the problematic a.s. for ground water used in sugar beet (chloridazon, dimethenamid-P, ethofumesate, lenacile and metamitron) the ISAC'eau is calculated and shown on Fig. 2. The maximal ISAC'eau reaches 1,772, and the minimal value is around 0,9. By comparing both graphs (Figs. 1 and 2) it can be seen that in most case, almost half of the herbicides uses has potentially an impact on water resources. The case of the parcel "P008" is particularly interesting. Because it's ISAC is the lowest one, but it's ISAC'eau is one of the highest. Which means that, even though this farmer has a low consumption of a.s., it's pressure on water resources remains important.

The use of the ISAC'eau allows a focus on practices having a negative impact on Walloon water resources. Therefore, it's easier to modify these practices and search for alternatives. By focusing on a specific a.s. and the reason of its application on the field, solutions can be proposed to reduce its impact on water resources. Along with this focus, an unbiased quantification of the diminution of PPP use is made by the ISAC'eau. The awareness-raising of farmers, commercials, and advisors to the problematics a.s. in Wallonia is a first important step to enhance the quality of water resources.

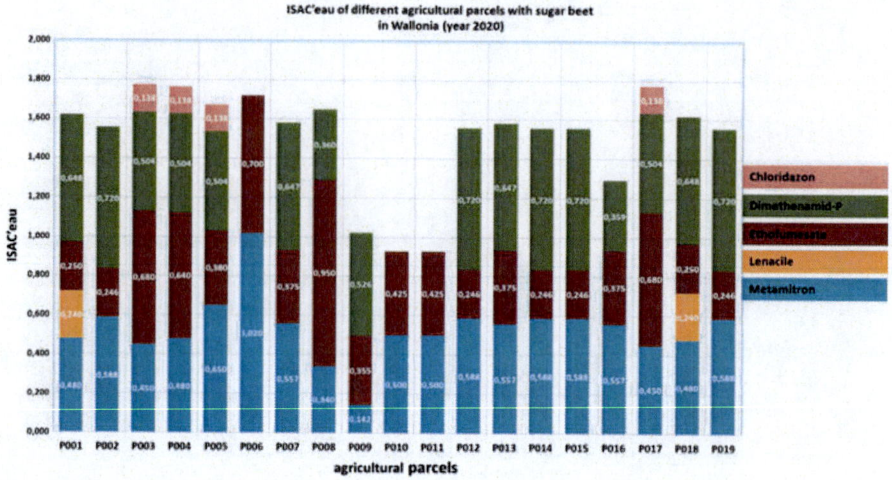

Fig. 2 Histogram showing the ISAC'eau for 19 different agricultural parcels of sugar beet in Belgium for the year 2020

4 Perspective

As a perspective to improve, the ISAC'eau could be associated to the hazard properties regarding water (e.i. groundwater) of each active substance. The association of hazard properties of the a.s. to ISAC'eau would allow to determine the Ground Water (GW) score. In this way the importance of each a.s. into the pesticides would change and be balanced regarding their hazard to groundwater. Practically, the ISAC'eau of each substance would be multiplied by its corresponding hazard to groundwater. The aim of the Ground Water risk score is mainly to raise farmer's awareness of problematic a.s. and focus on its application modalities. Topographic information are essential to complete the diagnosis.

This part is under development for the moment. The difficulty is to determine the hazard indicator of a.s. for the water resources (e.i. groundwater). For instance, several indicators and index have been created to characterize the lixiviation of pesticides into groundwater. These indicators and index can be simple cause index or based on operational model depending on their construction mode and input parameters. The Macro model being the most complex model which is based on mechanical processes. Amongst simple cause index, a certain variety exists in the literature.

Combining different physico-chemical properties of the a.s. [1] would be a solution. For that purpose, we propose to base the Ground Water (GW) score on the three most represented physico-chemical characterization of the a.s.: the water solubility in water at 20 °C (mL.g^{-1}), the half-life (DT50) in field conditions (days) and a soil adsorption coefficient (mL.g^{-1}). The M.LEACH index includes theses three parameters [1]. This index allows a ranking of lixiviation tendency of the a.s. This index would be mathematically modified in order to reduce the range of the values, thus allowing an easier use of it. The data used for calculation could be mobilized in the peer review's document from the European Food Safety Authority (EFSA). The half-life (DT50) of the a.s. were chosen amongst the data from the EFSA's peer review in order to match better to Belgian climate. The formula to calculate the Ground Water score could be expressed as below.

$$Ground\ water\ lixiviation\ index = log10\left[\frac{solubility * DT50}{\text{soil adsorption coefficient}}\right] + 10$$

As mentioned before, this process is still under development and need to be precise. We are currently collecting all the necessary data for the calculation of the ground water lixiviation index, in order to calculate finally the Ground Water Risk score of the active substances.

5 Conclusion

This new unbiased index provides an objective basis for raising farmers' awareness of water resource protection. The results presented highlighted contrasting PPP uses for the 2020 crop year. In addition, the ISAC'eau via the quantification of PPP use allows an unbiased comparison between the different applications of herbicides that can potentially be problematic for the water compartment. It also allows at the watershed level to identify possible improvements with a view to reducing the impact on water through the implementation of alternatives, particularly in terms of weed control such as: the promotion of mechanical weeding the reduction of doses or reflections concerning treatment schemes (chemical substitutions) used in sugar beet crop. By collecting each year the data of herbicides use on sugar beet crop in this region, we forecast to obtain referential guidelines that can objectively provide a tool to compare farmer's practices.

The ISAC allows the analyze of different agricultural practices and PPP use amongst farmers. It's essential to focus on problematic a.s., as the results have shown despite a low PPP use, the potential impact on water resources can remain important. Therefore, a consciousness-raising of the problematic a.s. and their likelihood behavior in the environment is essential to enhance PPP practices. By applying all solutions on the whole territory at national scale, and by focusing on "hot spot", the contaminations of water resources are expected to decrease [7]. Undoubtedly, farmers need a powerful and official agronomical support to operate the variety of existing solutions and to help drive its decisions according to the field's conditions [8]. Finally, it is clear that the effective pollution of groundwater will obviously depend on the use of the a.s. (including the conditions of application) [10]. Moreover, the pedoclimatic data and topographical conditions corresponding to the agricultural parcels are relevant and must be taken into account for estimating the transfer to water resources, thus including the factors influencing infiltration and potential leaching for each active substance.

Since not all a.s. have the same weight on water resource, a focus on more problematic a.s. among the selected ones would be possible, by coupling the ISAC'eau with an hazard index, resulting in the development of a Ground Water Risk score. The development of this score is under development.

Acknowledgements This research project was funded by the Service Public de Wallonie (SPW) and Société Publique de Gestion de l'Eau (SPGE).

References

1. Akay Demir AE, Dilek FB, Yetis U (2019) A new screening index for pesticides leachability to groundwater. J Environ Manag 231:1193–1202. https://doi.org/10.1016/j.jenvman.2018.11.007
2. Cioni F, Maines G (2010) Weed control in sugarbeet. Sugar Tech 12:243–255. https://doi.org/10.1007/s12355-010-0036-2
3. Durenne B, Weickmans B, Malice E, Huyghebaert B, Henriet, F (2021) ISAC, an unbiased indicator based on active substances for evaluating PPP use. In: Proceedings in 72nd international symposium on crop protection (ISCP). Ghent University, 18 mai 2021
4. European Commission, Directorate-General for Health and Food Safety (2017) Sustainable use of pesticides: overview report. https://doi.org/10.2875/846869
5. Guichard L, Dedieu F, Jeuffroy M-H et al (2017) Le plan Ecophyto de réduction d'usage des pesticides en France: décryptage d'un échec et raisons d'espérer. Cahiers Agricultures 26:14002. https://doi.org/10.1051/cagri/2017004
6. O'Driscoll JH, Siggins A, Healy MG, McGinley J, Mellander P-E, Morrison L, Ryan PC (2022) A risk ranking of pesticides in Irish drinking water considering chronic health effects. Sci Total Environ 829:154532. https://doi.org/10.1016/j.scitotenv.2022.154532
7. Kudsk P, Jørgensen LN, Ørum JE (2018) Pesticide load—a new Danish pesticide risk indicator with multiple applications. Land Use Policy 70:384–393. https://doi.org/10.1016/j.landusepol.2017.11.010
8. Latruffe L, Diazabakana A, Bockstaller C, Desjeux Y, Finn J, Kelly E, Ryan M, Uthes S (2016) Measurement of sustainability in agriculture: a review of indicators. Stud. Agr. Econ. 118:123–130. https://doi.org/10.7896/j.1624
9. Le Bellec F, Vélu A, Fournier P et al (2015) Helping farmers to reduce herbicide environmental impacts. Ecol Ind 54:207–216. https://doi.org/10.1016/j.ecolind.2015.02.020
10. Lefrancq M, Jadas-Hécart A, La Jeunesse I et al (2017) High frequency monitoring of pesticides in runoff water to improve understanding of their transport and environmental impacts. Sci Total Environ 587–588:75–86. https://doi.org/10.1016/j.scitotenv.2017.02.022
11. Lopez B, Ollivier P, Togola A et al (2015) Screening of French groundwater for regulated and emerging contaminants. Sci Total Environ 518–519:562–573. https://doi.org/10.1016/j.scitotenv.2015.01.110

Assessment of the Suitability of Apple Pomace for Food Purposes in the Context of Sustainable Agriculture

Rafał Nadulski, Paweł Sobczak, Kazimierz Zawiślak, and Zbigniew Kobus

Abstract The aim of the study was to determine the chemical composition of pomace from various apple varieties in terms of their suitability for the production of food products. The research covered pomace obtained from five apple varieties. The content of dry matter, ash, protein, fat, sugar, sorbitol, vitamin C, dietary fiber, polyphenols and antioxidant activity were determined. Significant differences in the chemical composition of the tested pomace were found, especially in terms of the content of health-promoting compounds. Therefore, when processing the pomace for food purposes, it is important to take into account the variety of apples they come from. Due to its chemical composition, pomace from the Szempion variety is particularly suitable for food purposes.

Keywords Pomace · Apple · Utilization · Technology · Chemical composition

1 Introduction

The main aim of apple processing is the production of juice and its concentrate. Apple pomace is a major byproduct obtained during apple juice processing. Its quantity depends primarily on the technical and technological conditions of the process, as well as the characteristics of the fruit variety, and may vary from 10 to 40% of the input material [20]. Currently, the processing of apples for juices is also carried out by local processing plants run by producer groups of farmers [12]. This is consistent with the modern concept of sustainable agriculture. In small processing facilities pomace obtained after pressing is characterized by, among others, higher moisture

R. Nadulski (✉) · P. Sobczak · K. Zawiślak
Department of Food Engineering and Machines, University of Life Sciences, Głęboka 28, 20-612 Lublin, Poland
e-mail: rafal.nadulski@up.lublin.pl

Z. Kobus
Department of Technology Fundamentals, University of Life Sciences, Głęboka 28, 20-612 Lublin, Poland

S. Pascuzzi and F. Santoro (eds.), *Farm Machinery and Processes Management in Sustainable Agriculture*, Lecture Notes in Civil Engineering 289,
https://doi.org/10.1007/978-3-031-13090-8_35

content as compared with the pomace obtained under industrial conditions after initial enzymatic treatment and pressed using basket press with drainage tubing.

The pomace is a problem for processing plants because it is unstable and susceptible to the development of microorganisms. The plant needs to get rid of or process them quickly. There are many methods of processing pomace for food and non-food purposes [4, 5, 17]. Pomace contains plenty of important compounds, such as carbohydrates, phenolic compounds, dietary fiber, vitamin C and minerals [7]. Apple pomace is of great nutritional value, providing health benefits, and can be used to produce functional food [8]. For this reason, apple pomace is a valuable raw material for the food industry [10]. In general, the pomace is processed without distinguishing what variety of apples it comes from. The aim of the study was to assess the chemical composition of pomace from various apple varieties in terms of their use in the production of food products with health-promoting properties.

2 Material and Methods

2.1 Material

The research was carried out on pomace obtained from five apple varieties (Szampion, Idared, Ligol, Red Jonaprince, Redchief). The apples were purchased from the company Zrzeszenie Producentów Owoców Stryjno-Sad with its seat in Kawęczyn near Piaski (Poland) (GPS coordinates: 51°9′56.0″ N, 22°50′49.1″ E) The tests were carried out on the raw material after two months of storage in a cold store with a controlled atmosphere (temperature 1.6–2.2 °C, oxygen content 1.6%, carbon dioxide 2.2% and nitrogen 96.2%).

The shredded pulp (mash) was divided into portions weighing 300 g and then pressed on a laboratory basket press of our own design with a capacity of approx. 1000 ml [13]. For each variety of apples, pressing was performed in 10 repetitions. Before further measurements, the pomace was stored in a refrigerator at a temperature of approx. 4 °C.

2.2 Methods

Methodology of the Chemical Research
The content of dry matter, ash, protein, fat, sugar, sorbitol, vitamin C, dietary fiber, polyphenols and antioxidant activity were determined. Each measurement was perfomed in three repetitions.

Dry Matter

The dry matter of the pomace was measured with a moisture analyzer (MAC 210/NH Radwag, Radom, Poland). Samples weighing 5 g were dried at 105 °C, after which, the changes in the water content of the samples were not recorded by the device.

Ash

The ash content was determined by the gravimetric method (AACC Method 08–01, 2000) [1].

Protein

The protein content was determined using the Kjeldahl method (AACC Method 46–08, 2000) [1].

Sugar and Sorbitol

The content of total sugar and sorbitol was determined using HPLC with a refractive index detector PN-EN 12,630:1999 [14].

Fat

The fat content was determined using the Soxhlet method (AACC Method 30–10, 2000) [1].

Dietary Fibre

The total dietary fibre content (TDF) was measured using the enzymatic–gravimetric method (AACC 32–05, 2000; AACC 32–07, 2000) [1].

Vitamin C

The vitamin C content was determined using the HPLC method (PN-EN 14,130: 2003, 2003) [15].

Polyphenols

The content of polyphenolic compounds was measured using a spectrophotometric de-termination in the presence of Folin-Ciocalteu reagent, in accordance with the modified Singleton method [18] presented in the work by Wilczyński et al. [20].

Antioxidant Activity

The antioxidant activity of the extracts obtained from apple pomace was evaluated using a free radical DPPH (2, 2-diphenyl-1-picrylhydrazyl) assay according to the method presented by Wilczyński et al. [20].

Statistical Analysis of the Results

Statistical analysis was performed by means of Statistica 12 software (StatSoft, Inc., Tulsa, OK, USA) [19] using one-way analysis of variance. The significance of differences be-tween mean values was determined using Fisher's test at $p < 0.05$.

3 Results

As a result of the research, the chemical composition of the pomace was determined, with particular emphasis on the compounds with health-promoting properties. The test results are presented in the charts.

In the case of the cultivars Szampion, Ligol, Idared and Red Jonaprince, the dry matter content was similar and no statistically significant differences were found between the examined pomace (Fig. 1a). The Redchief variety was characterized by the highest dry matter content in pomace. The most rich in minerals were pomace from the Red Jonaprince and Redchief cultivars, for which the ash content oscillated around 6% (Fig. 1b). For the cultivars Szampion, Idared and Ligol, the ash content was almost twice lower and ranged from 3.07% to 3.46%. A high ash content indicates a high minerals content. These minerals include: sodium, calcium, phosphorus, magnesium, iron, zinc, copper, manganese [12].

Figure 2a and b shows the content of dry matter and ash in the pomace of the tested apple cultivars. The pomace obtained from the Szampion and Red Jonaprince cultivars had the highest sugar content. According to Savatović et al. [16], the total sugar content of fresh apples is 12–14%.

Our research shows that the tested pomace is characterized by a lower sugar content compared to fresh apples. The highest content of sorbitol was recorded for pomace from the Szampion (0.67/100 g) and Redchef (0.48/100 g) varieties.

The content of sorbitol in the other varieties (Idared, Ligol and Red Jonaprince) was at a similar level and amounted to approx. 0.36/100 g of pomace. In fresh apples, the sorbitol content is higher, ranging from 0.87/100 g to 1.49/100 g [6].

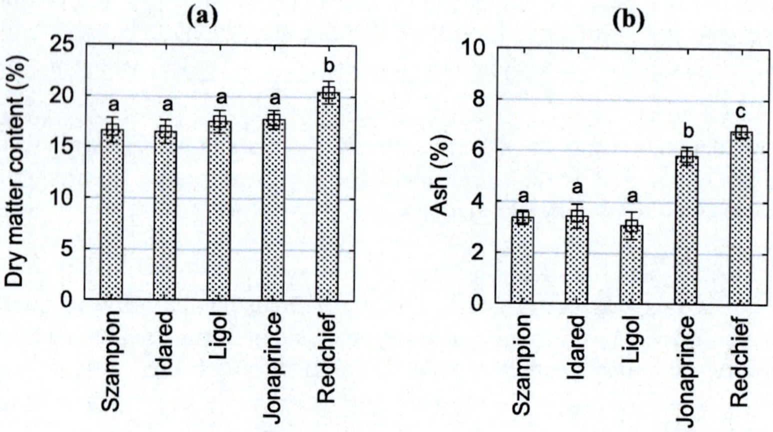

Fig. 1 Dry matter **a** and ash **b** content in the apple pomace. [a,b,c]Values in a column marked with the same letter are not statistically significantly different ($p > 0.05$)

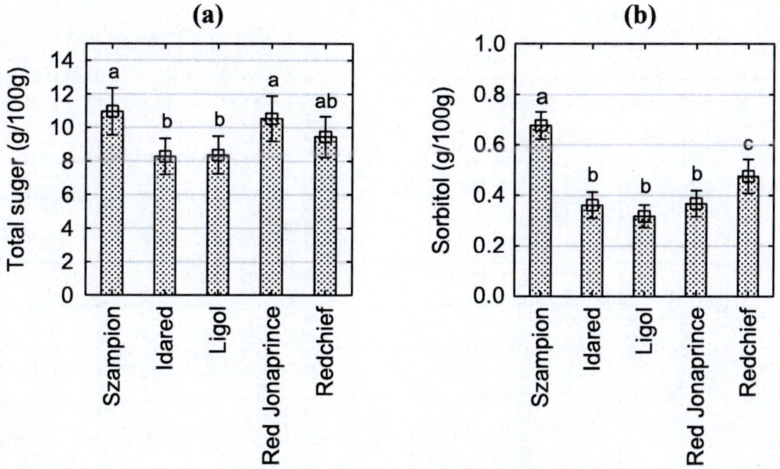

Fig. 2 Total sugar **a** and sorbitol **b** content in the apple pomace. [a,b,c]Values in a column marked with the same letter are not statistically significantly different ($p > 0.05$)

Apples are generally low in protein and fat. The obtained pomace differed in terms of protein and fat content. The pomace from the Ligol cultivar (3.35%) had the highest protein content, while the pomace from the Idared and Redchief varieties (approx. 2%) was the lowest in fat (Fig. 3a). In the case of the Szampion and Red Jonaprince varieties, the protein content in the pomace was similar and amounted to approx. 2.5%. Generally, a much higher protein content is observed in the pomace compared to the protein content in fresh apples [16]. The pomace of the Szampion variety (1.29%) had the highest fat content, while the pomace of the Idared variety (0.71%) was the lowest (Fig. 3b). In fresh apples, the fat content generally does not exceed 0.1% [16].

Among the components of the pomace, vitamin C, fiber, polyphenols and antioxidant activity are of particular importance from the nutritional point of view [2].

The pomace of the tested apple cultivars differed significantly in terms of vitamin C content (Fig. 4a). The highest content of vitamin C in the pomace was observed in the Szampion variety (5.92 mg/100 g), and the lowest in the Red Jonaprince variety (1.55 mg/100 g). The pomace from the Idared and Ligol cultivars had a similar vitamin C content (approx.4.5 mg/100 g), while in the case of the Redchief cultivar, the vitamin C content in the pomace was 1.84 mg/100 g. For comparison, the vitamin C content in fresh apples is approx. 7.5 mg/100 g [3]. The pomace from the Szampion variety was yellow in color and did not undergo enzymatic browning, which is probably related to the high content of vitamin C, which partially remained after pressing.

As reported by Morales-Contreras et al. [11] the apple pomace is an important source of pectin, and 14% of the worldwide-produced pectins comes from this byproduct. In our research, the highest content of dietary fiber was found in the

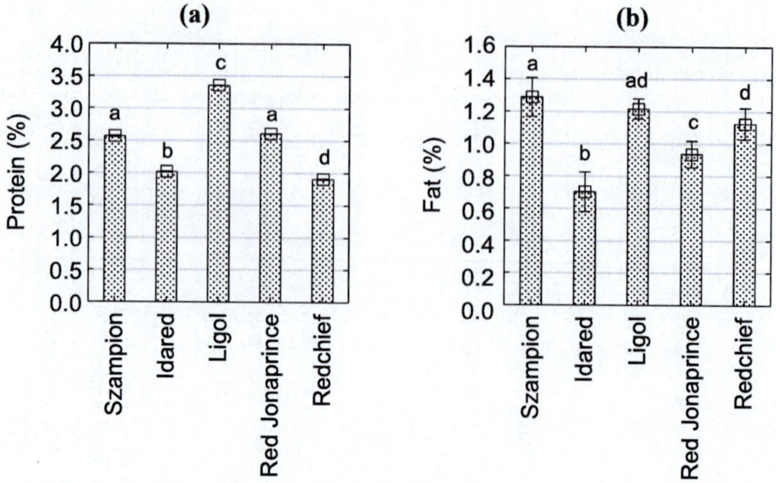

Fig. 3 Protein **a** and fat **b** content in the apple pomace. [a,b,c,d]Values in a column marked with the same letter are not statistically significantly different ($p > 0.05$)

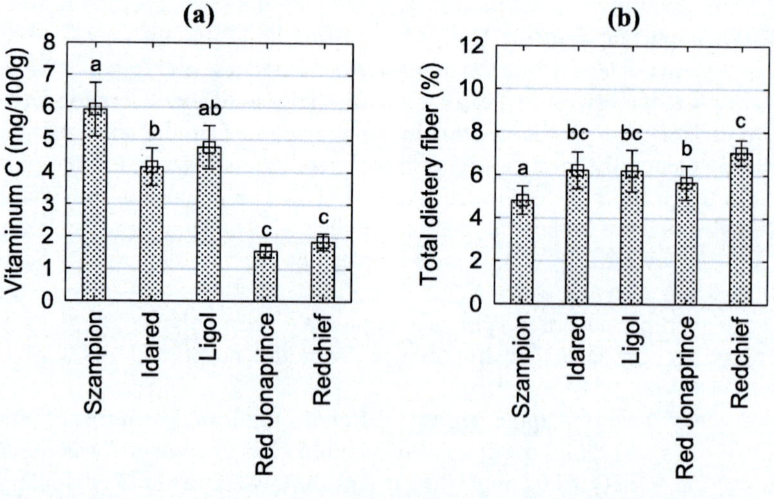

Fig. 4 Vitamin C **a** and total dietery fiber TDF **b** content in the apple pomace. [a,b,c]Values in a column marked with the same letter are not statistically significantly different ($p > 0.05$)

Redchief variety (6.75%), and the lowest in the Szampion variety, 5.28% (Fig. 4b). The other varieties were characterized by the fiber content at a similar level.

The tested apple pomace is characterized by a high content of polyphenols and shows high antioxidant activity. The red Jonaprince cultivar (65.24 mg GAE/100 mg) and cultivar (57.15 mg GAE/100 mg) had the highest content of polyphenols (Fig. 5a). For the others cultivars, the content of polyphenolic compounds was approximately

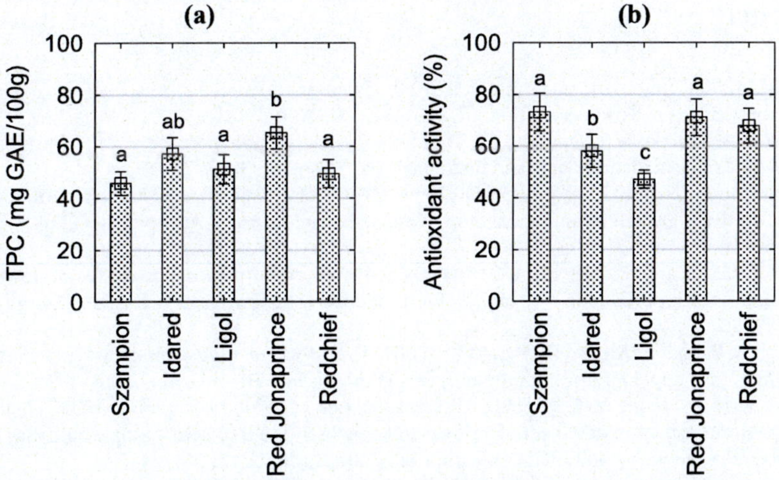

Fig. 5 Total polyphenols content TPC **a** and antioxidant activity AA **b** in the apple pomace. [a,b,c]Values in a column marked with the same letter are not statistically significantly different ($p > 0.05$)

50 mg GAE/100 mg. On the other hand, the extracts obtained from pomace cv. Szampion (72%), Red Jonaprince (70.8%) and Redchief (68%) were characterized by the highest capacity to scavenge free radicals (Fig. 5b). In the case of the Idared variety, the inhibition rate was 57%, and for the Ligol variety only 47%. Research by other authors also indicates that apple pomace has a broad content of natural antioxidants constituents that have strong antioxidant activity [9].

Due to the high content of dietary fiber, phenolic compounds, vitamin C and other nutrients, the pomace tested by us can be used as a component of food products with health-promoting properties.

4 Conclusion

The research shows that the obtained pomace from different cultivars differs significantly in the chemical composition. The previous concepts of apple pomace management did not take into account their chemical composition. The results of the research indicate that in further work on new food products based on apple pomace, their chemical composition should be taken into account. This will allow you to compose products with pro-health properties. Based on the conducted research, it was found that the pomace obtained from all tested apple varieties has a high content of nutrients and bioactive compounds. The pomace from the Szampion variety is characterized by the highest content of total sugars, vitamin C, and the highest free radical scavenging capacity, and is also resistant to enzymatic browning.

References

1. AACC (2000) International Approved Methods of the American Association of Cereal Chemists, 10th ed. The Association: St. Paul, MN, USA
2. Antonic A, Jancikova S, Dordevic D, Tremlova B (2020) Apple pomace as food fortification ingredient: a systematic review and meta-analysis. J Food Sci 85:2977–2985
3. Campeanu G, Neata G, Darjanschi G (2009) Chemical composition of the fruits of several apple cultivars growth as biological crop. Notulae Botanicae Horti Agrobotanici Cluj-Napoca 37:161–164
4. Dhillon GS, Kaur S, Brar SK (2013) Perspective of apple processing wastes as low-cost substrates for bioproduction of high value products: a review. Renew Sustain Energy Rev 27:789–805
5. Gullon B, Falque E, Alonso JL, Parajo JC (2007) Evaluation of apple pomace as a raw material for alternative applications in food industries. Food Technol Biotechnol 45:426–433
6. Kim I, Ku K-H, Jeong M-C, Kim SS, Mitchell AE, Lee J (2019) A comparison of the chemical composition and antioxidant activity of several new early- to mid-season apple cultivars for a warmer climate with traditional cultivars. J Sci Food Agric 99:4712–4724
7. Kołodziejczyk K, Markowski J, Kosmala M, Król B, Płocharski W (2007) Apple pomace as a potential source of nutraceutical products. Polish J Food Nutr Sci 57:291–295
8. Kruczek M, Gumul D, Kačániová M, Ivanišhová E, Marecek J, Gambuss H (2017) Industrial apple pomace by-products as a potential source of pro-health compounds in functional food. J Microbiol Biotechnol Food Sci 7:22–26
9. Lu Y, Foo LY (2000) Antioxidant and radical scavenging activities of polyphenols from apple pomace. Food Chem 68:81–85
10. Lyu F, Luiz SF, Azeredo DRP, Cruz AG, Ajlouni S, Ranadheera CS (2020) Apple Pomace as a functional and healthy ingredient in food products: a review. Processes 8:319
11. Morales-Contreras BE, Wicker L, Rosas-Flores W, Contreras-Esquivel JC, Gallegos-Infante JA, Reyes-Jaquez D, Morales-Castro J (2020) Apple pomace from variety "Blanca de Asturias" as sustainable source of pectin: composition, rheological, and thermal properties. LWT-Food Sci Technol 117:108641
12. Nadulski R, Kobus Z, Guz T, Strzałkowska K, Kosik D (2014) Usefulness of the selected apple cultivars for pressing in farm condition. Agric Eng 18:147–154
13. Nadulski R, Kobus Z, Wilczyński K, Zawiślak K, Grochowicz J, Guz T (2016) Application of freezing and thawing in apple (*Malus domestica*) juice extraction. J Food Sci 81:E2718–E2725
14. PN-EN 12630:1999 (1999) Fruit and Vegetable Juices—Determination of Glucose, Fructose, Sorbitol and Sucrose Contents—Method Using High Performance Liquid Chromatography. PKN, Warsaw, Poland
15. PN-EN 14130:2003 (2003) Foodstuffs—Determination of Vitamin C by HPLC. PKN, Warsaw Poland
16. Savatović SM, Tepić AN, Šumić ZM, Nikolić MS (2009) Antioxidant activity of polyphenol-enriched apple juice. Acta Periodica Technologica 40:95–102
17. Shalini R, Gupta DK (2010) Utilization of pomace from apple processing industries: a review. J Food Sci Technol 47:365–371
18. Singleton VL, Orthofer R, Lamuela-Raventos RM (1999) Analysis of total phenols and other oxidation substrates and antioxidants by means of Folin-Ciocalteu reagent. Methods Enzmol 299:152–178
19. Statistica 12 (2014) StatSoft Inc.: Tulsa, OK, USA
20. Wilczyński K, Kobus Z, Dziki D (2019) Effect of press construction on yield and quality of apple juice. Sustainability 11:3630